Practical Guide to Vegetable Oil Processing

AF791420

Practical Guide to Vegetable Oil Processing

Second Edition

Monoj K. Gupta
MG Edible Oil Consulting Int'l Inc.
Lynnwood, TX, United States

AOCS PRESS
ELSEVIER
ACADEMIC PRESS
An imprint of Elsevier
elsevier.com

Academic Press and AOCS Press
Academic Press is an imprint of Elsevier
125 London Wall, London EC2Y 5AS, United Kingdom
525 B Street, Suite 1800, San Diego, CA 92101-4495, United States
50 Hampshire Street, 5th Floor, Cambridge, MA 02139, United States
The Boulevard, Langford Lane, Kidlington, Oxford OX5 1GB, United Kingdom

Copyright © 2017 AOCS Press. Published by Elsevier Inc. All rights reserved.

Published in cooperation with American Oil Chemists' Society www.aocs.org
Director, Content Development: Janet Brown

No part of this publication may be reproduced or transmitted in any form or by any means, electronic or mechanical, including photocopying, recording, or any information storage and retrieval system, without permission in writing from the publisher. Details on how to seek permission, further information about the Publisher's permissions policies and our arrangements with organizations such as the Copyright Clearance Center and the Copyright Licensing Agency, can be found at our website: www.elsevier.com/permissions.

This book and the individual contributions contained in it are protected under copyright by the Publisher (other than as may be noted herein).

Notices
Knowledge and best practice in this field are constantly changing. As new research and experience broaden our understanding, changes in research methods, professional practices, or medical treatment may become necessary.

Practitioners and researchers must always rely on their own experience and knowledge in evaluating and using any information, methods, compounds, or experiments described herein. In using such information or methods they should be mindful of their own safety and the safety of others, including parties for whom they have a professional responsibility.

To the fullest extent of the law, neither the Publisher nor the authors, contributors, or editors, assume any liability for any injury and/or damage to persons or property as a matter of products liability, negligence or otherwise, or from any use or operation of any methods, products, instructions, or ideas contained in the material herein.

Library of Congress Cataloging-in-Publication Data
A catalog record for this book is available from the Library of Congress

British Library Cataloguing-in-Publication Data
A catalogue record for this book is available from the British Library

ISBN: 978-1-63067-050-4

For information on all Academic Press publications visit our website at
https://www.elsevier.com/books-and-journals

ELSEVIER
Book Aid International
Working together to grow libraries in developing countries
www.elsevier.com • www.bookaid.org

Publisher: Nikki Levy
Acquisition Editor: Nancy Maragioglio
Editorial Project Manager: Billie Jean Fernandez
Production Project Manager: Susan Li
Designer: Victoria Pearson

Contents

12. Insight to Oil Quality Management

13. *Trans* Fat Alternatives and Challenges

Preface

It was my desire to introduce the second edition of the book because of the introduction of certain newer techniques in vegetable oil processing. These are discussed in various chapters in this book.

The first edition of this book was received well by the readers. Many readers asked when the second edition of the book would be published. I also received requests from readers to include the processing practices for palm oil, coconut oil, cottonseed oil, and sesame seed oil, as these are important vegetable oils. Unfortunately, it was not possible to do so. The reason for their exclusion is that the basic principles and practices described in this book do apply to most vegetable oil processing operations. Additionally, the volume of information would have been too large to be included a single book.

Vegetable oil processing is an essential part of the food industry. Current unit operations have been developed over many years by processors and equipment manufacturers, with the assistance of universities and federal laboratories. Public universities have changed over time, resulting in the current emphasis on programs that meet the prevailing business needs. In today's market, the vegetable oil processing industry does not offer enough jobs to warrant a more detailed training of future technical personnel. The size of oil processing programs, where they exist at all, depends on local initiatives in attracting and maintaining sufficient numbers of students and external funding of research. For this reason, Texas A&M University, Cornell University, Purdue University, Iowa State University, University of Illinois, University of Florida, and Ohio State University are among the few exceptions, although most of these institutions have much stronger Food Technology and Food Engineering curriculum than programs on fats and oils.

The majority of these graduating students prefer food manufactures because of job availability, while only a few find employment in the vegetable oil refining industry.

Pioneers in the vegetable oil processing industry in the United States were Durkee, Procter & Gamble, Anderson Clayton, Hunt Wesson, Humko, Unilever, A.E. Staley Co., and Corn Products Co. These companies were very strong in their research and development activity. They maintained product and process development activities that trained fresh university graduates in chemistry and chemical engineering in processing and applications of vegetable oils and animal fats.

The oil companies in the United States were mostly stand-alone refiners, that is, they purchased crude oils from the crushers and processed them to make various products. They had their own pilot plants that facilitated the training programs in the area of oil processing. The fresh recruits could get hands-on experience in oil processing and product formulation. This was done primarily through project assignments to the newcomers. Some of these companies also had well-established training programs to provide the necessary tools to their technical recruits in oil processing and product formulation.

Numerous changes have taken place in the oil-processing industry in the United States since the 1970s:

1. The oil crushers, such as Archer Daniels & Midland Co., Cargill Co., and Bunge Corporation realized that it was more profitable to integrate their crushing operation with the refinery. They started to refine their own oil, in addition to selling the crude oil to the stand-alone refiners. They soon entered the market with packaged fats and oils products initially through acquisitions and later by building their own facilities. They expanded their R&D capabilities and now have become well established in the area where many stand-alone refiners filled the industry needs.
2. Oil prices soared in the mid-1970s during the Middle East oil embargo, causing a serious blow to the stand-alone refiners.
3. The stand-alone refiners started to see declining profit margins on their products because they could not match the production and reduced cost of production of the integrated crusher refiners.
4. As the competition grew from the crusher refiners, the R&D activity in the stand-alone refineries declined seriously due to lack of funds.
5. Some of these stand-alone refiners started to provide copacking services to the crusher refiners as they entered the consumer product market. Soon, some of these companies were bought out by the crusher refiners and subsequently either upgraded or disbanded.
6. Eventually, many of the stand-alone refiners either closed down or were bought out by the crusher refiners or other food companies.
7. Some of the stand-alone refiners switched their product lines to go into a niche market where the large crusher refiners were not competing.
8. Initially the crusher refiners were not up to speed with the R&D work.
9. The oil-processing equipment manufacturers picked up the slack and started to offer the technology needed for the oil refineries.
10. During this period the USDA laboratories remained active in the oil research field.
11. Universities, such as Texas A&M and Iowa State University, became active in providing pilot plant services to the oil industry.
12. Independent facilities, such as the POS pilot plant in Canada, became available as a source of basic, as well as applied research work in fats and oils.
13. A.C. Humko of Memphis, Tennessee, United States, offered pilot plant services to the oil companies.

These changes in the vegetable oil industry essentially eliminated opportunities for on-the-job training of fresh college graduates in fats and oils technology in the manner that was possible prior to 1970. Very few individuals from that era are still working for major oil refiners. A few are working as consultants, but a great majority of them have either retired or are deceased.

I am probably one of the few fortunate ones who received training in fats and oils at Procter & Gamble Co. and am still around to talk about my experience. The company hired fresh engineering graduates from the universities. Every new engineer hired received training through the assignment of projects. The new recruit had to go through the following steps:

1. The engineer was assigned a project.
2. The engineer prepared a project proposal that contained the following elements:
    a. project objective,
    b. experimental plan,
    c. data to be collected,
    d. analytical and product testing to be performed,
    e. duration of the project,
    f. list of all internal resources, and
    g. list of all internal R&D reports on related topics.
3. The proposal had to be approved by the immediate supervisor and the Director.
4. At the end of the project the engineer had to write a formal report that had to be approved by the Director.
5. A copy of the report was kept in the company archive for future reference.
6. The project could then be officially closed.
7. If the product required any plant trial, a completely new proposal had to be initiated by the engineer with all pertinent information of cost, besides the objective.
8. The product performance had to be proven through several tests, such as market sample data collection and analysis, customer complaint data, product storage study, and consumer tests.
9. At the end of the study another report had to be prepared, approved, and archived as before.

I am not aware of such a rigorous training program that might be available anywhere today.

In this book, I will make my best effort to explain why certain processing steps are considered necessary. I will also provide adequate theoretical explanations to the readers so they can appreciate the significance of the steps taken in a vegetable oil processing. It might not be possible to cover every detail or I might even leave certain material out of this book to protect any proprietary information that I have gathered during my tenure at various companies. I believe that the readers will find the information provided in this book to be useful.

In recognition, I would like to express my appreciation to Late Robert L. Wille and Cornelius Japikse, my original mentors at Procter & Gamble Co., for training me during the early days of my career. I also am indebted to Late Walter E. Farr and the Late Dr. Thomas H. Smouse for their support in advancing my career in oil processing and applications at Anderson Clayton Co. My sincere appreciation also goes to my wife, Mina Gupta, for her untiring encouragement to write this book. I also wish to express my sincere gratitude to the reviewers of the various chapters of this book in spite of their busy schedules.

Finally, a trend is developing in the area of technical communication, which serves as a reminder that we must all be critical thinkers. There are some trade journals, as well as some technical journals, that now publish editorial reviews of scientific and technical issues written by the editors or the assistant editors, who gather information and compile a presentation. While the information has been gathered through speaking with experts in the field, and references are clearly made to the persons providing the information, there are times when this information is not absolutely accurate. Although I am sure no publication intentionally publishes erroneous information, it runs the risk of misleading or confusing less-experienced readers. In my opinion, we should look carefully at once again relying on experts in the field to provide not only original research but also these critical reviews to ensure we are providing a solid scientific foundation for readers.

Chapter 1

Requirement for Successful Production and Delivery of the Refined Vegetable Oils

Vegetable oils are refined with care so the resulting oils as well as the products formulated with the oils are of high quality.

In the rest of the book the various processing steps, their operating conditions, corrective actions through troubleshooting, etc. have been discussed for the reader. All of the processing conditions described are to assist the oil processors to understand the principles of oil processing and produce the best quality refined oil at the plant.

It must be stressed that even after using the guidelines provided in this book, one may not be able to produce the best quality refined oil if the incoming crude oil is not of high quality. It may sound strange, but the success of obtaining the highest quality finished oil depends greatly on the quality of the crude oil received at the refinery.

1.1 CRUDE OIL

Crude oil quality can vary and it depends on various factors that are not directly under the control of the oil refiner. Poor quality crude oil creates certain difficulties in the refining process along with the oil quality issues. Several tips to procure the highest quality crude oil are discussed in this chapter so the refiner is aware of these factors and can take certain actions in the refinery to minimize the negative impact of some of these factors.

1.2 OILSEEDS

As mentioned earlier, good quality of the refined oil starts with the high quality oilseeds or oil-bearing fruits and nuts. The quality of the crude oil depends on various factors, such as:

- maturity of the oilseeds,
- harvest conditions (excessive rain or drought condition before harvest),
- handling of seeds,

Practical Guide to Vegetable Oil Processing. http://dx.doi.org/10.1016/B978-1-63067-050-4.00001-5
Copyright © 2017 AOCS Press. Published by Elsevier Inc. All rights reserved.

- seed storage conditions, and
- insect infestation of the seeds.

(In the subsequent discussions only oilseeds will be mentioned. Fruit palm and oil-bearing nuts will be discussed separately.)

1.2.1 Maturity

Immature soybean seeds can exhibit various deficiencies. The crude oil may exhibit some different fatty acid profile and also some variations in the other components in the seeds. This may slightly impact the processing conditions and performance of the refined oil in certain applications. There are numerous literature references that indicate the following:

- The immature seeds tend to have lower lipoxygenase activity, trypsin inhibitor, and urease activity compared to the mature seeds.
- The immature seeds tend to have higher contents of FFA (free fatty acids) and chlorophylls compared to the mature seeds.
- Oil content and total protein contents are not very different between the immature and mature seeds.
- There are minor differences in some individual protein contents between the immature and mature seeds.

Therefore, the oil refiner may receive crude soybean oil that contains high chlorophyll because of immature soybeans. This will require some additional degumming and bleaching steps. This will be discussed in the chapter on bleaching.

1.2.2 Harvest Condition

1.2.2.1 Wet Harvest Condition

Soybean, sunflower, cottonseed, and canola crude oils can exhibit higher than normal green color when the seeds are harvested before they reach maturity or the harvesting season is too wet. The crude oils will require extra steps to remove the excess chlorophylls from them in the degumming and bleaching steps. The refined oil may have lower stability if these steps are not followed properly.

1.2.2.2 Dry Harvest Condition

Dry harvest condition due to droughts can cause physical damage to the seeds resulting in higher than normal FFA and oxidation in the crude oil. The oil will exhibit lower than normal stability.

1.2.3 Handling of Seeds

The seeds, if damaged, during harvest and transport and storage, the crude oil can develop higher FFA and exhibit higher oxidation. This oil will require extra steps in the refining process and will typically exhibit lower stability than normal.

The seeds are dried to <10% moisture before storage. The drying condition requires controlled air temperature and flow around the seeds during the drying step. The seeds may develop case hardening if the air temperature is higher and or the airflow rate is higher than normal. This can develop surface cracks in subsequent handling of the seeds and the crude oil will exhibit higher than normal FFA and initial oxidation.

1.2.4 Seed Storage

It is important that the seeds are properly dried to <10% moisture and stored under 40°C (104°F) with proper air ventilation. At temperature of 45°C (114°F) or moisture content of 14% or higher, the oilseeds develop higher concentration of nonhydratable phospholipids. This makes degumming, refining, and bleaching processes more difficult and it also results in higher refining loss and also the refined oil quality is compromised. In addition, there is color fixation of the oil. The crude oil develops darker than normal color that cannot be reduced through the normal bleaching process. The crude oil in most cases has to be treated with stronger alkali solution to reduce the color.

1.2.5 Insect Infestation

Typically, dry growing season and drought condition tend to promote insect damage of the seeds. This results in higher than normal FFA and initial oxidation in the crude oil. As described in some of the previous conditions, the crude oil exhibits higher than normal refining loss and lower stability of the oil.

1.3 ADDITIONAL COMMENTS ON OILSEEDS

Oilseeds mature at a slight different rate between the top and lower parts of the plant. This tends to be more pronounced in case of cottonseed. Similarly the soybean pods can have different degree of maturity on the same plant and not all the seeds on the same sunflower would be identical in maturity. Therefore, a lot of oilseeds shipment may contain some seeds that are somewhat less mature.

The oilseeds in a lot will always have some damaged (broken) seeds, some with lesser degree of maturity. However, the various grades of seeds that are sold under USDA specification seem to perform in a uniform manner in producing the crude oil of desired quality.

Higher than normal level of diglycerides are formed whenever the crude oil is treated with stronger than the normal strength of alkali solution used in the process. Sometimes the crude oil is alkali treated more than once to meet the refined oil specification on FFA and/or color. The excess alkali or stronger alkali can attack the neutral triglyceride molecules in the oil (in addition to the

FFA), forming diglycerides. Diglycerides are emulsifiers. High concentration of diglycerides in the alkali treated oil makes it difficult to separate the aqueous phase from the oil phase in the soap separation stage. This tends to increase the loss of neutral oil in the soap causing higher oil loss in the refining process.

1.4 FRUIT PALM

The fruit palm is harvested from the tree when they reach maturity. Like in case of oilseeds, the fruit palm on the same bunch may have somewhat different degree of maturity. Usually, the very ripe ones get damaged or ruptured under the normal harvesting procedure.

Lipase and lipoxygenase activity begin in the oil inside the fruit palm when the skin of the fruit is damaged. The fruit is treated for enzyme deactivation and the oil is extracted as soon as possible after the harvest. However, most commercial crude palm oil (CPO) contains as much as 5% FFA and the diglycerides content is typically 5%.

Whenever a molecule of FFA is formed from hydrolysis of a neutral triglyceride molecule, a diglyceride molecule is formed. When the palm fruit is damaged during harvest, the enzyme lipase hydrolyzes the triglyceride molecule forming FFA and diglyceride.

The author studied the damaged fruit palm and the impact on the FFA of the oil in a palm plantation in Costa Rica. Following tests were performed:

1. The damaged fruits from a fruit bunch were collected and weighed.
2. The total weight of the fruit palm in the bunch was taken.
3. It was found that the ruptured and damaged fruit constituted 6% of the total weight of the fruit in the bunch.
4. FFA content in the oil extracted from the damaged fruit was found to be 50%.
5. Therefore, it was estimated that whatever the FFA of the oil from the fruit palm extracted from the whole bunch would be increased by 3% ($0.06 \times 50 = 3.0$).
6. Thus, the majority of the FFA in CPO would have come from the over ripe and damaged fruit.

Typical commercial production of CPO does not separate the damaged fruit from the rest for oil extraction.

There are companies, such as Sime Darby Jomalina that do separate the damaged fruit before extraction in order to produce low FFA and low diglyceride CPO and refined PO and palmolein. Sime Darby Jomalina can deliver palm oil and palm oil fractions with guaranteed quality (JGQ). There are other companies in Malaysia that are also capable of delivering low FFA and low diglyceride palm oil if a customer needs it.

High diglyceride content in the palm oil increases the FFA in a fryer faster and also slows down the rate of crystal formation in the shortening and margarine process.

1.5 GROUNDNUTS (PEANUTS) AND TREE NUTS

The same comments made in connection with the oilseeds also apply for these oil-bearing nuts. An important additional issue that can be experienced with nuts is mold that can produce aflatoxins. Aflatoxin is a type of mycotoxin produced by *Aspergillus* molds. Aflatoxins are very toxic and highly carcinogenic. There are three different types of aflatoxins that can be found in food. Short-term heavy ingestion of the toxins can cause even death. Long-term exposure can cause growth impairment and liver cancer. *Aspergillus* molds grow mostly on crops, such as grains and nuts. Under the right conditions, *Aspergillus* often grows on grain before it is harvested. But it can also grow on harvested grain if the grain is stored damp.

This is why nuts should be analyzed for aflatoxins in addition to the other tests that are normally done for accepting the raw material for crushing.

1.6 CRUDE OIL HANDLING, STORAGE, AND TRANSPORT

Most solvent extraction plants that produce crude oil do not cool and filter the crude oil after desolventization. This causes oxidation in the oil. In addition, if the crude oil is stored for extended period, it undergoes oxidation and a few other reactions that are discussed later in the book. These reactions degrade the quality of the crude oil, which, in turn, increases difficulty in refining and produces less than desirable quality in the refined oil. Excessive aeration of the crude oil during loading and transportation can increase oxidation of the crude oil.

Crude oil should be refined soon after it is made. Crude oil, if stored before refining, should be done at <40°C (104°F) for seed oils. Palm oil should be stored <50°C and preferably <45°C. Higher storage temperature causes oxidation to the crude oil. In addition, the FFA can rise, the color darkens and it can even have color fixation. The PV value goes up with higher storage temperature and longer time of storage. The PV breaks down during the refining process but the anisidine value (AV) goes up. PV measures the primary oxidative state for the oil. AV indicates the degree of the secondary oxidation state of the oil. Crude oil with higher AV indicates prior exposure of the crude oil to oxygen. This results in refined oil that would oxidize rapidly when heated (principally in frying and baking applications).

1.7 CONCLUDING REMARKS

It should be clear from the previous discussions that the quality of the crude oil is of utmost importance in obtaining good quality refined oil because all of the reactions discussed here negatively impact the refined oil quality as well as the products formulated with the refined oil.

Chapter 2

Basic Oil Chemistry

Man has used vegetable oils for centuries. Oil bearing nuts and animal fats were consumed as sources of energy long before nutrition concepts were envisioned. Oils also were used early for lighting, as medicines, as cosmetics in religious ceremonies, and applied to weapons and utensils. The ancient oils of the Middle East, sesame and olive, were valued because of their long stability. Sunflower was cultivated in the Arizona–New Mexico area before the time of Christ, and seeds from the Missouri–Mississippi river basins were among the early plants transposed to Europe by explorers. Invention of the cotton gin in the late 1700s led to a major cotton export trade in the United States in the early 1800s, and to development of cottonseed oil as the first new oil of the Industrial Age in the mid-1800s. The continuous screw press, and early methods of caustic refining, bleaching, deodorization, winterization, and hydrogenation, including development of the first all vegetable shortening "Crisco" (shortened name for crystallized cottonseed oil) are among innovations developed. Processing of soybean, a crop first developed in China, led to further oil industry innovations including development of continuous solvent extractors and steam distillation technologies to reduce or remove the original raw flavor in the crude oil were developed in the mid-1900s. As flavor and stability improved, man expanded use of oils to: (1) cooking, (2) frying, (3) baking shortenings, (4) salad dressings, (5) food lubricants (like release agents in baking and candy making processes), (6) flavor carriers, and (7) dust-control agents. Each of the application requires oils with specific physical and chemical properties.

Other oils, such as palm oil, regular canola oil, high oleic and low linolenic canola oil, high oleic sunflower oil, high oleic safflower oil, and so on were all commercialized much later than the animal fat and cottonseed oil.

2.1 COMPOSITION OF OIL

All of the world's matter is composed from approximately 108 elements. The smallest divisible stable particle of an element is called an "atom." Compounds consist of atoms of two or more elements, with the smallest divisible stable particle called a "molecule." Carbon (C), hydrogen (H), and oxygen (O) atoms are the principal building blocks of fats and oils.

Practical Guide to Vegetable Oil Processing. http://dx.doi.org/10.1016/B978-1-63067-050-4.00002-7
Copyright © 2017 AOCS Press. Published by Elsevier Inc. All rights reserved.

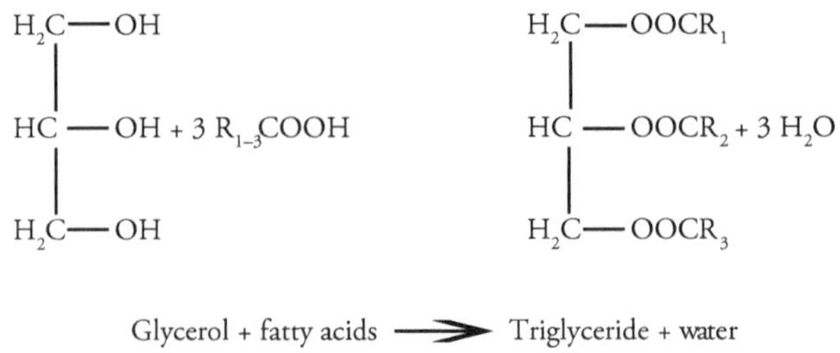

FIGURE 2.1 Formation of triglycerides.

Often, it is desirable to pictorially indicate relative positions of the elements in molecular structures. But, these must be carefully drawn by established convention, since the world exists in three dimensions, but only two dimensions are available for presentation on paper. In making such drawings, the knowledgeable chemist recognizes that some atoms only associate with others by extending links, while others only accept links. For example, each oxygen atom extends two links, while, each hydrogen atom accepts only one link. The chemistry of fats and oils is carbon chemistry, also known as "organic chemistry." The carbon atom is unique in that it can either extend or accept a total of four links, with link givers, link receivers, or even with other carbon atoms.

Oil is a mixture of 96–98% fatty acid triacylglycerols (commonly referred to as "triglycerides"), with the balance consisting of other fat-dispersible or fat-soluble compounds. Triglycerides consist of three fatty acids, which are substituted in the hydroxyl (alcoholic) sites of a glycerin (glycerol) backbone. The construction of a simple triglyceride is shown in Fig. 2.1, where each fatty acid is represented as a different "R."

Depending on the extent to which the three former hydroxyl groups of glycerol are replaced with fatty acids, the resulting compounds are known as follows.

Monoglycerides are formed when one of the three hydroxyl groups of glycerol is replaced by a fatty acid.

Diglycerides are formed when two of the three hydroxyl groups of glycerol are replaced by the same or different fatty acids.

Triglycerides are formed when all three of the hydroxyl groups of glycerol are replaced by fatty acids (also referred as neutral oil).

A molecule of water is formed each time a fatty acid molecule replaces a hydroxyl group. Fig. 2.2 further shows the structures of monoglyceride, diglyceride, and triglyceride molecules.

The major objective in refining and processing is to convert a shipment of purchased crude oil into the maximum possible amount of saleable "neutral oil" (triglycerides). Monoglycerides and diglycerides are formed when the neutral oil reacts with water molecules under undesirable storage and handling conditions. This reduces the yield of neutral oil in the refining process. It also creates poor quality refined oil. This will be discussed further in Chapter 11.

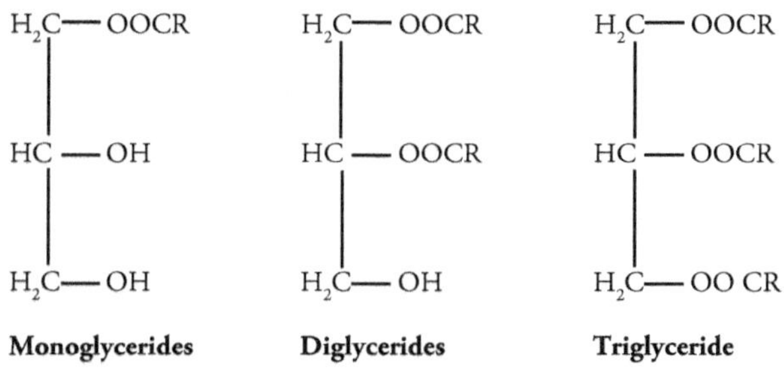

FIGURE 2.2 **Structures of mono-, di-, and triglycerides.**

2.2 DISTINCTIONS BETWEEN OILS AND FATS

A triglyceride molecule is called "oil" if it is liquid at ambient (room) temperature, and a "fat" if it is semisolid. Definitions of "room temperature" will vary greatly with the climate of the region. For example, "room temperature" in a tropical region can be >95°F (35°C), whereas that in a temperate region can be 68°F (20°C). A good example is coconut oil, which is liquid at room temperature in semitropical areas during the year except for the winter months when it becomes solid and might be called a "fat," although coconut oil is always referred to as oil. Similarly, partially hydrogenated oil, which might be semisolid or solid at room temperature, is commonly referred to as oil.

Products of reactions between hydroxyl groups and organic acids are called "esters" or sometimes "acyl- compounds." The broad variety of products includes waxes made by esterification of long chain alcohols and long chain fatty acids, various food and industrial emulsifiers, noncaloric sucrose-based frying oils, fatty acid methyl ester solvents, and biodiesel fuels.

2.3 FATTY ACIDS IN COMMON VEGETABLE OILS

Fatty acids are the building blocks of triglycerides. They generally contain 4–22 carbon atoms and are linear in structure. Sometimes, fatty acids are designated as "short chain" (4–8 carbon atoms), "medium chain" (10–12 carbon atoms), and "long chain" (14 or more carbon atoms). The following fatty acids are most common in vegetable oils:

Saturated	Unsaturated
Lauric (C_{12})	Oleic ($C_{18:1}$)
Palmitic (C_{16})	Linoleic ($C_{18:2}$)
Stearic (C_{18})	Linolenic ($C_{18:3}$)
Arachidic (C_{20})	
Behinic (C_{22})	

Oleic acid, which has one double bond, is called a "monounsaturated fatty acid" while linoleic and linolenic acids are called "polyunsaturated fatty acids" because they contain more than one double bond (2 and 3, respectively).

2.3.1 Saturated and Unsaturated Fatty Acids

A carbon atom with all four reaction sites of the carbon atom reacted with other elements is termed "saturated." The structure of a fatty acid with an end carboxyl group (—COOH) is shown below.

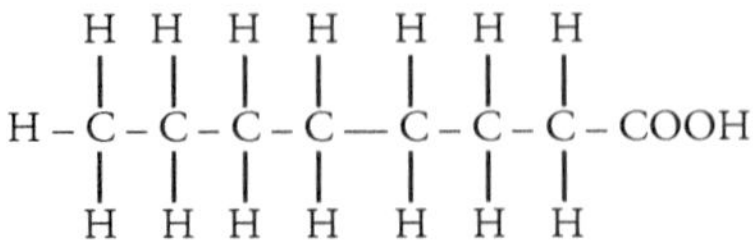

In this example, only single carbon-to-carbon bonds exist, and the fatty acid is called "saturated."

Unsaturated fatty acids contain fewer hydrogen atoms than required to fully satisfy the valence of each carbon atom in the molecule. Thus, some carbon atoms are connected to each other with a "double bond" as shown in the following.

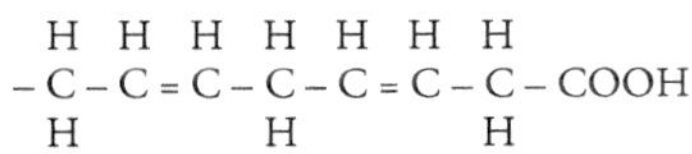

The double bonds in most vegetable oils (except for drying oils used in paints) contain two single bonds between the two double bonds in the chain.

Most of the hydrogen in double bonds of natural fatty acids is found on the same side of the double bond, indicating a "*cis* position" (or "*cis*-isomer"). But, some of the hydrogen atoms may move to the other side of the bond during hydrogenation process (chemical saturation of double bonds), to produce "*trans*-isomers." These structures are further clarified in the following.

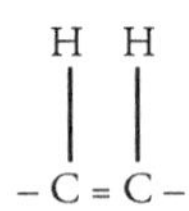

***Cis* Isomer**

***Trans* Isomer**

Both *cis* and *trans* isomers are "unsaturated," fatty acids. However, transformation of the *cis* to *trans* configuration raises the melt-point for the oil.

A small conversion of *cis* to *trans* forms also occurs when oils are heated to very high temperature as during hydrogenation and deodorization.

2.4 TYPICAL BEHAVIOR OF FATTY ACIDS

2.4.1 Unsaturated Fatty Acids

Unsaturated fatty acids are unstable and are very susceptible to oxidation even at ambient temperatures. They tend to:

1. readily oxidize when exposed to air or oxygen,
2. form aldehydes, ketones, etc.,
3. form primarily oxidative polymers, and
4. form cyclic compounds.

2.4.2 Saturated Fatty Acids

In contrast, saturated fatty acids are relatively stable. They do not oxidize in the presence of air or oxygen, but will decompose under high heat. They can produce:

- thermal polymers
- toxins, such as acroleins

2.5 OBJECTIVES OF PROPER OIL PROCESSING

The objective of proper oil processing is to obtain finished oil with the following traits:

1. long oxidative stability,
2. long thermal stability,
3. long flavor stability,
4. long storage stability, and
5. long shelf life of food products formulated with the oil.

It is critical that processors understand the basic constituents of oil, its properties, and how to maintain process conditions that deliver oil with the quality standards listed previously.

2.6 NONTRIGLYCERIDE COMPONENTS OF OILS

As mentioned earlier, crude vegetable oils generally contain 96–98% triglycerides. Although these components are present in small amounts, they can be very influential in determining overall stability and performance of the oil. They may be grouped as:

1. major nontriglyceride components
2. minor nontriglyceride components

2.6.1 Major Nontriglycerides

The following components generally are present at high levels in the crude oil and can be measured as percentages:

1. phospholipids
2. free fatty acids (FFA)
3. diglycerides
4. monoglycerides

2.6.1.1 Phospholipids

These compounds are also known as phosphatides or gums. Their levels are generally expressed in parts per million of phosphorus. The five major groups of phospholipids found in most vegetable oils are:

1. phosphatidylcholine
2. phosphatidylethanolamine
3. phosphatidylinositol
4. phosphatidylserine
5. phosphatidic acid

Typical phospholipids contents of common vegetable oils are shown in Table 2.1.

TABLE 2.1 Phospholipids Contents of Selected Vegetable Oils

Oil type	Phospholipids content (%)	Phosphorus[a] content (ppm)
Crude soybean oil	1–3	317–950
Degummed soybean	0.32–0.64	100–200
Crude corn oil	0.7–0.9	222–285
Crude peanut (groundnut) oil	0.3–0.6	95–190
Crude canola oil	1.8–3.5	570–1104
Superdegummed canola oil	0.13–0.16	41–51
Crude sunflower oil	0.5–0.9	159–285
Crude safflower oil	0.4–0.6	127–190
Crude palm oil	0.06–0.95	19–30

[a]*The relationship between phospholipids and phosphorus contents is: phosphorous (ppm) = [phosphatides (%) × 10^4]/31.7.*

2.6.2 Hydratable and Nonhydratable Phospholipids

Two types of phospholipids are present in crude oils from the standpoint of their affinity for water:

1. Hydratable phospholipids
2. Nonhydratable phospholipids

Treatment with water at 140–158°F (60–70°C) hydrates some of the phospholipids in crude oils, which settle out or can be separated by centrifugation. For example, 600–800 ppm phosphorus in crude soybean oil can be reduced to 200 ppm or less by simple water degumming. Phospholipids, which are not removed by water alone are considered "nonhydratable." The objective of acid-pretreatment of crude oil is to convert nonhydratable phospholipids into hydratable forms by sequestering (drawing away) absorbed bivalent cations (like calcium and magnesium metals) which interfere with their hydratability. Various methods for degumming crude oil are described in Chapter 3.

2.6.3 Free Fatty Acids

Fatty acids, separated from triglyceride molecules, are called "free fatty acids, "FFA" and dissociate into two moieties—a link-accepting hydrogen ion and the link-giving residual. Formation of FFA in the oil of stored oilseeds is a natural occurrence, initiated by "lipase" enzymes. A small amount of FFA also formed during seed crushing and subsequent handling and storage of the crude oil. Fatty acids bound in triglycerides are still reactive in oxidation and hydrogenation processes. Amounts of FFA in crude oil vary with the oil species and history of the sample. Typical FFA values in selected crude oils are shown in Table 2.2.

2.6.4 Monoglycerides and Diglycerides

Degradation of crude oils into FFA always is accompanied by formation of diglycerides and monoglycerides. These compounds have emulsifying

TABLE 2.2 Typical Free Fatty Acid (FFA) Content of Common Crude Vegetable Oils

Oil type	FFA content (%)
Most seed oils	0.5–1.5
Crude palm oil	1–4
Crude cottonseed oil	0.5–3
Extra virgin olive oil	<0.8
Virgin olive oil	<2

properties and can negatively impact on oil losses in refining and processing, and also on performance of the final oil. This will be discussed later in Chapter 11. Typical levels of monoglycerides and diglycerides in various fully processed oils are shown in Table 11.3.

2.6.5 Minor Nontriglycerides

Minor nontriglyceride components of crude oil, present in parts per million levels, include:

1. tocopherols and tocotrienols
2. sterols and sterol esters
3. volatile and nonvolatile compounds formed from decomposition of the triglycerides
4. color compounds
5. trace metals

2.6.6 Tocopherols

Tocopherols are naturally occurring antioxidants in vegetable oils, and one of nature's protections against oil oxidation. Four types of tocopherols are present: *alpha*, *beta*, *gamma*, and *delta*. Sometimes, these forms are identified by Greek letters α, β, γ, and δ, respectively. Alpha (α) tocopherol provides protection to the oil against photooxidation (oxidation under visible light). Functions of beta (β) tocopherol, found at very low concentrations in oils, are not fully known. Gamma (γ) and delta (δ) tocopherols protect oil against autoxidation. Autoxidation is the primary pathway for oil oxidation, with oil degradation occurring even in absence of light. This type of oxidation process occurs during processing, storage, distribution of oil as well as food ingredients containing oils and during food products manufacture and their storage. The reaction is initiated by formation of a free radical from the unsaturated oil by a metal initiator. The reaction propagates and continues until either oxygen or unsaturated fatty acids are exhausted in the oil.

Photooxidation can occur in unsaturated fatty acids when oil is exposed to ultraviolet rays and a metal initiator is present in the oil. This reaction is called photochemical reaction. This is a relatively slow reaction process like autoxidation. Photooxidation occurs to the oil in presence of a sensitizer like chlorophyll (or its oxidation products) when exposed to visible light. This reaction is very rapid and is 1500 times faster than autoxidation.

Tocotrienols, another group of natural antioxidants, have attracted strong attention to palm and rice bran oils, which contain 300–500 and 400 ppm of these compounds, respectively. Tocotrienols are especially effective against autoxidation. Autoxidation reaction mechanism is shown in Table 2.3. Rice bran oil and corn oil also contain ferulic acid, an excellent antioxidant at high

TABLE 2.3 Monoglycerides and Diglycerides Present in Fully Processed Oils

Oil type	Monoglyceride (%)	Diglyceride (%)
Most seed oils	0.2–0.4	<0.5
Palm oil	0.5–3	3–7

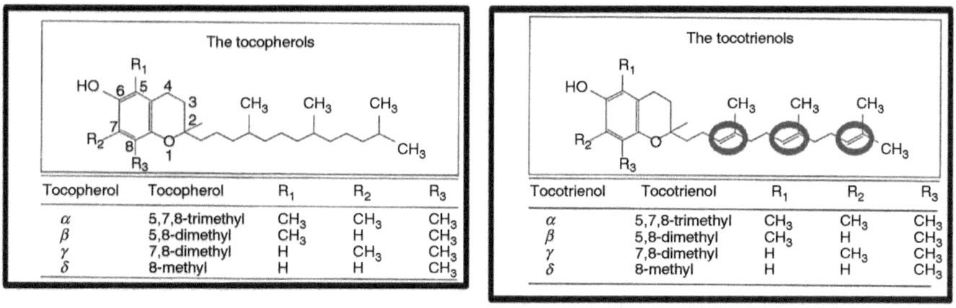

FIGURE 2.3 Structures of tocopherols and tocotrienols.

temperatures. Rice bran oil contains another group of antioxidants known as oryzanols, which are extremely effective as antioxidants at high temperature applications like frying and baking. Sesame seed oil contains sesamolin, sesamol, sesaminol, and episesaminol antioxidants, which are not present in other seed oils. Further, palm oil contains CO Enzyme Q-10 a unique antioxidant not present in the seed oils. The structures of tocopherols and tocotrienols in Fig. 2.3 and the typical tocol contents of various oils are shown in Table 2.4.

(RH) Unsaturated fatty acid + ⟶ Metal ion initiator — R· Free radical — Initiation (1)

R· Free radical + O_2 Oxygen ⟶ ROO· Peroxy radical — Autoxidation (2)

ROO· Peroxy radical + (RH) Unsaturated fatty acid ⟶ ROOH Hydro peroxide + R· Free radical — Propagation (3)

R· + R· (Free radicals) ⟶ RR (Dimer) — Termination (4)

TABLE 2.4 Tocols Contents in Crude Oils (ppm)

	Sunflower	Cottonseed	Soybean	Corn oil	Palm oil[a]	Canola
Tocopherols						
Alpha	403–935	402	90–120	191	129–215	290
Beta	ND–45	1.5	ND	—	22–37	—
Gamma	ND–34	572	740–1020	942	19–32	382
Delta	ND–7	75	240–300	42	10–20	13.4
Tocotrienols						
Alpha	ND[a]	ND	ND	23	44–73	ND
Beta	ND	ND	ND	—	44–73	ND
Gamma	ND	ND	ND	—	260–437	ND
Delta	ND	ND	ND	—	70–117	ND
Total tocols	440–1520	1050	1130–1450	1198	600–1000	685

ND, Nondetectable.
[a]*Palm oil contains +CO enzyme Q-10 = 15–30 ppm.*

Tocotrienols have three additional double bonds compared to tocopherols, which might be the reason for their improved antioxidant effects over tocopherols.

2.6.7 Sterols and Sterol Esters

Phytosterols and phytosterol esters are often present in low concentrations similar to tocopherols and other antioxidants mentioned previously. These compounds also have antioxidant properties, although this property has not been studied as extensively as with the tocopherols. However, sterols and their derivatives have been studied more extensively in human nutrition. Like tocopherols, different types of sterols and derivatives exist, with type and concentration varying with the oil species. Sterols and sterol ester contents of common vegetable oils are shown in Table 2.5.

2.6.8 Volatile and Nonvolatile Compounds

Autoxidation generates a large number of oil decomposition products, including:

1. primary oxidation products, for example, peroxide value (PV).
2. Secondary oxidation products, for example, aldehydes, ketenes, etc.
3. Tertiary oxidation products, for example, alcohols, acids, oxidation polymers, epoxides, cyclic fatty acids, and so on.

The majority of these compounds has low molecular weight and volatilizes as the oil is heated. But, some fatty acid derivatives are too large and do not

TABLE 2.5 Sterol Compounds and Their Levels in Common Crude Vegetable Oils (ppm)

Type of sterols	Soybean	Canola	Sunflower	Corn oil	Palm
Brassicasterol	ND–12.3	950	ND–9.2	ND–44.2	ND
Beta-sitosterol	918–2,460	3,600	1,348.8–2,990	4,384–14,718.6	150.6–434.7
Campesterol	284.4–992.2	1,900	177.6–593.4	1,488–5,326.1	56.1–192.5
Stigmasterol	268.2–783.1	35	168–529	344–1,701.7	25.5–97.3
Delta5 avenasterol	34.2–151.7	130	ND–317.4	336–1,812.2	ND–19.6
Delta7 stigmastenol	25.2–213.2	76	168–1104	80–928.2	0.6–16.8
Delta7 avenasterol	18–188.6	160	74.4–243.8	56–596.7	ND–35.7
Total sterols	1,800–4,100	6,900	2,400–4,600	8,000–22,100	300–700

ND, Nondetectable.

volatilize. These compounds have distinct effects on oil and product flavors and their stability, which will be discussed in Chapter 12.

2.6.9 Color Compounds

The main color compounds in vegetable oils are carotenes and chlorophylls, although other chromophoric compounds also are present. Among the vegetable oils, palm oil contains the highest amount of carotenes. On the other hand, soybean and canola contain the highest amounts of chlorophylls. Most of the carotenes are removed from the oil by heat bleaching in deodorization described later. Most of the chlorophylls are removed from the oil during the bleaching process using bleaching clay.

Most of the carotenes are retained in the deodorized palm oil called the "red palm oil," using a very special process. This oil is sold as a naturally rich-in-carotene oil. The carotene content of this oil is 500–600 ppm, compared to 600–800 ppm in crude palm oil (CPO). Benefits of carotenes for human eyesight have been demonstrated in human studies in India and the Far East, and red palm oil is promoted for this nutritional property. This oil also has higher tocopherol and tocotrienol contents than the conventionally processed palm oil or palm olein.

2.6.10 Trace Metals

Trace metals are undesirable in processed oils because they initiate the autoxidation reaction and shorten storage stabilities of oils and food products formulated with them. The most common metals found in the crude oil are: iron, calcium, magnesium, and sometimes very low levels of copper. Toxic "heavy metals" may also be present in very low concentrations in crude oils. Trace metals are removed from the crude oil by the bleaching clay, and bound by citric acid after the deodorization process. This will be discussed later in Chapters 6 and 12.

2.7 OIL ANALYSIS USED IN VEGETABLE OIL INDUSTRY AND THEIR SIGNIFICANCE

Vegetable oil is analyzed at various stages of processing. Each analysis provides specific information to the processor as well as to the users. The most commonly conducted analyses in oil processing plants are listed in the following with brief descriptions.

Analysis	*Method of Analysis (Version)* **AOCS Method**
Iodine value (IV)	
• Cyclohexane–acetic acid method	Cd 1d-92 (09)
• NIR method	Cd 1e-01 (09)
• Calculated from GLC	Cd 1c- 85 (09)
• Cyclohexane method	Cd 1b-87 (12)
FFA	
• Crude and refined fats and oils	Ca 5a-40 (12)
Acid value	
• Of fats and oils	Cd 3d-63 (09)
PV	
• Isooctane method	Cd 8b-90 (11)
• Chloroform method[a]	Cd 8-53 (03)
para Anisidine value (pAV)	Cd 18-90 (97)
Soap in oil	
• Titrimetric method	Cc 17-95 (09)
• Conductivity method	Cc 15-60 (89)
Conjugated dienes	Ti 1a-64(09)
Polar material (TPM)	Cd 20-91 (09)
Polymerized triglycerides	Cd 22-91 (09)
Solid fat index (SFI)	Cd 10-57 (95)
Solid fat content (SFC)	Ca 5a-40 (12)
Fatty acid composition (FAC)	
• Capillary GLC method	Ce 1e-91(01)
• Packed column method	Ce 1c-89 (95)
trans fatty acid (TFA)	
• *trans* of partially hydrogenated oils by GLC-IR	Cd 14b-93 (95)
• *cis, cis* and *trans* isomers by GLC[a]	Ce 1c-89 (95)
• Isomers isolated by FTIR	Cd 14-95 (09)
• By capillary GLC method	Ce 1f-96 (09)

[a] *Surplus method—either superseded or obsolete.*

Bleaching test	
• For refined cottonseed oil	Cc 8a-52 (12)
• For refined soybean oil	Cc 8b-52 (11)
• For refined sunflower oil	Cc 8b-52 (11)
Lovibond color	
Wesson (Lovibond) method	Cc 13b-45 (09)
Color (per ISO Standard)	Cc 13e-92 (09)
Color (automated method)	Cc 13j-97 (09)
Chlorophyll pigment	
Refined and bleached oils	Cc 13d-55 (09)
Crude vegetable oils	Cc 13i-96 (13)
Crude vegetable oils	Cc 13k-13 (13)
Trace metals	
By AAS (Cr, Cu, Fe, Ni)	Ca 15-75 (09)
By graphite furnace AAS (Cr, Cu, Fe, Ni, Mn)	Ca 18-79 (09)
By graphite furnace direct (Cu, Fe, Ni)	Ca 18b-91 (09)
By graphite furnace AAS (Pb only)	Ca 18c-91 (09)
By ICP-OES (all metals)	Ca 17-01 (09)
Phosphorus in oils	
By AAS	Ca 12b-92 (09)
By ICP-OES	Ca 20-99 (09)
By IO method	Ca 12a-02 (09)
Smoke point, flash point, and fire point	
Cleveland open cup method	Cc 9a-48 (09)
Melt point	
Capillary tube method	Cc 1-25 (09)
Mettler dropping point	Cc 18-80 (09)
Slip melting point	Cc 3-25 (09)
Slip melting point, ISO Standard	Cc 3b-92 (09)
Wiley method[a]	Cc 2-38 (91)

[a]*Surplus method—could be considered obsolete.*

Active oxygen method (AOM)[a]	Cd 12-57 (93)
Oil stability index(OSI)	Cd 12b-92 (09)
Refining loss	
• Degummed, expeller soybean oil	Ca 9a-52 (09)
• Degummed hydraulic and extracted soybean oil	
• Extracted and reconstituted prepressed cottonseed oil	
• Vegetable oils crude	
Neutral oil	
• Loss	Ca 9F-57 (09)
• In soap stock	G5 -40 (09)
Unsaponifiable matter	Ca 6a-40 (11)
Saponification value	Cd 3-25 (13)
Mono and diglycerides	
• By capillary GLC	Cd 11b-91 (09)
• By HPLC-ELSD	Cd 11d-96 (09)

Mono, di and triglycerides	
• By silica gel chromatography	Cd 11c-93 (09)
Alfa monoglycerides	Cd 11-57 (11)
Moisture and volatiles (butter fat, margarines, oils)	
• By hot plate method	Ca 2b 38-(09)
• Vacuum oven method (except coconut oil)	Ca 2d-25 (09)
• By distillation method	Ca 2a = 45 (09)
• By Karl Fischer method	Ca 2e-84 (09)
Alkalinity	
• Of fats and oils	Cd 3e-02 (09)
• In soda soap and products	Da 7-48 (09)
Acetone insoluble matter	
• In lecithin	Ja 4-46 (11)

[a]*Surplus method—could be considered obsolete.*

2.8 SIGNIFICANCE OF THE ANALYTICAL METHODS AND RESULTS

2.8.1 Iodine Value

This method determines the degree of unsaturation in the oil. The results are expressed as grams of iodine absorbed per 100 g of the oil sample.

Oils with higher unsaturation show higher IV values. Iodine values of most common crude vegetable oils are listed in Table 2.6.

2.8.2 Free Fatty Acids

This method determines the amount of FFA present in the oil. Generally, results are expressed as percent oleic acid for seed oils. It is expressed as percent palmitic acid for palm oil and palm oil derivatives, and as percent lauric acid for palm kernel or coconut oils.

TABLE 2.6 Typical Iodine Values of Common Refined Vegetable Oils

Oil type	Typical iodine value
Soybean	132
Canola	120
Sunflower oil	128
Cottonseed oil	110
Palm oil	50

2.8.3 Acid Value

The acid value for oil is the number of mg of potassium hydroxide required to neutralize the free acid in 1 g of the oil. For easy reference:

AV = 1.99 FFA (%)

2.8.4 Peroxide Value

This method measures the primary state of oxidative of the unsaturated fatty acids in oil. The fatty acid can be in the form of FFA or as part of a triglyceride molecule. This method measures all substances in the oil, which oxidize potassium iodide under conditions of the method as milliequivalents of peroxide per 1000 g of oil or fat. PV of freshly bleached as well as deodorized oil must be "zero."

2.8.5 para Anisidine Value

pAV is defined by convention as 100 times the optical density of a solution containing 1 g of oil and 100 mL of a mixture of solvent and reagents specified in the test method, measured in a 1-cm cuvette at 350 nm. This test measures some of the secondary oxidation compounds of oils and fats generated from the decomposition of the peroxides. Specifically, 2-alkenals and 2, 4-dienals are measured by this method. Freshly deodorized oil may have a pAV content of 2–6.

2.8.6 Soap in Oil

This titrimetric method determines alkalinity in the oil as parts per million sodium oleate. Presence of soap in bleached oil indicates poor bleaching. Properly refined and bleached oil must have zero soap content. Soap in bleached oil can cause numerous production and quality problems which will be discussed later.

2.8.7 Conjugated Dienes

This spectrophotometric method determines diene linkages of unsaturated fatty acids present in oil in terms of percent of oil. This is a measure to understand the onset of autoxidation reaction, and will be discussed later in Chapter 12.

2.8.8 Polar Material (TPM)

This method determines the total amount of polar materials present in the oil by column chromatography. It is used as a measure of oxidative degradation for oil, especially in frying processes.

2.8.9 Polymerized Triglycerides

This method determines polymerized triglycerides in fats and oils by a gel-permeation method, and indicates the degree of thermal and oxidative abuse of the oil.

2.8.10 Solid Fat Index

This is a dilatometric method which determines the combined volume of solid and liquid in the sample at specific temperatures. It is an empirical measure of solids fat content in a sample of oil at specified temperatures. The information is used in formulating shortenings, margarines, and spreads.

2.8.11 Solid Fat Content

This Nuclear Magnetic Resonance Spectrometry (NMR) method estimates the amount of fat solids present in the oil (fat) sample at specific temperatures. It is also used in formulating shortenings, margarines, and spreads, and originally was developed for the emerging modern palm oil industry.

2.8.12 Fatty Acid Composition

This capillary method identifies the fatty acids in a fat or oil by analysis of the sample's fatty acid methyl esters by capillary gas–liquid chromatography. The fatty acid methyl esters are prepared according to AOCS Method Ce 2-66 (09). This method does not identify *cis* or *trans* isomers.

2.8.13 Fatty Acid Composition

The packed column method is especially suitable for analyzing hydrogenated fat because it is capable of providing (1) fatty acids identities and compositions, and (2) TFA and *cis–cis* methylene-interrupted unsaturation. This method yields slightly lower *trans* values as compared to the infrared spectrophotometric method (AOCS Method Cd 14-61).

2.8.14 *trans* Fatty Acid

TFA in hydrogenated fat is becoming increasingly critical for the vegetable oil industry.

2.8.15 Refined and Bleached Color Test

These methods are available applicable to refined cottonseed oil soybean and sunflower oils. These tests are particularly helpful to predict the color of the deodorized oil that could be obtained from a given crude oil.

2.8.16 Lovibond Color

Method Cc 13b-45 (09) compares the oil color by comparing against colored glasses. This method can be used to measure color of all normal oils provided there is no turbidity in the sample.

Method Cc 13e-92 (09) is preferred by the British Standard Lovibond International trade.

Method Cc 13j-97 (09) is suitable for measuring colors of all refined, bleached, and deodorized vegetable oils and also filtered and deodorized tallow. The Automated method gives results in the AOCS-Tintometer (Wesson method) or the Lovibond color scale.

Lovibond color can be used to track degree of removal of color bodies present in the original crude oil. Each type of oil has a characteristic Lovibond Red color. A higher color indicates problems either with the oil or the process. These will be discussed in detail later in Chapters 6, 8, and 12.

2.8.17 Chlorophyll Pigments

This spectrophotometric method determines the concentration of chlorophyll in expelled, refined, and bleached oils by measuring absorption at 630, 670, and 710 nm wavelengths. This method is not applicable to hydrogenated oils, deodorized oils, or finished products.

2.8.18 Trace Metals (ICP)

The ICP method, or Inductively Coupled Plasma Optical Emission Spectroscopy (ICP-OES), is used for quantitative determination of calcium, copper, iron, magnesium, nickel, silicon, lead, sodium, and cadmium in oil, when these impurities are present in the solubilized form in the oil. Suspended material, such as bleaching clay or nickel catalyst cannot be detected by this method. The detection level by this method is extremely low and precise.

2.8.19 Trace Metals (Atomic Absorption Method)

This method is suitable for crude oil and partially refined oil. It can determine copper, chromium iron and nickel as low as 0.1 ppm in the oil.

2.8.20 Phosphorus (Graphite Furnace)

This method determines the phosphorus content in parts per million. It involves vaporization of the oil in a suitable graphite furnace and an atomic absorption spectrophotometer for reading.

2.8.21 Phosphorus (ICP)

This method quantitatively determines the phosphorus level in oil by using Inductively Coupled Plasma Optical Emission Spectroscopy (ICP-OES).

2.8.22 Smoke Point, Flash Point, and Fire Point (Cleveland Open Cup method)

Smoke point is directly related to the amount of FFA in the oil, and also to the amounts of monoglycerides and diglycerides present in the oil. The flash point of solvent-extracted crude oil must be checked at receipt to make sure it is higher than 300°F (149°C). The smoke point for the degummed soybean oil or crude sunflower oil is 250°F (121°C) maximum, according to the Trading rules of NIOP.

2.8.23 Melt Point (Capillary Tube Method)

The complete melting point of fat is determined by this method.

2.8.24 Melt Point (Mettler Drop Point Method)

The temperature at which the fat sample becomes soft and flows under the specific conditions of the test is measured by this method. This is an approximate method for melt point because one can see higher melting solids in the melted sample even at a temperature higher than the melt point determined by this method.

2.8.25 Active Oxygen Method (AOM)

This method measures the time in hours needed for the PV of a sample to reach 100 mEq when tested under the conditions specified. This is a measure of the primary oxidative stability of oil. AOM provides good information oil stability for salad dressing and applications that do not require high temperature treatment for the oil. Most oil processors and end users stopped using this method because the following method is found to be more useful to determine the oxidative stability of the oil.

2.8.26 Oil Stability Index (OSI)

This method provides the tertiary oxidative state for the oil. In oil applications, OSI is a better measure of oil stability while processing of foods formulated with the oil and is subjected to high temperature.

The apparent basic difference between OSI and AOM is that OSI estimates the time required to exhaust antioxidants present in the sample and begins accumulating peroxides, while AOM measures the total time required for the sample to degrade to the 100 mEq PV. The OSI method provides better information about the secondary and tertiary oxidation of the oil. This will be discussed further in Chapter 12.

2.8.27 Refining Loss

There are several methods for the test that apply to different oils as listed previously.

2.8.28 Neutral Oil Loss

2.8.28.1 In Refining

The total natural oil of natural fats and oils consisting essentially of triglycerides and unsaponifiable matter is determined by this method. The fatty acids and miscellaneous nonfat substances are removed by passing through a column of activated alumina. The loss is the difference between the amount of feed and that recovered, expressed as percent.

This method is satisfactory for cottonseed, soybean, peanut (groundnut), linseed, coconut, and sunflower oils.

2.8.28.2 In Soap Stock

There is always a certain amount of neutral oil that remains in the soap stock that is discharged from the primary separator in caustic refining process. This method allows one to determine the amount of neutral oil in the soap stock.

2.8.29 Unsaponifiable Matter

Unsaponifiable matter include those substances frequently found dissolve in fats and oils, which cannot be saponified by the usual caustic treatment, but are soluble in ordinary fat and oil solvents. Included in this group are high aliphatic alcohols, sterols, pigments, and hydrocarbons.

This method is applicable to normal animal and vegetable oils and fats but is not applicable to marine oils and the feed-grade fats.

2.8.30 Saponification Value

This method is defined as the number of milligrams of KOH required to saponify 1 g of the oil sample. This method is important in the soap making industry.

This method is applicable for vegetable oils, deodorizer distillates, and sludges.

BIBLIOGRAPHY

Food Codex Alimentarius Commission Standard 210.(1999).

Bockisch, M., 1993. Fats and Oils Handbook. AOCS Publication, USA.

Gupta, M.K., 2017. Practical Guide to Vegetable Oil Processing. AOCS Publication, Peoria, IL, USA, 2007.

Muller-Mulot, W., 1976. J. Am. Oil Chem. Soc. 53, 732.

Ooi, C.K., Choo, Y.M., Yap, S.C., Basiron, Y., Ong, A.S.H., 1994. Recovery of carotenoids from palm oil. J. Am. Oil Chem. Soc. 71, 423–426.

Slover, H.T., Lehmann, J., Valis, R.J., 1969. Vitamin in foods: determination of tocols and tocotrienols. J. Am. Oil Chem. Soc. 46, 417–420.

Chapter 3

Crude Oil Receiving, Storage, and Handling

Crude oil is the starting raw material for making refined, bleached, and deodorized oil produced at vegetable oil plants. Some smaller operations receive refined oils, which they process further to make finished products. Major segments of palm and palm olein oils are sold worldwide as "neutralized" oils. This means that the FFA in the oil is partially neutralized with caustic (NaOH). The specification for the neutralized oil allows a maximum FFA content of 0.25% for the oil and soap content of 8 ppm. The soap in the oil comes from the neutralization step.

Soybean oil is sold as crude or degummed oil. Canola oil is normally sold as super degummed oil. Cottonseed oil in the United States is mostly sold as PBSY (prime bleached summer yellow). A certain amount is also sold as crude nondegummed oil.

The author has come across some cases in the Third World countries, where the oil processor receive partially neutralized seed oils for further processing. The author recommends against this practice because the presence of some soap remaining in the oil can increase the FFA in the oil and also can cause darkening of the oil during further processing. Incidentally, FFA in neutralized palm or palm olein does go up, when the oil is shipped overseas and it is stored in large storage tanks at the terminals.

All crude oil is received by various means, such as tank trucks, rail cars, ISO tanks, flexi bags/bladders, or by barge as listed in the following sections.

3.1 CRUDE OIL RECEIVING

Mode of crude oil receipt	Quantity per receipt (metric tons)
By tank trucks	5–30 tons/load (11,000–66,000 lb)
By rail cars	30–82 tons/load (66,000–180,000 lb)
ISO tanks	Typically 30 tons
Flex-Tanks/Flex-Bags	Variable sizes
By barge or ship	Several hundred tons/load

The mode of receipt depends on the size of the refinery or whether or not railroad facility is available at the plant. Small batch refiners receive oil by

Practical Guide to Vegetable Oil Processing. http://dx.doi.org/10.1016/B978-1-63067-050-4.00003-9
Copyright © 2017 AOCS Press. Published by Elsevier Inc. All rights reserved.

trucks. Large refiners use rail cars for oil receiving. Large refiners also receive oil by barge or by ship. In case of receipts by barge or ship, the oil received is stored at sites near the port. These are commonly referred to as *terminals* and are typically managed by independent companies. The oil refiners rent or lease certain amount of storage space for their incoming crude oils to be stored for a given time. The terminals are responsible for maintaining the temperature and nitrogen in the tanks (where applicable) and in some cases check the FFA and PV of the oil in storage. The terminals are used only when large shipments of oil are received as stated previously. (See Fig. 3.1 for the representative shipping containers.)

3.1.1 Crude Oil Quality in Trade

In the United States of America the quality of the crude oil shipped and received are defined by the rules set by the National Institute of Oilseed Products (NIOP). The NIOP issues trading rules govern the following:

1. the quality standard for the incoming oil whether it is crude, refined, or refined and bleached oil,
2. the procedure for settling the difference between the shipper's analyses and that analyzed by the client at the time of receipt, and
3. discounts or allowances permitted to the recipients based on the oil analysis and the rules of settlements.

Another trade organization in USA called National Oil Processors Association (NOPA) is participated and managed by the oil processors who define various guidelines regarding the quality of oil and trading rules for soybean meals and soybean oil. Customers wanting oil quality other than those in the NIOP or NOPA guidelines need to negotiate directly with the oil vendor to receive the specific quality of the oil they need.

Some of the pertinent NIOP oil specifications are shown in Tables 3.1–3.3.

The trading rules outlined in the NIOP and NOPA guidelines are quite extensive and comprehensive. The reader is recommended to obtain the latest trading rules (NIOP-2013-2014; NOPA October 2013) from both organizations and get familiarized with them as needed.

There are some critical areas that the oil refiners need to pay attention to as the oil on its arrival. The specific areas for the truck and rail cars are different from those for oil received by barge.

3.2 FOSFA INTERNATIONAL (HEADQUARTER—LONDON, UK)

The Federation of Oils, Seeds and Fats Associations (FOSFA) is a professional international contract issuing and arbitral body concerned exclusively with the world trade in oilseeds, oils and fats with over 1000 members in 84 countries.

FIGURE 3.1 Typical shipping vessels and containers for vegetable oils.

Internationally, 85% of the global trade in oils and fats is traded under FOSFA contracts. These members include producers and processors, shippers and dealers, traders, brokers and agents, superintendents, analysts, ship owners, and others providing services to traders.

FOSFA has an extensive range of standard forms of contracts covering goods shipped either CIF, C&F or FOB, for soybeans, sunflower seeds, rapeseed, and

TABLE 3.1 NIOP Specifications on Some Selected Oils

Specification	Crude coconut oil	Crude palm oil	Neutralized palm oil (unbleached)	Crude palm olein	Crude palm stearin
NIOP rule no.	6.4	6.7	6.8	NIOP Guidelines	NIOP Guidelines
FFA[a] (%)	4 Max.	5.0[b] Max.	0.25[b] Max.	5.0[b] Max.	5.0[b] Max.
Lovibond color (Y/R) (AOCS Cc 13b-45)	100/15				
Moisture and impurities, M&I (%)	1 Max.	1 Max.	0.1 Max.	0.5 Max.	0.5 Max.
Iodine value	10 Max.		50–55	57 Min.	43–48
Melting point °F (°C)			91.4–102.2 (33–39)		
Softening point, °F (°C)				75.2 (24) Max.	111.2–122 (44–50)
Cloud point °F (°C)					50 (10) Max.

All analyses are at the time of shipment.
[a]*Expressed as lauric acid.*
[b]*Expressed as palmitic acid.*

TABLE 3.2 NIOP Specifications on Some Selected Refined Oils

Specification	Refined, bleached, and deodorized (RBD) palm olein	Refined, bleached, and deodorized (RBD) palm stearin	Refined, bleached, and deodorized (RBD) palm oil
NIOP rule no.	NIOP Guidelines	NIOP Guidelines	NIOP Guidelines
FFA (%)	0.1[a] Max.	0.1[a] Max.	0.1[a] Max.
Lovibond color (Y/R)	3.0 Red, max.	3.0 Red, max.	3 Red, max.
Moisture and impurities, M&I (%)	0.1 Max.	0.1 Max.	0.1 Max.
Iodine value	57 Min.	43–48	51–55
Slip point (°C)		44–50	91.4–102.2 (33–39)
Softening point (°C)	24 Max.		

All analyses are at the time of shipment.
[a]*Expressed as palmitic acid.*

TABLE 3.3 NIOP and NOPA Specifications on Some Selected Oils

Specification	Crude canola (LEAR) oil	Crude degummed LEAR oil	Crude super de-gummed LEAR oil	Degummed soybean	Fully refined soybean oil
NIOP rule no.	6.17	6.17	6.17	NOPA Rule 103	NOPA Rule 103
FFA[a] (%)	1 Max.	1 Max.	1 Max.	0.75 Max.	0.05 Max.
PV (mEq/kg)					2.0 Max.
Refined and bleached Lovibond color (Y/R)	15/1.5 Y/R max.	15/1.5 Y/R max.	15/1.5 Y/R max.		20/2.0 Y/R
Cold test (h) AOCS method Cc 11-53 (97)					5.5 Min.
Fat stability (h)					35 mEq/kg @8 h
Moisture and volatile impurities, M&I (%)	0.5 Max.	0.3 Max.	0.3 Max.	0.3 Max.	0.1 Max.
Phosphorus (ppm)		200 Max.	50 Max.	200 Max.	
Chlorophyll (ppm)	30 PPB max.	30 Max.	30 Max.		
Flashpoint °F (°C)	302 (150) Min.	302 (150) Min.	302 (150) Min.	121.1 Min.	
Erucic acid, % by wt.	2 Max.	2 Max.	2 Max.		
Sulfur (ppm)	10 Max.	10 Max.	8 Max.		
Unsaponifiable matter				1.5 Max.	
Marine oil				Absent	Absent
Preservative				Absent	Absent
Appearance at °F (°C)					Clear and brilliant at 77–85 (21–29)
Settlings					Absent
Taste/odor					Bland/free of foreign odor
FDA GRAS status					Yes

All analyses are at the time of shipment.
[a]*Expressed as oleic acid.*

others, vegetable and marine oils and fats, refined oils and fats, from all origins worldwide, for different methods of transportation and different terms of trade.

The Federation's contracts incorporate a dispute procedure involving arbitration by experienced individuals from within the trade.

3.3 MEMBERSHIP

Membership of the Federation is open to any individual, company, or organization actively involved in the trade in oilseeds, oils and fats, or in the supply of services related to these trades.

Being in membership provides direct benefits to your business as the "FOSFA" terms will be familiar to you and, more often than not, your counterparty and service provider.

The Federation has over 1000 members, which are categorized under the following types.

3.3.1 Trading Members

Companies, firms, sole traders, organizations, or others conducting business as principals to contracts in the commodities covered by FOSFA.

3.3.2 Broker Members (Full or Associate)

Companies, firms, sole traders, organizations, or others which do not act as principals to the Federation's contracts and only receive remuneration in the form of commission from either or both of the contracting principals concerned.

3.3.3 Nontrading Members (Full or Associate)

These are companies, organizations, or individuals providing various services.

3.3.4 Superintendent Members

Independent superintendents recognized by the Federation for the purposes of superintending under the terms of FOSFA contracts (often also referred to as surveyors).

3.3.5 Analyst Members (Full or Associate)

Independent laboratories and analysts, providing laboratory services.

3.3.6 Kindred Associations

National or international organizations affiliated to FOSFA commodity trades.

3.3.7 Benefits of Membership

The Federation offers many services to members, including:

- a voice in the way international trade is conducted,
- direct involvement in committee work, influencing trade practices,
- FOSFA standard form contracts and contract referred documents,
- arbitration facilities and electronic access to awards (publication service),
- superintendents and analysts schemes,
- FOSFA Handbook containing member contact details, and
- advisory services covering contractual, legal, and technical areas.

It also provides:

- trade training and professional development programs,
- a wide range of free or discounted publications and services,
- access to FOSFA.net (private members website),
- a quarterly newsletter of interest to all members, and
- information on FOSFA and other trade events, providing networking opportunities.

3.3.7.1 Truck or Rail Car Receipt

3.3.7.1.1 Checking the Truck or Rail Car

1. Check the outside of the truck or rail car for cleanliness.
2. Do not accept the shipment if the outside of the truck, or the hose is dirty.
3. If the pump provided on the truck is to be used for oil unloading, one must make sure that the pump and the fittings are clean.
4. The truck driver must carry a cleanliness certificate, which indicates the material carried by the vessel as the prior load, showing who cleaned the vessel, when, and how.
5. Make sure that the previous load carried was a safe material for the vessel to carry food products (NIOP provides the guidelines). Many countries have implemented regulations for the shippers (FOSFA provides a list of previous cargo for shipping vessels) to use the dedicated containers for transporting food material only.
6. Check the shipping document. This must clearly indicate the following items:
   a. type of oil,
   b. a certificate of analysis (COA),
   c. the seal numbers,
   d. neutral oil loss (or refining loss) data,
   e. flash point (for solvent extracted oil), and
   f. for shipping oil into the EC countries, the oil must be certified to contain <50 ppm of mineral oil.
7. Make sure the supplier sent the right oil according to the order. If it is different, have the purchasing department or the person responsible for the purchase resolve the issue with the supplier.

8. Check the seals. These must not be broken.
9. If the seals are broken or missing, do not unload the oil.
10. The numbers on the seals must match those reported in the shipping document.
11. Do not unload the oil if the numbers on the seals do not match those on the shipping paper.
12. Notify the purchasing department or the individual responsible for purchasing the oil and have him resolve the issue with the supplier before the oil is unloaded. Remember, the supplier must declare the oil to be safe for processing for human consumption.
13. Prepare to unload the oil only if the purchasing department received complete information regarding the broken or missing seals as well as non-matching seal numbers between the vessel and the shipping document and the receiving personnel are satisfied with the supplier's explanation that no tampering has taken place.
14. Record the seal numbers on the oil-receiving document.
15. Review the certificate of analysis and attach it to the receiving document for Quality Control.
16. Open the hatch of the truck if all of the aforementioned requirements are met. Look for the following:
    a. foreign matter in the truck,
    b. foreign or unusual odor in the oil, and
    c. unusual color of the oil.
17. Do not unload the oil if:
    a. there is any foreign matter in the oil,
    b. the oil has an unusual odor, and/or unusual color.
18. Identify the foreign matter in the truck. Unload the oil only if the material is found to be safe as food contact material.
19. In case of an unusual odor in the oil, do not unload the oil until the source of the odor is identified and is determined to be safe to process the oil for human consumption.

3.3.7.1.2 Checking Crude Oil Quality Prior to Unloading

Once the shipping container and the oil are found to be acceptable for unloading, collect oil sample from the top, middle, and bottom of the vessel using a zone sampler (AOCS method: C 1-47), mix the three samples in a clean container and analyze it for:

1. FFA
2. PV
3. para anisidine value
4. Lovibond color
5. appearance
6. odor

7. sediments
8. moisture and impurities
9. fatty acid composition
10. flash point for solvent extracted oils
11. lab-refining loss—chromatographic, or cup loss depending on the type of crude oil
12. presence of mineral oils if the oil is received in nondedicated container

The proper practice for US processor would be to check for hydrocarbons (mineral oils) if the oil is received by ship or barge that is not dedicated to only vegetable oils. This is because these vessels can also carry crude petroleum. If the vessel is not properly cleaned, the crude petroleum may contaminate the vegetable oil from time to time. The new regulation requires dedicated vessels for vegetable oil or food material transport only. The US FDA (Food and Drug Administration) allows a maximum level of 10 ppm of mineral oil in US Foods (21 CFR SEC.178.3570 lubricants with incidental food contact). The European Council (EC) requires that all oils imported by the EU countries must have a mineral oil content <50 ppm. Therefore, any country shipping oil to any EC country must make sure that the 50 ppm limit for mineral content is met.

3.3.7.2 Barge Receipt (NIOP, NOPA)

1. This is quite different from receiving oil by truck or rail car directly from the crusher.
2. The barge is sent with the shipper's analyses.
3. An independent third party, surveyor, checks the inbound receipt, samples the oil, analyzes the sample to determine its suitability for unloading, supervises oil unloading, and certifies the quantity of the oil received.
4. There are guidelines regarding the resolution of any difference between the shipper's analysis and the analysis of the receipt, governed by the Trading Rules. This is resolved with the help of the surveyor. In extreme cases a mediator (Arbitrator) gets involved in resolving the dispute between the buyer and the seller.
5. Once the oil is unloaded in the tank, it becomes the buyer's property and the buyer then bears the responsibility to maintain the quality of the oil. In most cases, the terminal-staff generally perform the job under contract and are not responsible for maintaining the oil quality, unless there is specific contract drawn between the managements of the terminal and the buyer.

3.4 CRUDE OIL UNLOADING (TRUCK OR RAIL CAR)

Prepare to transfer the crude oil into dedicated storage tanks after the vessel inspection and crude oil analysis are found to be satisfactory. Following steps are recommended:

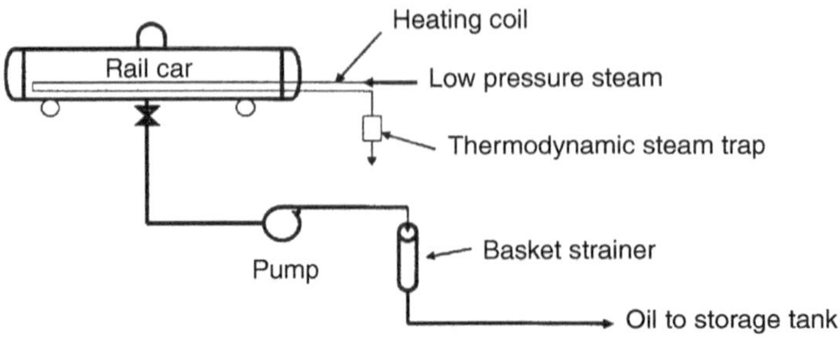

FIGURE 3.2 Schematic diagram for oil unloading.

1. Melt the oil in the truck or rail car if it is solidified (before sampling for analysis). A schematic diagram for oil unloading is shown in Fig. 3.2.
2. Apply heat carefully using low pressure steam <15 PSIG steam, (or 1 kg/cm^2) and using a thermodynamic steam trap to regulate the condensate temperature to 140°F (60°C) or lower. Absence of a thermodynamic trap, or the use of higher steam pressure, can scorch the oil causing oxidation and several other undesirable reactions in the oil as listed in the following:
   a. High oil temperature can cause oxidation of the unsaturated fatty acids in the crude oil. This reduces the flavor stability of the finished oil.
   b. At higher temperature, the neutral triglyceride can get hydrolyzed in presence of the phospholipids and the moisture present in the crude oil, causing a rise in FFA in the oil during storage.
   c. Some of these free fatty acids are unsaturated (linoleic and linolenic). These fatty acids can oxidize rapidly forming undesirable oxidation products in the oil. This may even have an impact on the shelf life of the final deodorized oil made from the crude oil.
   d. May cause flavor reversion in the deodorized soybean oil under this situation.
   e. Cottonseed oil or corn oil may become darker, requiring stronger caustic treatment at the refinery. This can cause additional refining loss and quality degradation in the deodorized oil.
   f. Carotenes get oxidized and the oil can develop fixed color. This is especially common for palm oil. This can produce high yellow color in plasticized shortening.
   g. It can darken the crude oil, requiring extra caustic refining and extra bleaching. This can reduce the stability of the finished oil.
3. The oil must pass through a bag filter or a stainless steel basket strainer at the discharge of the unloading pump with an opening of 150–200 μm in order to remove any foreign objects from the oil.
4. Clear the oil in the unloading line after each oil unloading. Do not use steam or air. Use nitrogen to blow the line.

5. In order to minimize air entrainment into the oil use either bottom loading line going into the oil storage tank or extend the oil discharge line to the floor of the tank.

3.4.1 Impact of Steam Blowing for Line Clearing

Use of steam blow can hydrate the phospholipids present in the crude oil. This can cause the following damages to the oil:

1. Can hydrate the phospholipids in the crude oil.
2. Hydrated phospholipids tend to settle to the bottom of the tank. This tends to increase the refining loss.
3. Can hydrolyze the oil, producing higher FFA in the crude oil in storage, which results into increased refining loss.
4. Hydrolysis of the triglyceride molecules produces diglycerides and sometimes monoglycerides in the crude oil in storage. This can cause higher refining loss and subsequent issues during processing and stability of the finished oil.
5. Can produce reverted flavor in deodorized soybean oil.

3.4.1.1 Avoid the Use of Air Blow or Improper Oil Discharge Into the Tank

Air blow causes aeration of the crude oil, which can cause the following reactions in the oil:

1. Increase PV in the oil.
2. The higher PV increases the secondary and tertiary oxidation reactions in the oil during subsequent processing and storage. This results in poor flavor stability in the finished oil as well as the product formulated with it.
3. It can also increase the probability for developing reverted flavor in soybean oil.

3.5 CRUDE OIL STORAGE

1. Crude oil storage tanks must have side entering mechanical agitators. Without agitation, the gum settles to the bottom of the tank causing hydrolysis of the neutral oil. This increased refining loss and increases the potential flavor reversion (in soybean oil) as discussed earlier. Fig. 3.3 shows the schematic diagram for crude oil storage.
2. The agitator must have a low level cut-off switch to avoid whipping of the oil by the impeller.
3. The oil must be stored at ambient temperature of 68–95°F (20–35°C) for all seed oils and palm olein.

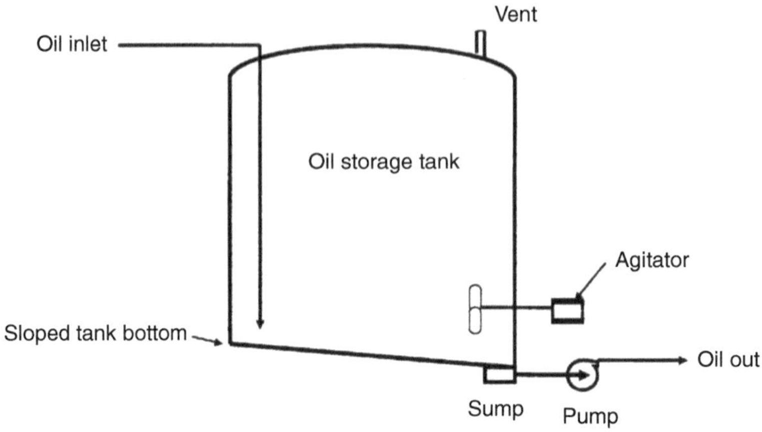

FIGURE 3.3 A typical crude oil storage tank.

4. Palm oil may be stored at 104–113°F (40–45°C).
5. Use hot water heating if heating is needed during storage.
6. Do not melt palm or cottonseed oil (if solidified) until it is time to refine it.
7. Use low-pressure steam to melt the oil to prevent overheating. It is advisable to use hot water at 176–179°F (80–82°C) for melting the oil.
8. Do not store the crude oil for long before refining it. The best results are obtained if the crude oil is refined within 3–4 weeks after it is received.
9. Clean the crude oil storage tank every 3–4 months to remove the sediment from the bottom of the tank.
10. Normally, physical cleaning by manual means can be adequate in most cases.
11. Caustic cleaning is needed when the deposits cannot be removed by manual means. Make sure the tank is washed thoroughly, neutralized with phosphoric or citric acid and the tank is washed once more after the final rinse with the acid neutralization.
12. Dry the tank immediately with dry hot air and coat the tank interior including the ceiling of the tank with fresh deodorized oil to prevent rust formation.

Table 3.4 lists the critical control points and their implications in crude oil receiving and storage and Table 3.5 lists the troubleshooting procedure for crude oil receiving and storage.

3.5.1 Special Notes on Oil Stored at Terminals

The clients of terminals must make sure that the oil temperature, oil FFA, and the PV is maintained properly to ensure the oil is usable for processing. The client needs to write a contract with the terminal's management and monitor the terminal's compliance to the terms of the contract.

TABLE 3.4 Critical Control Points in Crude Oil Receiving and Storage

Operation	Recommended limits	Implications/consequences
Vessel inspection	Cleanliness—must be clean Smell/odor—no foreign odor Foreign objects—none	The oil may not be suitable for human consumption
Blowing the unloading line with steam	Do not use steam	• Hydration of the phospholipids • Increased FFA in the crude oil during storage • Increased refining loss
Blowing the unloading line with air	Use nitrogen	• Increases oil oxidation • Can reduce flavor stability of the finished oil • May cause flavor reversion in soybean oil
Melting solidified oil for refining	Low pressure steam or hot water for heating the oil in the storage tank	Using high pressure steam can: • increase oil oxidation • increased hydrolysis • increased refining loss • produce poor flavor stability in the oil
Using mechanical agitation on the tank	Use a mechanical agitator with low level cut off	• Prevents settling of the phospholipids to the bottom of the tank • This prevents increased refining loss and also the other negative issues associated with the steam blowing during unloading as discussed before
Oil temperature	68–95°F (20–35°C) for seed oils and palm olein 104–113°F (40–45°C) for palm oil	A higher storage temperature increases: • oil oxidation • hydrolysis in the oil • the refining loss • color fixation
Storage time	3–4 weeks	A longer storage time can cause increased: • oil oxidation • hydrolysis in the oil • refining loss

TABLE 3.5 Troubleshooting Oil Receiving and Storage

Symptom	Probable cause/causes	Recommended solutions
FFA goes up rapidly in storage	• Steam blow was used during oil unloading • Incoming oil had high moisture • High storage temperature • Excessive sludge (gum) accumulated at the bottom of the tank • Leaks in the heating coil • The oil has been stored for long • Storage temperature is high	• Use nitrogen for blowing oil unloading line • Check moisture in the oil at receipt • Turn off heat, refine the oil as soon as possible if the FFA in the oil is rising • Use side mounted mechanical agitator designed for the tank to reduce sedimentation of the gum • Transfer the oil to another tank immediately and repair the leak • Do not store the crude oil for long. A maximum storage time of 3–4 weeks is ideal • Turn off heat, refine the oil as quickly as possible, transfer the oil to another tank and refine it if possible
Heavy layer of gum accumulated at the bottom of the tank	• Blowing the line with steam during unloading • No mechanical agitation in the tank • Agitator not working properly • Tank needs cleaning	• Stop using steam, use nitrogen • Install proper mechanical agitator on the tank with low level cut-off with automatic or manual restart • Check the amps on the agitator motor. One or more impeller blades might have fallen off if the amperage is low on the motor. • Clean the tank every 3–4 months
PV in the crude oil is rising rapidly	• Air entrainment into the oil occurred during unloading • High oil temperature in the tank	• Use proper piping design as suggested earlier also use nitrogen to clear the line after oil unloading • Follow suggestions listed above

Chapter 4

Degumming

4.1 INTRODUCTION

Phospholipids (also referred to as phosphatides) are essential components of vegetable cell structure. The phospholipids are commonly referred to as gum. These compounds, being oil soluble, get extracted from the seeds along with the crude oil and they remain in the crude oil.

In the oil processing industry, the phospholipids content in the oil is expressed in terms of parts per million of phosphorus. This is because there is a relationship between the phospholipids content and the corresponding phosphorus level in the oil.

Table 4.1 lists the phospholipids and phosphorus contents in common crude vegetable oils.

Crude oil is degummed to reduce the phosphorus content of the oil to a very low level because the phosphorus content of the oil has a profound influence in the flavor, color, and hydrolytic and oxidative stability of the refined, bleached, and deodorized oil, as well as it causes reduced refining yield. Several process-related issues are experienced when the oil contains high phosphorus in the finished oil.

Crude oil has to be degummed before storage or shipment for several reasons:

1. It deposits or settles to the bottom of the storage tank, which reduces refining yield if the crude oil is stored for long without mechanical agitation.
2. It deposits at the bottom of the shipping container, increasing the oil loss.
3. Increases the refining loss.

The oil after initial degumming must be further treated to reduce the phospholipids (or phosphorus) to a much lower level otherwise numerous process and oil quality issues are encountered, as follows:

1. The refined oil may have less than satisfactory quality and performance.
2. Increases bleaching clay consumption.
3. Deactivates bleaching clay, hydrogenation catalyst, interesterification catalyst, and enzymes for degumming and interesterification.
4. Creates impurities in the hydrogenation reaction.

Practical Guide to Vegetable Oil Processing. http://dx.doi.org/10.1016/B978-1-63067-050-4.00004-0
Copyright © 2017 AOCS Press. Published by Elsevier Inc. All rights reserved.

TABLE 4.1 Typical Phospholipids and Phosphorus Contents of Most Common Vegetable Oils

Oil type	Phospholipids content (%)	Phosphorus content (ppm)
Crude soybean oil	1–3	400–1200
Degummed soybean	0.32–0.64	79–158
Crude corn oil	0.7–2.0	250–800
Crude cottonseed oil	1.0–2.5	400–1000
Crude peanut (groundnut) oil	0.3–0.7	100–300
Crude canola oil	0.5–3.5	200–1400
Superdegummed canola oil	0.13–0.16	52–64
Crude sunflower oil	0.5–1.3	200–500
Crude safflower oil	0.4–0.6	160–240
Crude palm oil	0.03–0.1	12–40
Crude coconut oil	0.02–0.05	8–20

5. Prematurely plugs the filters in bleaching, hydrogenation catalyst removal, and filter to remove waxes in winterization.
6. Flavor regression in soybean oil.

4.2 PURPOSE OF DEGUMMING

Crude oil is degummed for various reasons. Soybean oil is water degummed to recover the phospholipids to manufacture lecithin. Canola oil is degummed as a general practice because the oil is sold as degummed or superdegummed crude oil. The gum recovered from canola oil is dark and does not have the same commercial value as the soy lecithin. It is used primarily in the animal feed. Sunflower oil is water degummed to make sunflower oil lecithin, which is less versatile in use compared to the soy lecithin. Corn oil, made from wet-milling process, is degummed to deliver higher quality crude oil that is easy to refine.

Discharge of plant effluents from the seed oil degumming process needed to be curbed when strict environmental regulations were imposed on in many countries. This forced the oil processors to look into refining methods other than chemical refining to reduce plant effluents that increased the biological oxygen demand (BOD) and chemical oxygen demand (COD) loadings in the plant discharge. Besides water degumming, various acid degumming processes evolved for treating crude oils that are either more difficult to refine or the phosphorus content in the oil had to be reduced to a very low level so physical refining process (described in Chapter 5) could be applied to these oils. Some experts claim that crude soybean can be refined through the physical refining process if the total phosphorus content of the oil is reduced to 15 ppm or less. The author believes that to physically refine soybean oil the phosphorus content in the crude oil must be <5 ppm. There are several other factors that are critical for obtaining stable soybean oil flavor besides low phosphorus content prior to physical refining. These are discussed in Chapter 11.

4.3 HYDRATABLE PHOSPHOLIPIDS AND NONHYDRATABLE PHOSPHOLIPIDS

Phospholipids in the crude oil can be broadly classified into two categories based on their removability with water and are called (1) hydratable phopholipids (HPs) and (2) nonhydratable phospholipids (NHPs). The HPs can be separated from the crude oil by treating the crude oil with deionized water. The water hydrates the HPs and forms a heavy phase, which is then separated from the oil with the help of a centrifuge. The NHPs require further treatment with acids, like phosphoric acid, citric acid, maleic acid, citric anhydride, etc. The acid reacts with the metal complex of the NHPs and makes them water soluble, so they can be removed from the crude oil using a centrifuge. Normally, the gums removed by the acid process cannot be used for making lecithin because of its very dark color. It is usually blended into the animal feed or can be used for industrial applications other than food manufacturing.

The integrated crusher refiners in the United States found that it is advantageous to degum the fresh soybean and corn oils immediately after extraction. Degummed oil produces lower amount of soap in the refining process. This reduces the load on the soap acidulation plant. This reduces the biological oxygen demand and chemical oxygen demand in the plant effluent as compared to that from the plant processing nondegummed crude oils. Cottonseed oil is mostly processed through micelle-refining process. This generates less soap in the oil refinery.

4.4 METHODS FOR DEGUMMING

Various degumming processes are used in the vegetable oil industry. They are:

1. Water degumming
2. Acid conditioning
3. Acid degumming
4. Deep degumming
   a. Superdegumming
   b. Combined superdegumming and neutralization
   c. TOP degumming (total degumming)
   d. Organic refining process (ORP)
   e. Soft degumming
5. Enzymatic degumming

4.4.1 Water Degumming

HPs absorb water and settle out from the rest of the oil. This property of the HPs is utilized in the water degumming process.

Water degumming removes most of the HPs from the crude oil. A very small amount of NHP lipid is also removed as they get entrained in the HP fraction in this process. The hydratable gum from the crude oil is easily separated from the crude oil when it is fresh and especially when made from good quality soybeans using proper seed crushing and extraction process. Crude oil derived from poor quality soybeans contains higher proportion of NHPs. The oil in this case needs acid degumming or acid pretreatment for better reduction of phospholipids before the crude oil can be refined.

Fig. 4.1 shows the Alfa Laval process for water degumming. Following steps are involved in the process:

1. Determine percent phospholipids in the crude oil by checking parts per million phosphorus.
2. Crude oil is heated to 140–150°F (60–65°C).
3. Deionized water, equal to percent phospholipids is added to the oil through an in-line high shear mixer.
4. The oil and water are gently mixed in a hydration tank for 30–40 min. The hydrated gum separates and agglomerates.
5. The oil is then gently pumped out of the tank and separated in a centrifuge.
6. The heavy phase contains most of the HPs. A very small amount of NHPs is also removed from the oil as entrainment.
7. The temperature of the oil leaving the centrifuge is 140–150°F (60–65°C).
8. Water-degummed oil contains 0.4%–0.8% moisture. The oil is heated to 185–190°F (85–90°C) in a heater and then pumped into a vacuum dryer, maintained at an absolute pressure of 50 mm of mercury (maximum). Heating of the oil is necessary because the temperature of the oil leaving the centrifuge is too low for vacuum drying at this level of vacuum.

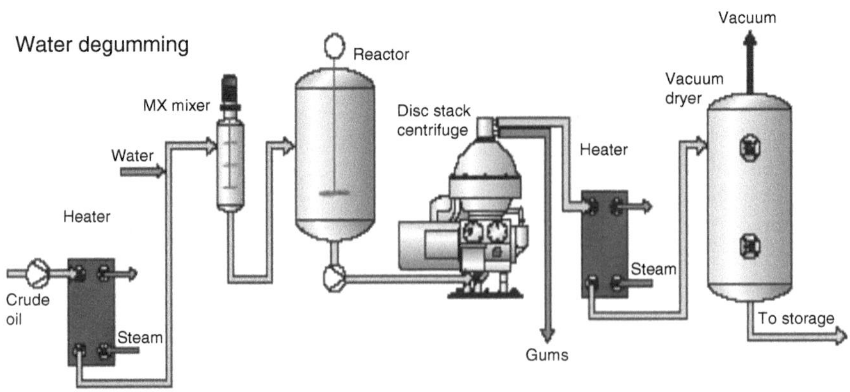

FIGURE 4.1 Schematic diagram for Alfa Laval water degumming process. *(Courtesy of Alfa Laval)*

9. Vacuum-dried oil should have moisture content of 0.05% or less.
10. The oil is cooled to <130°F (55°C) before storage.
11. Hydrated gum is dried in a wiped-film vacuum dryer if the gum is to be used to make lecithin. The vacuum dryer is equipped with specially designed mist eliminator to prevent any product loss due to excessive foaming in the vessel.
12. The vacuum system must be designed with noncontacting condensers to prevent any mixing of the phospholipids into the condenser water and forming an emulsion in the water tower basin.

4.4.1.1 Critical Control Points in Water Degumming

The critical control points in water degumming are:

- oil temperature
- amount of deionized water
- residence time in the hydration tank
- agitation in the hydration tank
- vacuum drying of oil
- vacuum drying of the gums

4.4.1.2 Oil Temperature

Temperature of oil at hydration is critical. At temperature <130°F (55°C) the degree of hydration is better but the viscosity of the oil is higher which makes the separation of oil and gum difficult. Less hydration of the phospholipids occurs at temperature above 149°F (65°C). This reduces the efficiency of gum removal and higher phosphorus is found in the degummed oil.

4.4.1.3 Amount of Deionized Water

The amount of water is normally equal to that of the total phospholipids content of the crude oil in percent. At lower water addition, the hydration of the phospholipids is incomplete, causing a reduction in the removal of the phospholipids. At higher water addition, the difference of density between the oil and the gum is reduced, causing poor separation in the centrifuge. This may leave more phospholipids in the oil and also increased oil loss in the gum phase.

4.4.1.4 Residence Time (Contact Time) in the Hydration Tank

A certain minimum amount of contact time is necessary for the hydration of the phospholipids in the crude oil. Lower than 30 min of contact time may not allow sufficient hydration of the HPs. This reduces the efficiency of separation of the oil and the gum. A longer than 40 min contact time is not harmful but is not required. Besides, a longer contact time requires a larger hydration tank that makes the process more costly.

4.4.1.5 Agitation in the Hydration Tank

The agitation in the tank must be gentle to prevent any emulsion formation.

Any amount of emulsion in the hydration tank will reduce separation efficiency of the process.

4.4.1.6 Vacuum Drying of the Oil

The oil must be dried if it is to be stored. The moisture in the vacuum-dried oil should be <0.05% and not >0.1%. The oil does not have to be dried if it is sent immediately for bleaching before chemical refining or for physical refining.

Storing wet oil can cause the following issues:

- The gum in the oil can hydrate and cause hydrolysis of the neutral oil. This will increase the FFA in the crude oil in storage.
- The hydration of the gum can increase the refining loss.

4.4.1.7 Vacuum Drying of the Gum

The gum goes through several steps before it becomes lecithin. It is dried to <0.1% moisture. Crude lecithin contains 65%–70% oil. Higher moisture in the lecithin makes the lecithin very viscous and unacceptable in most applications. In addition, high moisture can produce higher FFA in the lecithin during storage.

4.4.1.8 Target Water-Degummed Oil Quality

Phosphorus content	50–200 ppm (max)
Moisture	<0.1% (after vac. drying)

4.4.1.9 Target Quality of Dried Lecithin

Acetone insoluble (AI)	70%; (minimum)—65%

4.4.2 Acid Conditioning

This method is generally used to treat high quality palm oil and coconut oil. This is quite a very simple method but it has its limitations. The oil is treated with 400–1000 ppm of phosphoric acid using a high shear mixer. The oil is then directly sent to bleaching for physical refining. The steps of procedure for acid conditioning are listed as follows:

1. Good quality palm oil or coconut oil is heated to 130–140°F (55–65°C).
2. 400–1000 ppm of phosphoric acid is added to the crude oil and the two liquids are mixed in a high shear mixer. In some cases citric acid works better and is considered to be the preferable over phosphoric acid. For example, citric acid or maleic acid is preferred for treating canola oil.
3. The oil/acid mixture enters a retention tank called conditioning tank. This tank is provided with gentle agitation.

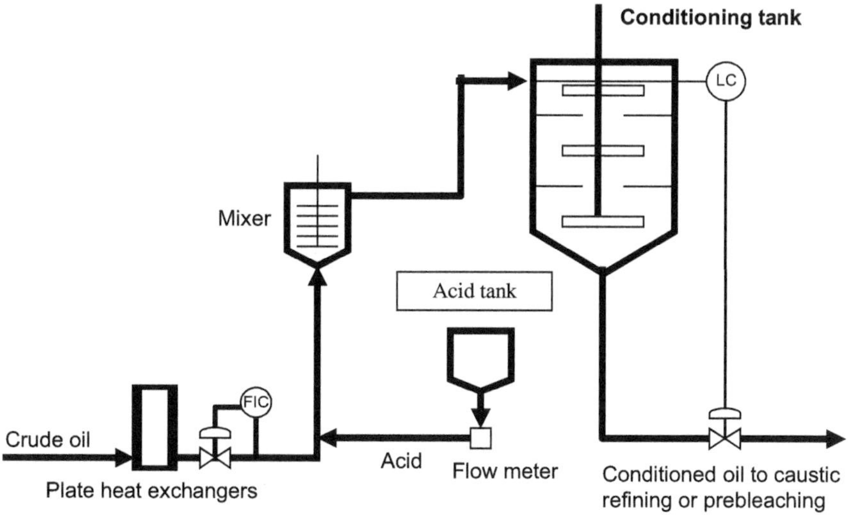

FIGURE 4.2 Schematic flow diagram for Alfa Laval acid-conditioning process.

4. The oil stays in the retention tank for 15–30 min. This hydrates the phospholipids and the hydrated phospholipids separate out of the oil because the metal salts of phosphoric acid are insoluble in oil.
5. The hydrated phospholipids are not separated from the crude oil before it goes to bleaching or caustic neutralization.

The precipitated salts and the hydrated gums have a tendency to plug up the filter screens. This is because the metal salts of citric acid or other acids are insoluble in the oil and precipitate out. This tends to plug up the filters in bleaching. It is always recommended to precoat the filter screens with diatomaceous earth. This procedure will be described under prebleaching (Chapter 6).

Many standalone refiners tend to use this technique to treat the crude oil to hydrate the NHPs in the crude oil before it is refined.

Many standalone refiners use this technique to pretreat the crude oil from poor quality or hard-to-refine seed oils. This is inexpensive but not as efficient as acid degumming process.

Fig. 4.2 shows the schematic flow diagram for the acid-conditioning process.

4.4.2.1 Critical Control Points in Acid Conditioning

Following are the critical control points in the acid-conditioning process:

1. oil temperature
2. amount of acid added
3. mixing of oil and acid
4. holding time after acid treatment (in the conditioning tank)
5. agitation in the conditioning tank

4.4.2.2 *Oil Temperature*

- At temperatures <55°C (130°F) the reaction between the acid and phospholipids is slow.
- At temperatures >60°C (140°F), some of the neutral crude oil may get hydrolyzed by the acid, causing an increased FFA level in the bleached oil. This will increase neutral oil loss in the distillate in the deodorizer.

4.4.2.3 *Amount of Acid*

As stated earlier, most refiners use 400–1000 ppm of phosphoric acid based on the crude oil flow. This may or may not be the right amount depending on the NHPs in the crude oil. At low acid dosage, the NHPs may not be adequately hydrated. With high acid dosage there may be some hydrolysis of the neutral oil, resulting in higher refining loss.

A better way to determine the required amount of phosphoric acid is outlined next:

For crude soybean oil:

1. Analyze the parts per million of calcium in the crude oil.
2. Estimate the amount of acid needed for treatment using the following example.

Crude soybean oil with calcium content of 100 ppm

$$\begin{aligned}\text{Atomic weight of calcium} &= 40\\ \text{Molecular weight of phosphoric acid} &= 98\\ \text{Theoretical ratio of phosphoric acid/calcium} &= 2\text{ Moles of phosphoric}\\ &\quad\ \text{acid/3 atoms of calcium}\end{aligned}$$

Theoretically, 120 parts of calcium will react with 196 parts of phosphoric acid.

Therefore, the amount of phosphoric acid needed if the crude oil that contains 100 ppm of calcium:

$$\{(98\times 2)/40\times 3\}\times 100 = 163\,\text{ppm}$$

In actual practice, the amount of phosphoric acid needed for this crude oil could be as much as 5–10 times of the calcium content in the crude oil. This is because:

1. In addition to calcium the acid reacts with magnesium and iron present in the crude oil.
2. There is also the lack of intimate mixing between the oil and the acid even with a high shear mixer.

The exact amount of acid must be estimated by the actual analysis of the oil for parts per million calcium in the crude oil. This amount will vary from lot to lot of the crude oil and also between different types of crude oil.

4.4.2.4 *Mixing of Oil and Acid*

Oil and acid must be intimately mixed. Phosphoric acid and oil are not miscible liquids. Therefore, a high shear mixer is needed to disperse the acid into the oil in microsize droplets to maximize the interfacial area between the oil and the acid. The importance of mixing is listed as follows:

- Hydration of the NHPs is incomplete if the dispersion of the acid in the oil is poor, resulting in high phosphorus in the refined oil.
- Inadequate dispersion of the acid in crude oil may cause the acid to settle out causing eventual corrosion in the tank, piping, etc., even if they are made of 304 stainless steel.

4.4.2.5 *Conditioning Time (Retention Time)*

- A minimum retention time of 15 min is needed for the reaction between the acid and the NHPs in the crude oil when a high shear mixer is used.
- A retention time longer than 30 min does not improve the hydration process.
- A very long retention time may allow some of the acid and the separated hydrated material to settle to the bottom of the tank making it difficult to filter the bleached oil.

4.4.2.6 *Agitation in the Conditioning Tank*

The agitation of the acid/oil mixture must be gentle to prevent emulsification in the oil.

4.4.3 Acid Degumming

Acid degumming provides more complete removal of phospholipids from the crude oil. This method is especially used to treat seed oils like corn and cottonseed oils and oils that are difficult to refine because of damaged seeds or because the crude oil is old or has been heat abused or oxidized. Fig. 4.3 shows

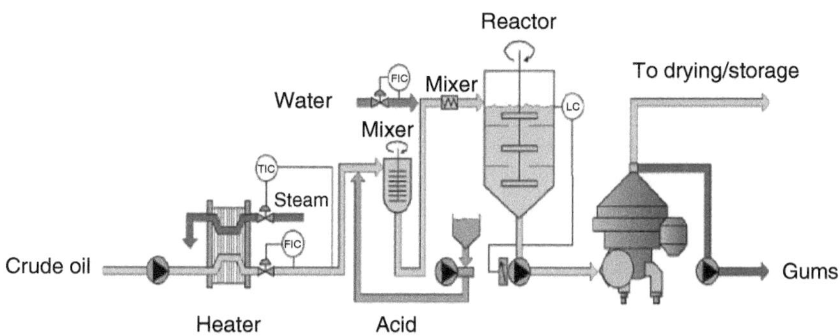

FIGURE 4.3 Schematic flow diagram for Alfa Laval acid degumming process. *(Courtesy of Alfa Laval)*

the schematic flow diagram for the acid degumming process by Alfa Laval. Degumming of oil works best in this process when water-degummed oil is used as the feedstock instead of crude oil.

This is an improvement over the acid-conditioning process. Following steps of procedure are involved in this process:

1. Deionized water (approximately 2% of the oil flow) is added to the acid/oil mixture leaving the high shear mixer.
2. The composite liquid stream passes through a static mixer before it enters the hydration tank.
3. The hydration tank is very similar to that in the water degumming process in terms of construction.
4. The oil/acid/water mixture is gently agitated in the hydration tank for 20–30 min.
5. The oil is then gently pumped out of the hydration tank and centrifuged.
6. The heavy phase (acid-treated gums) and the light phase (treated oil) are separated.
7. For high quality palm oil or coconut oil, the water can be added just before the acid/oil mixture enters the centrifuge.

One can see that the acid degumming process contains several aspects of water degumming and acid-conditioning processes.

4.4.3.1 Critical Control Points

The critical control points for this process are the same as described under water degumming and acid-conditioning processes.

4.4.3.2 Target Acid Degummed Oil Quality

Phosphorus content	20–50 ppm (max.)
Moisture	<0.1% (after vac. drying)

4.4.4 Deep Degumming

Several degumming processes that are capable of reducing the phosphorus content of the degummed oil to 5–10 ppm are called deep degumming process. These are:

1. Superdegumming (Unilever)
2. Superdegumming (Alfa Laval)
3. Combined superdegumming and neutralization (Alfa Laval)
4. TOP degumming
5. Organic refining process
6. Soft degumming

In the deep degumming process, acids like phosphoric acid or citric acid is used to make the acid react with the metal (calcium, magnesium, iron) of the

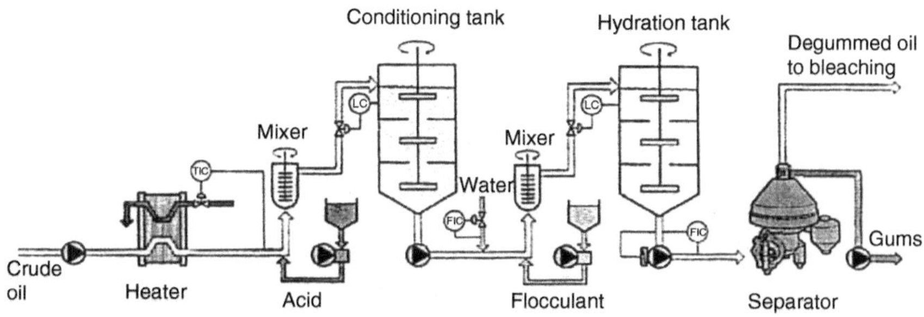

FIGURE 4.4 **Flow schematic diagram for superdegumming process (Unilever).**

NHPs. Chemically, this process is also referred to as chelation of the metals. These complex compounds between the acid and the NHP then become hydratable and are removed from the rest of the oil.

4.4.4.1 Superdegumming (Unilever Process)

This process was developed by Unilever to process soybean crude oil for physical refining and also treating other seed oils that are hard to refine. This is also known as the "special degumming" process. This process works better on crude oil as compared to water-degummed oil. It is believed that the HPs help the separation of the NHPs by agglomerating them, making it easy to remove them in a centrifuge. Fig. 4.4 shows the schematic flow diagram for the superdegumming process.

The process is carried out through the following steps:

1. The oil is heated to 35–40°C (95–104°F) and mixed with 0.1% citric acid in a knife mixer.
2. The acid-treated oil is held in a conditioning tank for 30 min with gentle agitation.
3. The conditioned oil is pumped out of the conditioning tank and mixed with 2% deionized water in a second knife mixer. So far the process looks very similar to a combination of acid conditioning and water degumming.
4. Along with the water a small amount of caustic is added (as flocculant) before the mixture enters the second knife mixer.
5. The mixture is allowed at least 30 min of retention time in the hydration tank. The flocculation step is a distinct step that differentiates it from the other degumming methods.
6. The oil from the hydration tank is gently pumped out and sent to a self-cleaning type centrifuge to separate the oil from the gum. It is important to use a self-cleaning centrifuge for this process. With a standard type of centrifuge, the bowl will get dirty too soon requiring frequent shut downs for cleaning of the bowl.

This process separates more gums compared to acid degumming. The gum separated is also very sticky. Therefore, in addition to the use of a self-cleaning centrifuge, it requires the hydrated oil flow to be about 60% of what is used in acid degumming. Thus, to obtain the same oil flow a larger centrifuge and the corresponding pumps, etc. are needed.

Even with the special care mentioned earlier, the crude oil loss in this process is higher than that in the acid degumming process. This is because of lower operating temperature, which increases the viscosity of the fluid and the separation of the two phases is more difficult.

The gum produced by this method contains higher percentage of neutral oil compared to acid degumming or water degumming process. This produces lower AI level in the gum.

Like in acid degumming process, the gum produced by the superdegumming process is not suitable for human consumption. Therefore, its outlet is limited to animal feed.

This process is used to make superdegummed canola oil. Some processors claim that maleic acid produces better results in canola oil degumming than citric or phosphoric acid.

4.4.4.1.1 Critical Control Points in Superdegumming Process

The critical control points for the superdegumming process are:

1. Oil temperature.
2. Retention time in the conditioning tank.
3. Retention time in the hydration tank.
4. Addition of flocculant.
5. Self-cleaning centrifuge.
6. Oil flow rate (reduced).

Items # 1–3 are similar to those in the acid-conditioning process.

4.4.4.1.2 Addition of Flocculant

A small amount of caustic is added to form soap with the FFA in the crude oil. The soap reduces the interfacial tension between the oil and the aqueous phase of the acid solution. This improves the reaction between the acid and the NHPs. Thus, a little amount of soap acts as an aid to hydrate the phospholipids and precipitate them. Addition of large amount of caustic would produce more soap causing emulsion formation. This will reduce the efficiency of the degumming process.

4.4.4.1.3 Self-Cleaning Centrifuge

A self-cleaning centrifuge maintains the bowl clean of the sticky gum and provides longer operation cycle improving productivity.

4.4.4.1.4 Oil Flow Rate

At increased flow rate the centrifuge bowl gets clogged too soon. This makes separation of the heavy and the light phases less effective. This requires more

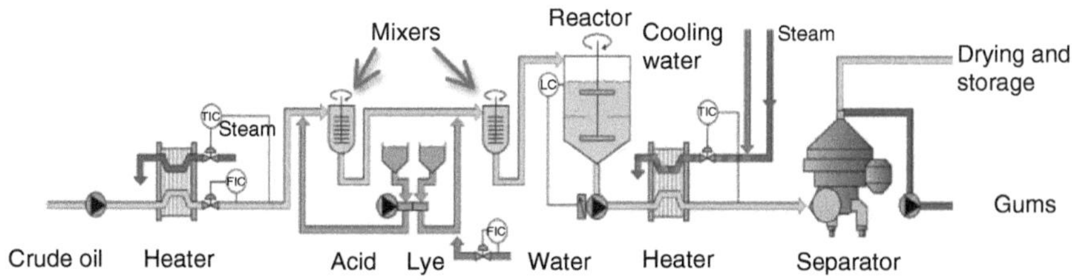

FIGURE 4.5 Schematic diagram for Alfa Laval superdegumming process. *(Courtesy of Alfa Laval)*

frequent shut downs to clean the centrifuge bowl and, thus, reduces productivity of the process.

4.4.4.1.5 Target Oil Analysis

Phosphorus in degummed oil (without water washing)	20–30 ppm
Phosphorus in degummed oil (with water washing)	<10 ppm

4.4.4.2 Superdegumming (Alfa Laval Process)

This process is similar to the Unilever process except that it does not use an acid-conditioning tank and a hydration tank. The schematic flow diagram is shown in Fig. 4.5.

4.4.4.2.1 Process Description

In this process:

1. Water-degummed oil is heated to 140°F (60°C).
2. Approximately 0.05–0.2 phosphoric acid is mixed into the oil stream with a high shear mixer.
3. A very small amount of caustic is added to the oil stream along with roughly 2%–2.5% deionized water to react with the FFA in the oil through a high shear mixer. The little amount of soap formed in this process helps hydration of the phospholipids.
4. The mixture is placed in a reaction tank for 60 min with gentle agitation to aid hydration of the phospholipids.
5. The hydrated gums are separated from the oil in a centrifuge.
6. The gums are sent to storage for further processing.
7. The degummed oil can be mixed with deionized water, heated to 185°F (85°C) sent through a separator to remove the water and then the oil is sent to vacuum drying and storage.

4.4.4.2.2 Target Oil Analysis

Phosphorus in degummed oil (without water washing)	20–30 ppm
Phosphorus in degummed oil (with water washing)	<10 ppm

The gums produced in both of these processes are unsuitable for food application and have lower AI value.

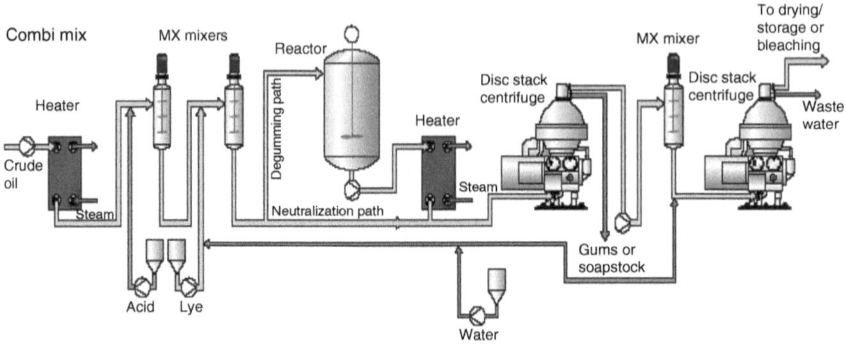

FIGURE 4.6 Alfa Laval combined special degumming and neutralization (Combi Process).

4.4.4.2.3 Critical Control Points

The critical control points in both superdegumming processes are very similar.

1. This is basically a parallel system to provide flexibility to plants that do not strictly process palm or coconut oil (or other lauric oils) but also seed oils that may be of poor quality.
2. The centrifuge placed in parallel can be used for water washing neutralized oil or can be used for degumming oils as needed. The system offers the flexibility of both physical and chemical refining along with degumming.

4.4.4.3 Combined Special Degumming and Neutralization

Fig. 4.6 shows the schematic flow diagram for combined superdegumming and neutralization. The system offers the flexibility of both:

1. Straight neutralization of high quality crude oil.
2. Use of a combination of superdegumming and neutralization capability for poor quality or hard-to-refine crude oil.

4.4.4.4 TOP Degumming

TOP is an acronym for the Dutch name for the process "Totaal Ontslijmings Process." This process is sometimes referred to as total degumming process. This process was developed and patented by Dijkstra and Van Opstal for water-degummed oil.

An acid (0.4–2.0 wt.%) was added to previously water-degummed oil and was dispersed with a high shear mixer, producing 100 million droplets of aqueous acid per gram of oil. It was claimed that the acid is added to decompose the NHPs present in the oil.

Following steps were used in this process:

1. Water-degummed oil was used as the feed material.
2. Approximately 0.4%–2.0% of phosphoric acid (or citric acid) was added.

3. The acid was dispersed in the oil in extremely fine droplets using a high shear mixer. This brought the acid and the metal in the NHPs in intimate contact for chelation to render the NHP hydratable.
4. The contact time between the acid and the oil was approximately 5 min.
5. A very small amount of base was added to bring the pH to 2.5 with the total amount of water not exceeding 5%.
6. The gums were then separated in a centrifuge.
7. The phosphorus level in the degummed oil was not low enough for physical refining.
8. Subsequently, an acid retention tank was added to the process before the alkali addition step.
9. A caustic treatment and a water-washing step were added to the process to treat hard-to-refine oils and obtain low phosphorus in the refined oil. This required a total of four centrifuges and an acid retention tank to achieve the reduction of phosphorus in the oil.

4.4.4.5 Organic Refining Process

Ag Processing, USA developed the ORP. Following steps are involved in this process.

1. Crude or water-degummed oil is heated to 197–208°F (92–98°C).
2. Citric acid (1%–5% solution) is added to the oil with the dosage of 3%–20% by weight.
3. The mixture passes through a knife mixer. The residence time in the knife mixer is approximately 16 min.
4. The gums are separated in a centrifuge.
5. The citric acid is recovered and reused.
6. The phosphorus level in the refined oil is <10 ppm.

This process generates high level of lysophospholipids compared to all other acid degumming processes.

Water-degummed oil was further degummed by this patented method by Unilever. Following steps are involved in this process:

1. The acid (50% acid in deionized water) was added to the oil at temperature of 149–194°F (65–90°C) mixed in a high shear mixer.
2. 2.5% deionized water is added and mixed into the acid treated oil in a static mixer.
3. The mixture is cooled to 104°F (40°C).
4. The cooled mixture is allowed to remain in a vessel for 30–60 min with gentle agitation for the phospholipids to form semicrystalline structure. At this temperature the semicrystals of phospholipids are insoluble in the oil.
5. The cold mixture can be centrifuged to separate the oil from the semicrystalline phospholipids at 104°F (40°C) or the mixture can be heated rapidly to 158°F (70°C) and then centrifuged.

6. At 104°F (40°C), the rate of flow of the liquid through the centrifuge will have to be significantly reduced or use a larger centrifuge.
7. The oil from the first centrifuge can be mixed with deionized water, mixed through an inline mixer and centrifuged in another centrifuge to get further reduction in phosphorus content in the oil.

Target phosphorus content of the oil from the previously mentioned process:

- Oil from the first centrifuge <30 ppm
- Oil from the water wash centrifuge <10 ppm

The entire process requires a total of two or three centrifuges:

- Two centrifuges without water washing
- Three centrifuges with water washing

4.4.4.6 Soft Degumming

In this process, the water-degummed or nondegummed oil is treated with a water-soluble chelating agent, ethylenediaminetetraacetic acid (EDTA). The chemical structure in two dimensions is shown in Fig. 4.7.

This method was promoted by Fractionnement Tirtiaux, B-6220 Fleurus, Belgium and applied at several oil refiners between 2000 and 2010.

The compound contains two amino groups and four acid moieties. These reaction sites provide very unique capability for the EDTA to form complexes with the metal ions with the Ca^{+2}, Mg^{+2}, and $Fe^{+2,\,+3}$ present in the NHPs. The chelated metal complexes are then extracted from the lipid phase into the aqueous phase and then removed by centrifugation to get the best results.

Soft Degumming process works on the following principles when water-degummed or nondegummed oil is treated with an aqueous solution of EDTA or its salts:

1. EDTA forms complexes with the metals present either in the calcium, magnesium, or iron salt of the phosphatidic acid or those of phosphatidylethanolamine.
2. These EDTA/metal complexes are very stable compounds and are water-soluble. Therefore, a very effective high shear mixer is needed to bring the

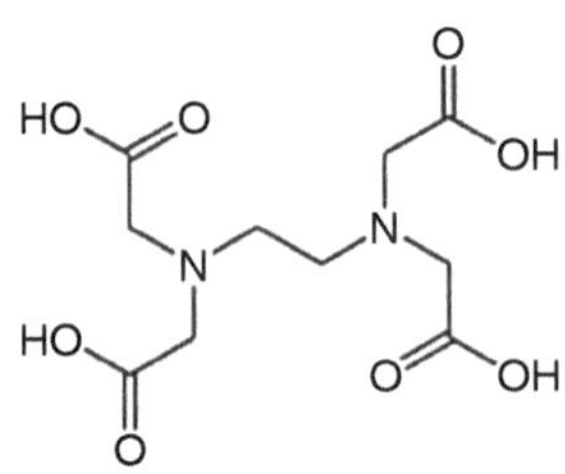

FIGURE 4.7 EDTA structure in two dimensions.

water and the oil phase to be in intimate contact to solubilize the metal complexes of the EDTA in a high-speed homogenizer.

3. An emulsifier, sodium dodecyl sulfate (SDS), is used to improve the solubility of the aqueous and the oil phases for the extraction of the metal complexes into the aqueous phase.
4. Sometimes HPs are used as emulsifier in this process.

4.4.4.6.1 Soft Degumming Process

Following steps are normally carried out in this process:

1. Crude oil is water degummed and water-degummed oil is used as feed to the process.
2. The oil is heated to 140°F (65°C).
3. Five weight percent of heated water at 140°F (65°C) or slightly higher, is added with EDTA dissolved in it. Concentration of EDTA is typically 100–150 mM.
4. Emulsifier, SDS is also added at a concentration of 50 mM.
5. The mixture is homogenized at the temperature in a high speed homogenizer for 20 min.
6. The degummed oil is then separated from the aqueous phase in a centrifuge.

4.4.4.6.2 Critical Control Points

The critical control points in this process are:

1. Oil/water mix temperature
   The mix must be maintained at a temperature 140°F (65°C) or higher. At lower temperature the chelation reaction is less complete resulting in higher phosphorus content in the degummed oil.
2. Intimate dispersion of the EDTA solution into the oil phase
   It is necessary to bring the EDTA and the metal ions in the NHPs into intimate contact for the metals to react with the EDTA and form a complex, which is hydrated and removed into the aqueous phase.
3. EDTA concentration
   Concentration of EDTA has to be high and most cases it is 100 mM or higher for effective reduction of phospholipids.
4. SDS concentration
   The emulsifier helps stabilize the emulsion phase created in Step 3. Like the EDTA, the emulsifier also requires a threshold level to get effective reduction of phospholipids. Typically the emulsifier level required is 50 mM or higher.
5. Contact time (hydration time) in the homogenization step
   An optimum contact time is required. It is normally 20 min. Below 20 min phosphorus content of the oil remains higher. Around 20 min it reaches the lowest value. Above 20 min the phosphorus level in the degummed oil

increases. It is believed that the phospholipids are redispersed into the emulsion and it becomes difficult to separate it from the oil.

6. Water/oil ratio in the emulsification step
   A higher proportion of water helps phospholipid reduction. Water to oil ratio of 70/100 the phospholipid reduction improves and it is better at the water/oil ratio of 1:1. This is because the chelated metal complexes of the EDTA are easier to hydrate at higher water level in the mixture.
   The soft degumming process is capable of producing low phosphorus (<5 ppm) but a high level of EDTA has to be added, especially for water-degummed oils with high phosphorus content near 200 ppm. High cost of EDTA and its use at fairly high level made the process costly and its application could not expand as expected.

4.4.5 Enzymatic Degumming

Enzymatic degumming process has gained popularity because it can provide higher oil yield. The increased yield comes from:

1. Increased yield of diglycerides (DAG).
2. Reduced neutral oil loss in the gums.
3. Reduced oil loss in the refining process.
4. Oil content in the meal is reduced (for feed). This increases the protein value of the meal that brings higher revenue.

Specific phospholipase enzymes are selected to simply cleave the bonds between the glycerol backbone and those of the FA and the phosphate ester as shown in Fig. 4.8.

Fig. 4.9 represents the concept showing the enzyme degumming process.

Lecitase PLA_1 is selective to all phospholipids. The process produces high level of FFA by cleaving the FA in the position 1 of the glycerol backbone of the phospholipids and releases lysophospholipid. The FFA remains in the oil phase while the latter is discharged along with the heavy phase in the centrifuge.

Purifine PLA_2 is selective to PC, PE, and PA. It cleaves the FA from position 2 of the glycerol backbone and the end products are FFA and lysophospholipid.

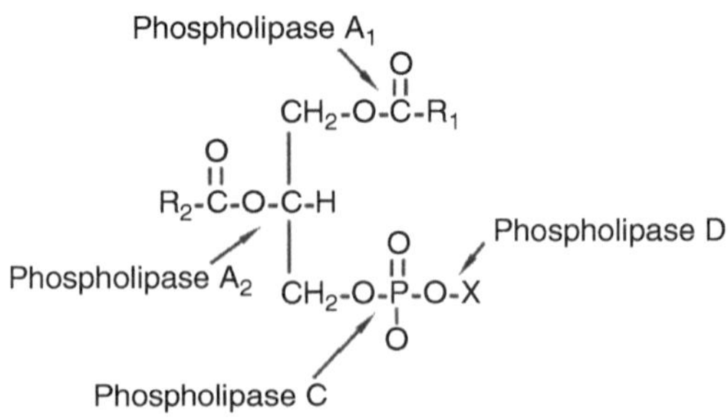

FIGURE 4.8 Phospholipase enzymes react on the phospholipase molecule.

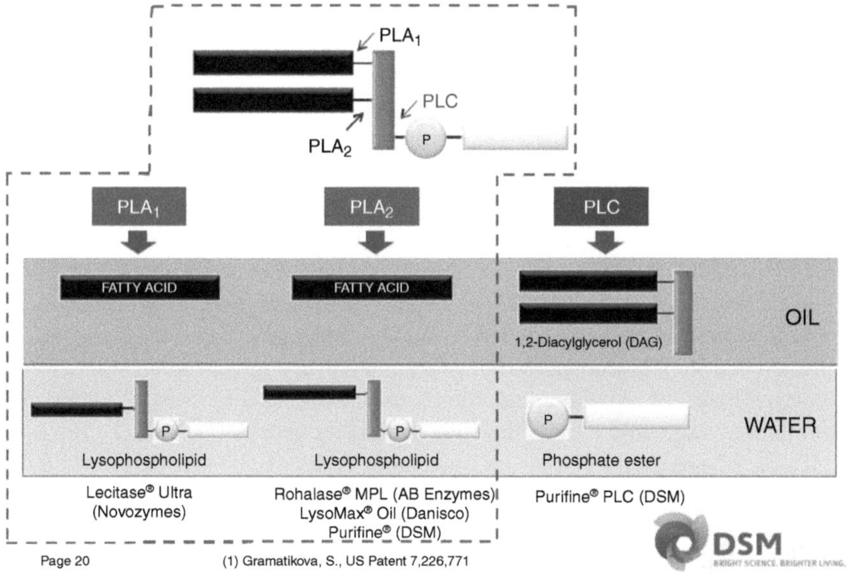

FIGURE 4.9 Example of enzymatic degumming with PLA_1 and PLA_2. *(Courtesy of DSM)*

They also appear in the oil phase and the heavy phases as described earlier for the reaction with PLA_1.

Originally, PLC was specific to PC and PE. However, the latest generation of Purifine 3G from DSM is specific for all four common phospholipids, PC, PE, PA, and PI. It is a mixture of PLC (effective on PC and PE) + PLA2 (effective on PC, PE, and PA) + PI-PLC (effective on PI). This makes this enzyme (DSM 3G) very effective in the degumming process.

Before engaging into the discussion on enzymatic degumming, it would be appropriate to recapitulate some of the previous discussions to properly illustrate the benefits as well as the challenges of enzymatic degumming.

It has been mentioned earlier that there are two categories of phospholipids in the crude vegetable oils based on their ability to be hydrated with water. They are referred to as:

1. HPs that can be removed from the crude oil by hydrating with water and separating the degummed oil and the hydrated gums through a centrifuge. Therefore most of the HPs can be removed through water degumming alone.
2. NHPs cannot be removed from the crude oil by simple water treatment because they do not hydrate under normal water treatment. These phospholipids can only be removed by treating the crude oil with acids, such as citric, phosphoric, maleic, acid anhydride, etc. or by using chelating agents, such as EDTA, and separating the degummed oil and the gums via centrifuging.

Hydratability and nonhydratability of the phospholipids depend on their chemical structures and their complexes with metals, such as Ca, Mg, and Fe.

This can vary from crude to crude, age and condition of the seeds, harvest conditions, drying and storage conditions for the seeds (such as rate of heating during drying, physical damage of the seeds, storage temperature and humidity for the seeds), and finally extraction process and postextraction conditions for the crude oils. In the oil industry, soybean and canola oils are the most important seedoils in terms of their usage and, at the same time, for the difficulty of reduction of their phospholipids to low levels. Low phospholipids in the refined oils are important. It is also important to consider the process of degumming so the phosphorus contents of the degummed oils are very low so the degummed oils could be physically refined. This is because the waste generated in chemical refining process is being regulated in many countries and the land and water pollution is a very big concern today in almost every industrial and advancing country. Therefore, enzyme degumming with minimum amount of chemicals for refining can greatly reduce the environmental concern.

Table 4.2 shows the list of the four major types of phospholipids and their metal compounds, their averages and the ranges observed in the soybean gums produced in the typical solvent extraction plants.

The specific types of phospholipids found in the crude soybean oil and crude canola oil bear some resemblances as well as there are some differences. Table 4.3 lists the phospholipids from crude soybean oil from a plant that used 100% expander process and the same from crude canola oil from the standard canola process. There are certain similarities between the phospholipids in the two oils. Crude canola oil contained much higher phosphatidic acid (PA).

TABLE 4.2 Phospholipids Found in Soybean Lecithin

	Range (%)		
Phospholipids	Low	Intermediate	High
PC	12.0–21.0	29.0–39.0	41.0–46.0
PE	8.0–9.5	20.0–26.3	31.0–34.0
PI	1.7–7.0	13.0–17.5	19.0–21.0
PS	0.2	5.9–6.3	—
Ca salt of phosphatidic acid	0.2–1.5	5.0–9.0	14.0
Lyso-PC	1.5	8.5	—
Lyso-PI	0.4–1.8	—	—
Lyso-PS	1.0	—	—
Lyso-PA	1.0	—	—

Source: Cherry, J.P., Kramer, W.H., 1989. Plant sources of lecithin. In: Szuhaj, B. (Ed.), Lecithins—Sources, Manufacture & Uses. AOCS, pp. 16–31.
PC, Phosphatidylcholine; PE, phosphatidylethanolamine; PI, phosphatidylinositol; PS, phosphatidylserine.

TABLE 4.3 Phosphorus and Phospholipid Make-Up in Crude Soybean and Canola Oil

	Soybean oil		Rapeseed oil	
Components	(ppm)	(%)	(ppm)	(%)
Phosphorus	850–1200		200–900	
Total phospholipids		2.0–2.9		0.5–2.3
PC		47		27
PI		24		17
PE		20		17
PA		9		39

PA, Phosphatidic acid; PC, phosphatidylcholine; PE, phosphatidylethanolamine; PI, phosphatidylinositol.

Each of the phospholipid components can be hydrated. The degree of hydration of each type of phospholipid varies widely. Table 4.4 lists the relative hydratability of the phospholipid components and some of their metal salts. As a general rule, the metal salts are far less hydratable than the corresponding phospholipids themselves.

One can see from the hydratability comparison data in Table 4.4 that not all of these phospholipids will be removed from the crude oil through water degumming. This is why the chemical degumming procedures were introduced. These procedures have been described earlier in this chapter. Citric or phosphoric acid or metal chelator EDTA, added in the chemical degumming process, reacts with

TABLE 4.4 Relative Rates of Hydration of Phospholipid Components and Their Metal Salts

Phospholipids and their salts	Relative hydration rate
PC	100
PI	44
PI (calcium salt)	24
PE	16
PE (calcium salt)	0.9
PA	8.5
PA (calcium salt)	0.6

PA, Phosphatidic acid; PC, phosphatidylcholine; PE, phosphatidylethanolamine; PI, phosphatidylinositol.

TABLE 4.5 Approximate Levels of Acetone Insoluble (AI) and Lipid Material in Gums

Process	AI (%)	Lipid material (%)
Water degumming	62–70	30–38
Acid degumming	50–60	40–50

the metal complexes in the PA and other phospholipid metal complexes, chelating the metals. A small amount of added NaOH and high shear mixing helps the hydration process to remove the additional phospholipids from the crude oil. In normal degumming process, there is always some neutral oil loss because the oil remains occluded in the gums. A very rough way to measure the oil in the gums is the method called AI method. The neutral oil, FFA, and diacylglycerol (DAG) are soluble in acetone while phospholipids and lysophospholipids are insoluble in acetone. Therefore, a very approximate estimate of oil loss in the gums can be related to the AI value for the gums. The typical AI value of the gums in the water-degummed soybean oil is 62–70%, and that in the gums in acid-degummed soybean oil is 50%–60%.

Table 4.5 shows there is 30%–38% of lipid material in the gums from the water-degummed process. This lipid material consists mostly of neutral oil. The other oil soluble components are FFA, DAG. The acid-degumming process loses more neutral oil in the gums than the water-degumming process.

Enzyme degumming process can release the FFA and the DAG from the phospholipid molecules. With citric acid pretreatment and simultaneous enzyme and low level of caustic treatment of the crude oil, one can achieve low NHP in the degummed oil along with increased yield of oil (due to DAG and less entrained neutral triglyceride), instead of a loss as in the chemical degumming process. Higher FFA increases the cost in the deodorizer system but the higher DAG yield and reduced neutral oil loss provide an overall economic advantage to the crusher. It is important for the prospective processor to take into account of the additional costs of equipment, such as the high shear mixers, reactors, plus the extra capacity for FA collection system at the deodorizer against the savings from the additional yield in the process.

It must be made clear that enzymatic treatment can reduce the NHP in the oil to a very low value (5 ppm) needed to use physical refining process on soybean or canola oil, but it requires proper pH control of the treatment media. This is because the enzymes are most active at certain pH and very small amounts of acid and base are needed to maintain the specific pH buffer required. On the surface this may appear that both enzymatic and deep degumming processes have got some commonality, such as the use of acid and base. However, the relative amounts of the acid and alkali used in the enzymatic degumming process are very little compared to the deep degumming process and the acid and the alkali are used only to maintain certain pH required for optimum reaction from the enzymes.

Ideally, it would be highly desirable to have an increased yield of DAG and no increase in FFA. This is because the higher FFA in the degummed oil requires larger fatty acid stripper and the handling system for the increased volume of FFA. In addition, there is also a small additional oil loss in the deodorizer when the FFA level increases. However, an overall better yield is obtained when a mixture of PLA and PLC are applied in degumming crude oil. This is believed to be due to the following:

1. Enhanced ability of the enzymes to release the metals (Ca, Mg, and Fe) from their complexes with the phospholipids.
2. Reduced neutral oil loss in the heavy phase due to the reduced viscosity of the gums and less emulsification by the phospholipids that occur in the conventional degumming processes.
3. The amount of gums produced in the enzymatic degumming process is lesser than the conventional degumming processes.
4. Also the gums produced in the enzymatic degumming process are far less viscous so they trap less neutral oil. Therefore, there is less entrained neutral oil in the gums.

Figs. 4.9 and 4.10 show the specific reaction sites of the phospholipase enzymes, PLA_1, PLA_2, and PLC with the phospholipids.

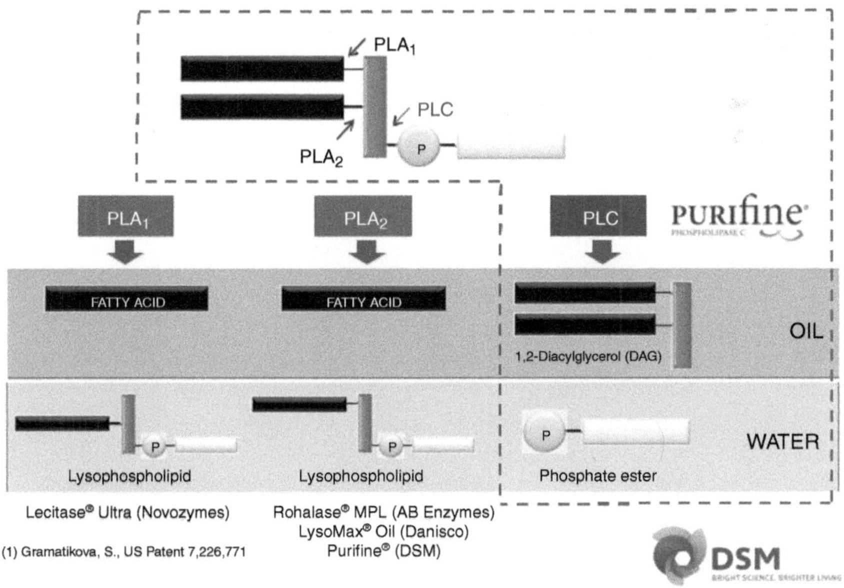

FIGURE 4.10 Example of enzymatic degumming with PLC. *(Courtesy of DSM)*

Fig. 4.9 represents the concept showing how the enzyme Lecitase (PLA_1, which is selective to all types of phospholipids) hydrolyzes the FA moiety in position 1 of the glyceride part of the phospholipid molecules. The remainder of the phospholipid molecules, which are called lysophospholipids remain in the oil phase. Purifine (PLA_2, which is selective to PC, PE, and PA), hydrolyzes the FA moiety from position 2 of the glycerol backbones and the end products are FFA and lysophospholipids. Fig. 4.10 shows Purifine (PLC) which produces diglycerides and phosphate esters.

All lysophospholipids and the phosphate esters are water soluble, while the FFA and DAG are oil soluble. The older versions of PLC were specific for PC, PE, and PA but Purifine 3G from DSM is specific to all four common phospholipids, PC, PE, PA, and PI.

Table 4.6 lists the currently used phospholipase enzymes for degumming.

The basic premise of enzyme degumming is based on the following critical steps:

1. Converting the PL micelle with citric acid with the help of high shear mixing. This forms reverse micelle.
2. Enzymatic hydrolysis of the reverse micelle by addition of water, high shear mixing and using small amount of caustic, if needed, and followed

TABLE 4.6 Currently Used Commercial Phospholipase Enzymes

Phospholipase category	Commercial names	Manufacturer	Effective on phospholipids	End products
PLA 1	Lecitase Ultra Lecitase NOVO A-PLA	Novozymes Novozymes Novozymes	All	FFA and lysophospholipids
PLA 2	Rohalase PL-Xtra	AB Enzymes	All	FFA and lysophospholipids
	LysoMax	Danisco	All	
	Purifine PLA_2	DSM	All	FFA and lysophospholipids
	Gumzyme	DSM	PA and PE	FFA and lysophospholipids
PLC	Purifine PLC (1G)	DSM	PC and PE	DAG and phosphate esters
	Purifine PLC (2G)	DSM	PC, PE, and PA	DAG and FFA
	Purifine PLC (3G)	DSM	All	DAG and FFA

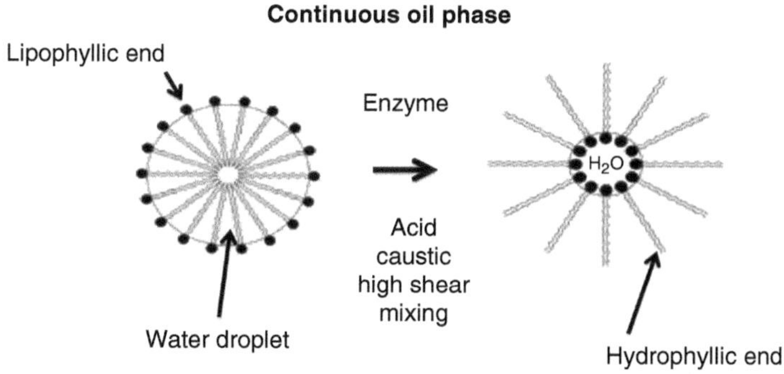

FIGURE 4.11 **Breaking up of the micelle.**

by incubation and separation of the lipid and water phases by a high speed centrifuge.

Fig. 4.11 shows the schematic view of the basic premise of enzyme degumming process.

The functions of the PLA's and the PLC's have been discussed earlier. These enzymes have undergone improvements over the years. The previous generations of the enzymes are already proven to be effective in the degumming process. Their latest generation of products has been already tested and the manufacturers are ready for their large-scale industrial applications in 2016.

Three leading entities that are actively developing and applying enzymes in oil degumming and numerous other food and industrial applications are:

- Novozymes
- AB Enzymes
- DSM N.V.

4.4.5.1 Novozymes

Chronological development of Novozymes' PLA enzyme is illustrated as follows:

1. A US Patent No. 5,264,367 was received by Rohm GmbH, Darmstadt, 1993.
2. The first enzyme, Lecitase 10L, a pancreatic PLA_2 enzyme was released in 2000.
3. Lecitase Novo, a microbial PLA_1 enzyme was released in 2001.
4. Lecitase Ultra, a microbial PLA_1 enzyme with higher temperature stability was released in 2003.
5. A-PLA, with higher temperature and higher acidity (lower pH) stability was released in 2016.

Numerous laboratory and industrial application reports available on the first three versions of the Lecitase enzymes can be found in literature. Some recent test data were received from Novozymes on A-PLA, which will be discussed.

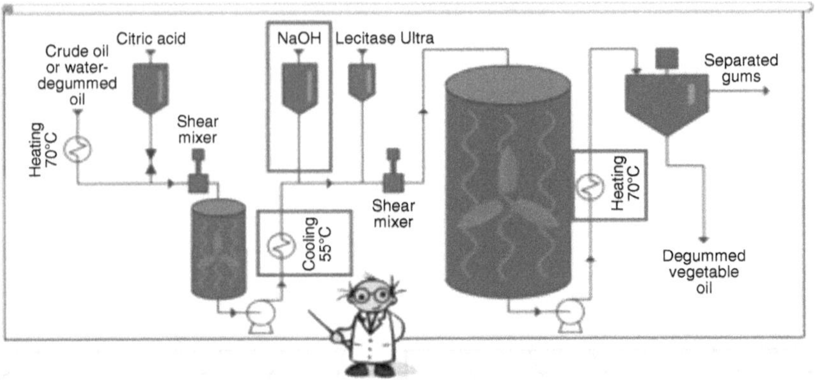

FIGURE 4.12 Novozymes process using Lecitase Ultra. *(Courtesy of Novozymes)*

Novozymes schematic process diagram for degumming oil with Lecitase Ultra is shown in Fig. 4.12. This process requires cooling down the oil to 55°C after citric acid treatment and before adding NaOH and Lecitase Ultra.

Novozymes schematic process diagram for degumming oil with A-PLA is shown in Fig. 4.13. Here one can see that there is no oil cooling required as well as no NaOH is added to the process.

Table 4.7 shows the comparison between the performances of Lecitase Ultra and A-PLA.

Fig. 4.14 shows the oil yield results on degumming crude rapeseed and crude soybean oils comparing the results obtained in acid degumming, Lecitase Ultra degumming, and A-PLA degumming. Crude oil P was 475 ppm. Citric acid was added at 650 ppm and the NaOH addition was at 1.5 molar equivalent and the temperature was 55°C.

Fig. 4.15 shows the relative yields from crude soybean oil using Lecitase Ultra and A-PLA enzymes. The crude soybean oil had phosphorus content of 595 ppm. The degummed oil in both cases were <10 ppm.

FIGURE 4.13 Novozymes process using A-PLA. *(Courtesy of Novozymes)*

TABLE 4.7 Comparisons Between the Performances of Lecitase Ultra and A-PLA

Operating parameters	Lecitase ultra	A-PLA
Optimum pH	5.5	4.1
Denaturation temp. at optimum pH	63°C	80°C
NaOH, molarity in gum phase	1.5	0
pH	5	3.7
Optimum operating temperature	55°C	75°C
P in degummed oil at optimum temp.[a]	—	<3 ppm
Fouling of heat exchanger and centrifuge	Yes	No
Oil yield	Higher than acid degumming process	Higher than Lecitase Ultra process

[a]*Crude oil P (595 ppm); citric acid added (650 ppm); NaOH (none); pH (3.8).*

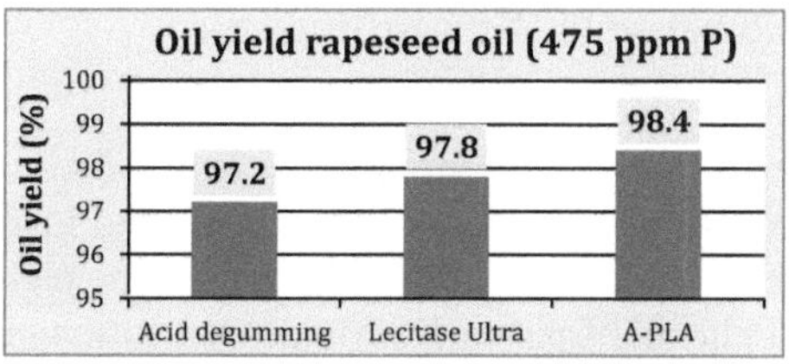

FIGURE 4.14 **Oil yield data comparison.** *(Courtesy of Novozymes)*

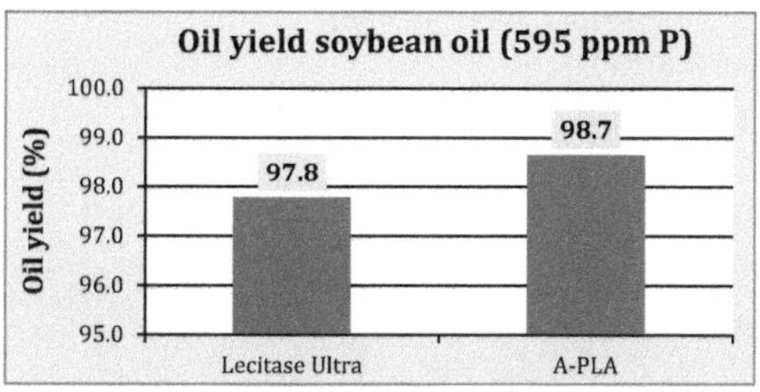

FIGURE 4.15 **Oil yield data comparison.** *(Courtesy of Novozymes)*

4.4.5.2 AB Enzymes GmbH (Feldbergstrasse, Darmstadt, Germany)

Rohalase PL-Xtra is the latest PLA_2 enzyme from AB Enzymes (Fig. 4.16). This enzyme is capable of reducing NHP from all types of phospholipids in one stem without NaOH. The general required reaction conditions are as follows:

1. Reaction temperature	55°C or lower
2. Dosage	25–50 g/ton of crude oil
3. Reaction time	2–4 h
4. Optimum pH	4.0

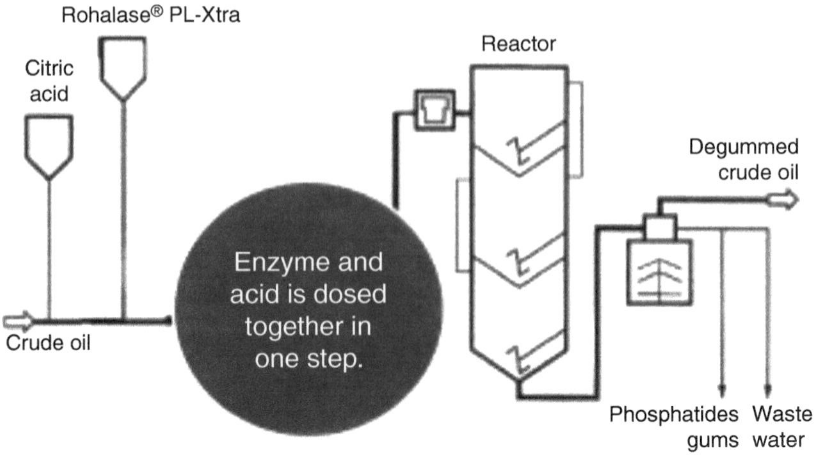

FIGURE 4.16 Schematic diagram for Rohalase PL-Xtra degumming process. *(Courtesy of AB Enzymes)*

The benefits of Rohalase PL-Xtra are listed as:

1. Optimum pH is 4.0 which results in the following:
   a. A one-step enzymatic degumming process that does not require NaOH
   b. Increased solubility of calcium citrate because of lower pH stability
   c. Less fouling in the equipment downstream of the high shear mixer
2. It hydrolyzes all phospholipids and provides the following:
   a. Improved removal of phospholipids
   b. Less entrained neutral oil in the gum phase
3. No lipase activity resulting in following:
   a. Less FFA formation
   b. Decreased neutral oil loss

4.4.5.2.1 Results of Degumming of Crude Soybean Oil With Rohalase PL-Xtra

Reaction Conditions:

1. Temperature	55°C
2. Enzyme dosage	50 ppm
3. Citric acid dosage	Not available
4. Reaction time	4 h

Results:

1. Crude oil P	800 ppm
2. Degummed oil P	5.8 ppm[a]
3. FFA increase	0.74%
4. DAG formation	None

[a]*P from nonenzymatic degumming 14.6 ppm.*

4.4.5.2.2 Results of Degumming of Crude Canola Oil With Rohalase PL-Xtra

Reaction Conditions:

1. Temperature	55°C
2. Enzyme dosage	50 ppm
3. Citric acid dosage	Not available
4. Reaction time	2 h

Results:

1. Crude oil P	477 ppm
2. Degummed oil P	4.2 ppm[a]
3. FFA increase	0.44%
4. DAG formation	None

[a]*P from nonenzymatic degumming 25.4 ppm.*

4.4.5.2.3 Effect of Reaction Time

The average oil yield was found to be higher at longer reaction time.

Degumming process	Yield (%)	
	2-h reaction	4-h reaction
Nonenzymatic	90.25	90.68
Rohalase PL-Xtra	94.52	94.68
Yield gain	0.17% over 2-h reaction time	

4.4.5.3 DSM N.V.

DSM has gone through the evolution of the enzyme Purifine over the years. On Dec. 7, 2011, Verenium Corporation, a leading industrial biotechnology company announced the introduction of Purifine phospholipase C (PLC) enzyme for use in pretreatment of oil for biodiesel production to reduce phospholipids in vegetable oils. Verenium's Purifine PLC provided improved overall economics of edible oil and biodiesel production and was in the works since 2007.

In 2012 DSM N.V. acquired Verenium's oilseed processing business and its IP portfolio, licenses for certain food enzymes, and access to biodiversity libraries that Verenium will create using proprietary technology.

Purifine PLC enzyme has been further modified by DSM. There are three generations of this enzyme that are commercially available today. The brief introduction to the three different generations of Purifine PLC is listed in Table 4.8.

4.4.5.3.1 Degumming Crude Oil With Lecitase Ultra PLA

The schematic flow diagram for the Alfa Laval enzymatic degumming process is shown in Fig. 4.17.

This system can be used for degumming with enzyme PLA or PLC or combination of PLA and PLC.

Enzyme, Lecitase Ultra PLA_1 reacts with all four types of phospholipids, PC, PE, PI, and PA. The reaction time varies with the type of phospholipid.

TABLE 4.8 Three Generations of Purifine Phospholipase C (PLC)

Purifine PLC generation	Comprised of	Active on	Function
Gen. 1(1G) (2008–2013)	PLC	PC and PE	Increased DAG yield Increases total yield 100 ppm added caustic further improves yield
Gen. 2 (2G) (2013—Argentina; 2014—USA)	PLC + PLA_2	PC, PE, and PA	DAG: same as 1G Total oil yield: higher NHP: more complete conversion to HP
Gen. 3(3G) (2016)	PLC + PLA_2 + PI-PLC		DAG: higher than 2G Total oil yield: significantly higher Overall economics: highly favorable

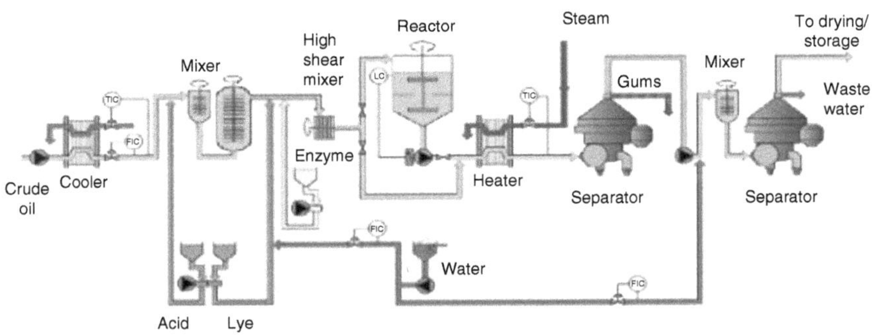

FIGURE 4.17 Schematic diagram for enzymatic degumming of crude oil. *(Courtesy of Alfa Laval)*

The steps of procedure for the PLA_1 degumming process is outlined below:

1. Crude or water-degummed oil is heated in a plate and frame heat exchanger to a temperature of 104–140°F (40–60°C) and not to exceed 158°F (70°C).
2. Citric acid is added to the oil stream continuously through a mixer (preferably a high shear mixer) and held in a retention tank with constant agitation.
3. Caustic in water solution is added into the oil stream.
4. Total amount of water added is up to 2 times of the phospholipids content of the oil.
5. PLA enzyme is added @ 30–200 ppm of the oil flow.
6. The stream of liquid then passes through an ultra high shear mixer to completely emulsify the enzyme and the oil and then is discharged into the reactor where the reaction is allowed to proceed to completion. It can take up to 6 h to complete the reaction with the PLA enzyme. PI and PA require longer reaction time with this enzyme.

7. The oil from the reaction tank is then heated to 167–185°F (75–85°C) through a plate and frame heat exchanger.
8. The heated mix is then centrifuged in a self-cleaning centrifuge.
9. The degummed oil can be physically refined with or without water washing.

4.4.5.3.2 Critical Control Points for Degumming With Lecitase Ultra PLA

The critical control points for the Lecitase Ultra PLA process are:

1. Reaction temperature
2. pH
3. Amount of water added
4. Enzyme dosage
5. Reaction time
6. Centrifuging temperature

1. Reaction temperature
   With the older generation of enzymes the oil temperature is maintained at 104–140°F (40–60°C) because at this temperature range the enzyme exhibits the optimum performance. In actual practice the reaction temperature is maintained <128°F (50°C) to keep the enzyme active and not letting it lose its activity due to high temperature and also maintain the viscosity of the mix low enough to get good reaction. However, the newer generation of enzymes are active at higher temperatures. Novozymes A-PLA enzyme has an optimum operating temperature of 75°C.
2. The pH
   The Lecitase Ultra PLA requires an optimum pH of 5.5. Citric acid is used as source of acid. The acid in this case produces calcium citrate, which creates hard deposits in the heat exchanger and inside the centrifuge. A pH < 4.5 reduces the amount of hard deposits in the process but the enzyme loses some of its activity and the process becomes less efficient. The newer enzyme A-PLA requires an optimum pH of 4.1. This makes the enzyme more robust both from the standpoint of temperature and pH.
3. Amount of water
   Amount of water added is up to twice the amount of total phospholipids in the oil. For example, if the total phospholipid content is 2% of the oil feed the amount of added water is 4% of the oil flow.
4. Enzyme dosage
   Phospholipase enzyme dosage is variable (Typically, 30–200 ppm) on the oil feed rate basis. The manufacturer of the enzymes recommends the dosage to their client plant.
5. Reaction time
   The time needed for complete hydrolysis can be 4–6 h. Therefore the reaction tank must have the capacity to hold at least 6 h of production.

6. **Centrifuge temperature**
   The centrifuge temperature must not be allowed to drop below 167°F (75°C). At lower temperature more lysophospholipids may appear in the oil phase. At temperature higher than 185°F (85°C), there can be more neutral oil in the heavy phase.

4.4.5.3.3 Expected Results

1. The phosphorus content of the degummed oil is <5–10 ppm.
2. Amount of FFA produced in the degumming process is directly proportional to the phospholipid content of the oil. This can be calculated as shown below: Assuming the phospholipid content of the oil is 2.4%

   - Avg. mol. wt. of FA in soybean oil = 282
   - Avg. mol. wt. of phospholipid = 750
   - Amount of FFA formed = (282/750) × 2.4
     = 0.902%

3. Half of the estimated FFA generated is expected to remain in the gum and the remainder in the degummed oil. The amount of FFA increase in the degummed oil in the previous example would be 0.5 × 0.902 = 0.45%.
4. The AI is 60%–65%. A higher AI value is indicative of the following:
   a. Incomplete reaction because of
   - insufficient enzyme dosage
   - not enough high shear mixing
   - inadequate reaction time
   - insufficient agitation
   - nonoptimal pH
   - high reaction temperature
   - water feed is inaccurate

   b. Inappropriate centrifuge operation, which can be
   - improper setting of the back-pressure for the heavy phase discharge
   - lower feed temperature to the centrifuge
   - dirty centrifuge bowl
   - lack of centrifuge cleaning and maintenance
5. Phospholipids in the oil are almost totally converted into lysophospholipids by the Lecitase Ultra PLA. With complete reaction, there should be no PA and PE in the gums, and no PA in the degummed oil. Incomplete reaction caused by any of the factors listed under #4 would show the presence of PA and or PE in the degummed oil.
6. The appearance of the gums from this process is much lighter and less viscous than those from the water-degummed oil.

4.4.5.3.4 Degumming Crude Oil With PLC

PLC degumming produces DAG and small amount of increased FFA in the oil. The basic flow diagram for the process is shown in Fig. 4.18.

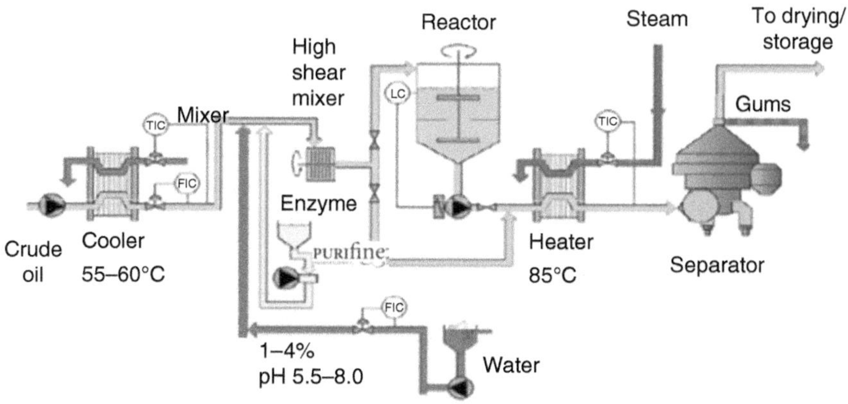

FIGURE 4.18 **Schematic flow diagram for PLC degumming.** *(Courtesy of DSM/Alfa Laval)*

Following steps are involved in this process:

1. Crude oil is cooled to 130–140°F (55–60°C) in a cooler.
2. 1%–4% water is added to the oil stream.
3. The enzyme supplier based on phospholipids analysis of the crude oil recommends the amount of PLC dosage.
4. The pH of 5.5–8 is maintained.
5. The mixture is passed through a high shear mixer and then enters the reactor.
6. The reaction mixture is held for 2 h with agitation.
7. The hydrated oil from the hydration tank is then heated to 167–185°F (75–85°C) through a plate and frame heat exchanger.
8. The heated mix is then centrifuged in a self-cleaning centrifuge.

DSM reported the following results based on the test conducted at a manufacturing plant in 2008 using Purifine (1G):

• Crude oil P content	1151 ppm
• Total phospholipids	2.05%
• DAG yield gain	1.03%
• Degummed oil P content	41 ppm

The P level in the degummed oil can be as high as 100 ppm according to some oil processors.

PLC-degummed oil has a residual P content very similar to that of water-degummed oil. However, there is a yield gain and lower loss of neutral oil in the heavy phase as shown in Table 4.9. PLC degumming shows significantly higher in yield and it is believed to be due to the production of DAG and the destruction of the emulsification properties of the PC and PE present in the oil.

Caustic refining and PLA degumming of the water-degummed oil produced very different results as shown in Table 4.10. PLA processed oil showed significantly less gums, less oil loss, and higher yield.

TABLE 4.9 Yield Differences Between Water Degumming and PLC Degumming of Crude Soybean Oil [Dayton and Gelhardo, 2014. Green veg. oil processing. AOCS Press, Urbana, IL (Chapter 6)]

Analysis	Water degummed	PLC degummed
P in crude oil (ppm)	600	600
P in degummed oil (ppm)	150	150
Dry heavy phase generated	20.3 kg	11.2 kg
Lipid lost	20.3 kg	6.2 kg
Oil yield	980 kg	991 kg

Note: Starting oil weight: 1000 kg; crude oil FFA: 0.6%; 2% water was used for both degumming operations.

TABLE 4.10 Yield Differences Between Caustic Refining and PLA Treatment of Degummed Soybean Oil [Dayton and Gelhardo, 2014. Green veg. oil processing. AOCS Press, Urbana, IL (Chapter 6)]

	Caustic[a] refining	PLA[b] treatment
Initial P in the water-degummed oil	150 ppm	150 ppm
P in the final oil	1 ppm	1 ppm
Dry heavy phase generated	17.3 kg	5.0 kg
Lipid lost	16.3 kg	3.7 kg
Oil lost in bleaching	0.8 kg	0.8 kg
Fatty acid	1 kg	6.7 kg
Unsaponifiables	2 kg	2 kg
Oil yield	981 kg	988 kg

Starting oil weight: 1000 kg; crude oil FFA: 0.6%.
[a] *500 ppm phosphoric acid, 0.03% excess caustic.*
[b] *500 ppm citric acid, 90 ppm sodium hydroxide (pH 4.5), 2% hydration water.*

4.4.5.3.5 Degumming Crude Oil with PLA and PLC (US Patent No. 8,460,905 B2 Authors: Christopher L.G. Dayton & Flavio da Silva Galhardo)

Degumming of crude oil is more efficient and produces better oil yield when the enzymes, PLC and PLA are used in the same process. The results are very similar whether the crude oil is treated with the enzymes. The patent shows the following results.

4.4.5.3.6 PLC Degumming of Crude Soybean Oil, Followed By PLA_1 Degumming

The authors followed the steps as described:

1. 2110.5 g of crude soybean oil containing 560.1 ppm of phosphorus was used.
2. The oil was heated to 141°F (60°C).
3. 63 g of deionized water and 123 g of Diversa Purifine PLC was added to the oil.
4. The mixture was agitated for 1 h under normal agitation maintaining a temperature of 131–141°F (55–60°C).
5. The mixture was centrifuged and the wet gum and the degummed oil were collected.
6. To the PLC-degummed oil, 2.0 g of 50% w/w water solution of citric acid was added.
7. The mixture was mixed in a high shear mixer for 1 min and then mixed with an overhead mixer for 1 h.
8. 1.8 mL of 4 molar Na (OH) solution was added to the oil mixture and the agitated through high shear mixer for 10 s.
9. 59 g of deionized water was added to the mixture and then mixed in high shear mixer for 1 min.
10. 0.1061 g of Novozyme's Lecitase Ultra PLC was added and the mixture was high shear mixed for 1 min.
11. The mixture was agitated under normal speed for 4 h at a temperature of 45°C (113°F).
12. The oil was then centrifuged.
13. The oil and the wet gum samples were collected.
14. The residual phosphorus (P) in the final oil was 3.2 ppm.

4.4.5.3.7 PLC and PLA_1 Degumming of Crude Soybean Oil Together

The authors used the following steps of procedure:

1. 2021.4 g of crude soybean oil containing 547.9 ppm of phosphorus was used.
2. The oil was heated to 167–177°F (75–80°C) under normal mixing using an overhead mixer.
3. 2.0 g of 50% w/w citric acid solution was added and mixed in high shear mixer for 1 min.
4. The oil mixture was agitated at normal speed for 1 h.
5. The oil was allowed to cool until the temperature reached 104–113°F (40–45°C).
6. 1.8 L of 4 molar NaOH solution was added.
7. The mixture was high sheer mixed for 10 s.
8. To this mixture the following were added:

   a. 61 g of deionized water
   b. 0.1184 g of Diversa Purifine PLC enzyme
   c. 0.1038 g of Novozyme's Lecitase Ultra PLA_1 enzyme
9. The mixture was high shear mixed for 1 min.
10. The mixture was then agitated under normal speed for 1 h at 113°F (45°C).
11. The oil and wet gums were separated in a centrifuge and the samples were collected.
12. The residual phosphorus in the degummed oil was 2.4 ppm.

The combined PLC and PLA_1 degumming produced essentially the same residual phosphorus as in the sequential degumming process with only 1 h of mixing.

4.4.5.3.8 Degumming Crude Oil With PLA and PLC Mixture [Using DSM Purifine (3G), Which is a Mixture of PLC + PLA_2 + PI-PLC]

The new generation of Purifine (3G) from DSM is active on all PL's (PC, PA, PE, and also PI). The previous generations of Purifine (2G) were not active on PI. The schematic diagram from DSM is shown for Purifine (3G) degumming with deep degumming in Fig. 4.19.

The basic process parameters are:

1. Purifine PLC dosage is 1/6th of the total P level in the oil feed.
2. The operating pH is 5–8.
3. The temperature is 104–158°F (40–70°C).
4. Water added is 2%–4% of the oil feed.
5. The optimum pH is 5.5 and the optimum temperature is 140°F (6°0C).

The yield improvement of the Purifine (3G) process has been illustrated in Fig. 4.10 and reduced fat in the meal in Fig. 4.11 earlier.

The enzyme activity is significantly reduced at lower pH but the P in the final oil is lower and the yield is greatly reduced. Therefore an optimum condition is maintained to maximize the yield in the crushing operation. If the oil is sold as degummed crude the crusher can make a higher profit.

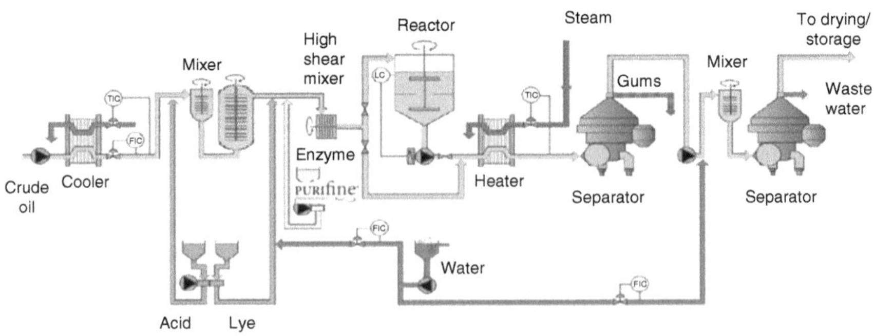

FIGURE 4.19 Schematic diagram for PLC deep degumming using Purifine (3G) from DSM. *(Courtesy of DSM/Alfa Laval)*

4.4.5.4 Concluding Remarks on Enzymatic Degumming Process

Enzymatic degumming of vegetable oil offers numerous benefits. However, it comes with an added cost. Therefore, the refiner must evaluate the cost and benefit of the process and determine the overall economic advantage. Following are some of the key points to consider in enzymatic degumming process:

1. Enzymatic degumming reduces the NHP in the crude soybean oil essentially to the same level as water degumming.
2. It does not reduce the level of NHP to a level low enough to be ready for physical refining.
3. However, the NHP in the oil can be reduced to a very low level through combined enzyme degumming and deep degumming as described before.
4. Enzymatic degumming can significantly increase the total oil yield by increased DAG and reducing neutral oil loss in the gum.
5. PLA treatment increases the FFA in the oil that requires additional fatty acid stripping and handling system, depending on the phospholipid content of the oil. There is also a small amount of oil loss through entrainment in the FFA removal process.
6. Use of a mixture of PLC and PLA improves the process yield, NHP reduction and the oil yield.
7. The process requires a Reaction tank that must have 2–6 h of residence time, depending on the quality of the incoming oil.

4.4.5.5 Real Benefit of Enzymatic Degumming Process

Enzymatic degumming process is superior to the chemical degumming process as described at the beginning of this section. This will be more evident with the following example.

Soybeans from some of the Latin American countries arrive with very high FFA and phosphorus contents. Conventional methods of degumming and refining would produce very high losses and relatively inferior quality oil (simply because of the quality of the beans). Normally, the crude oil produced from these beans has been chemically degummed and caustic refined with heavy bleaching of the refined oil. In some cases it would be next to impossible to process these crude oils under conventional method. Enzymatic degumming has opened the door to many processors who have access only to these beans.

4.4.5.6 Establishing the Yield Gain From Enzymatic Degumming

The enzymatic degumming principle is sound. The process is well established through the work of various researchers and the enzyme manufacturers. The oil processor had to carefully establish the economic advantage. The key is for the processor to establish the increased yield and low P level where physical refining is intended.

Establishing refining gain on daily basis will require the following:

1. A ^{31}PNMR to measure the different types of the phospholipids in the feed oil for the enzyme manufacture recommend the ratio as well as the of PLC/PLA to be used
2. An HPLC to measure the phospholipids in the crude oil as well as the DAG in the enzyme degummed oil
3. Precise mass balance and inventory system for:
   a. oil feed
   b. degummed oil produced
   c. amount of gums produced
   d. neutral oil content of the gums

The first two items will require expensive instrumentation, highly skilled analytical chemists, and besides that, the use of the expensive instruments will be minimal. It is not possible to have these in an oil processing plant. However, the oil processing plants can establish very good yield information by doing the following:

1. Maintain good control over the feed oil phosphorus content. For this the plant needs an ICP and trained personnel
2. Stay in touch with the enzyme manufacturer if the feed oil changes and have them guide the oil processor on enzyme dosage
3. An excellent mass balance around the:
   a. crude or water-degummed oil feed
   b. degummed oil produced
   c. gums produced
   d. monitoring the neutral oil content of the gum
4. Monitoring inventory:
   a. daily
   b. weekly
   c. monthly
   d. quarterly
   e. annually

Daily and weekly inventory checks can be done from the loss monitor instruments. Monthly and quarterly inventories can be done with complete production, receipts, measured losses, and looking into the damages and variations (D&V). In reality, monthly and quarterly inventory measurements can provide most important information and the plant might not have to spend on ^{31}PNMR and HPLC. However, an ICP instrument might be very helpful for all-purpose plant operation including measurement of total P in the crude, degummed and the refined, and bleached oils.

Chapter 5

Refining

5.1 PURPOSE OF REFINING VEGETABLE OIL

The term "refining" implies a processing step that reduces or removes certain undesirable impurities in the crude oil. The primary goal is to produce high quality edible oil that provides satisfactory results in all oil applications.

The crude oil contains several major and minor impurities. The refining step either reduces or removes the major and minor nontriglycerides (impurities) from the crude oil in order to make it suitable for food use.

5.1.1 Major Nontriglycerides

The major impurities in the crude oil are present in relatively high levels. These are expressed as percentage of the crude oil. Examples of the major impurities are:

- free fatty acid (FFA)
- phospholipids
- diglycerides
- monoglycerides

In addition, the crude oil contains some suspended material, such as meal, fiber, dirt, etc. These suspended materials must be removed from the crude oil via filtration before the oil enters the refining process. These impurities tend to plug the centrifuge bowl prematurely requiring frequent process shut down to clean it.

All of the major impurities are undesirable in the finished oil above certain levels. This is discussed later in this chapter and also in Chapter 12. The refiner should maintain the process conditions within the prescribed limits in order to obtain finished oils with the lowest levels of all of the aforementioned impurities.

5.1.2 Minor Nontriglycerides

These impurities are present in the crude oil in parts per million (ppm) levels. The minor nontriglycerides present in the oil are:

- tocopherols
- tocotrienols in palm oil and rice bran oil
- sterols and sterol esters

Practical Guide to Vegetable Oil Processing. http://dx.doi.org/10.1016/B978-1-63067-050-4.00005-2
Copyright © 2017 AOCS Press. Published by Elsevier Inc. All rights reserved.

- squalene and ferrulates in corn oil
- color compounds
- oil decomposition products, such as:
    - polar compounds
    - polymers
    - aldehydes
    - ketones
    - other volatile and nonvolatile compounds

Tocopherols, tocotrienols, sterols, sterol esters, and squalene are natural antioxidants. They also provide nutraceutical value to the oil and foods made with it. It is desirable to retain these components in the refined, bleached, and deodorized (RBD) oil at their highest levels possible in order to maintain high oxidative stability as well as higher nutritional value of the oil. The color compounds and the oil oxidation components must be reduced to very low levels. Some of the color compounds are reduced in order to improve the appearance of the oil, while others are reduced to improve the oxidative stability of the oil.

For example, the red color in crude palm oil is contributed by a number of carotene compounds. In most commercial operations, the carotenoids get bleached at high temperature under vacuum. However, the "red palm oil," which is being marketed for its nutritional value contains high level of carotenes. This oil is processed under careful bleaching and deodorization conditions in order to maintain most of the original carotenoids in the oil.

5.2 METHODS OF OIL REFINING

There are two primary methods for refining vegetable oils. They are:

1. physical refining process
2. chemical refining processes:
    a. batch refining process
    b. continuous refining process

In addition, there is another refining process, known as the cold refining process, used primarily for sunflower oil and other wax-bearing crude oils.

There are also other processes that are slight modifications of the physical or chemical refining methods. These are known as:

- modified physical refining (MPR) process
- modified chemical refining process
- semiphysical refining process

Selection of a specific method of crude oil refining process depends on several factors, such as:

1. type of crude oil
2. quality of the crude oil to be refined, especially its phospholipids (phosphorus) content

5.3 PHYSICAL REFINING PROCESS

The chemical refining process, which dominated the oil-processing field, came under scrutiny when the environmental requirements in many parts of the world became stringent. The chemical refining process produces soap as a by-product, which is treated with sulfuric acid to reclaim the fatty acids. This generates a plant effluent that is very rich in COD and BOD as discussed in Chapter 4. In order to alleviate this situation, the oil processing industry developed what is widely known today as the physical refining process. This process has been found suitable for the types of crude oil that are low in nonhydratable phospholipids (NHP) and those with high FFA.

The process is especially beneficial for refining crude palm oil and coconut oil because of their low phospholipids and high FFA contents. Chemical refining of either of these oils produces a very large amount of soap and that leads to excessive loss of neutral oil in the soap produced. In addition, physical refining process is suitable for the super degummed canola oil (with a phosphorus content of <10 ppm or less). This process is unsuitable for seed oils because of the higher levels of phospholipids in these oils. Sunflower or safflower oil could be refined via this process when the phosphorus in these oils are reduced to a very low level (<10 ppm or less) by using deep degumming or enzymatic degumming of the crude oil. Many processors are refining soybean oil via the physical refining process, but the outcome of this is highly questionable. Soybean oil can be successfully refined via the physical refining process if the phosphorus content is reduced to <10 ppm, preferably <5 ppm) via deep degumming or enzymatic degumming process.

Physical refining process has been difficult for any oil, which contains high levels of nonhydratable phospholipids, until the enzymatic degumming process was improved with the help of the new and more effective enzymes and fine-tuning of the enzymatic degumming process that took place in recent years.

Following is the list of oils where the physical refining process can be used to produce high quality refined oil:

1. super degummed canola oil with phosphorus content of <10 ppm
2. sunflower oil
3. safflower oil
4. palm oil
5. coconut oil
6. palm kernel oil
7. deep degummed or enzyme degummed soybean oil (phosphorus content <10 ppm, preferably <5 ppm)

Physical refining of the crude oil is carried out using the following steps:

1. The crude oil is pretreated with phosphoric acid or citric acid to hydrate the nonhydratable phospholipids and chelate trace metals impurities. This helps reduce the total phospholipids in the oil to a lower level.

2. The oil is bleached in a vacuum bleacher using acid activated clay. This removes the remaining phospholipids, trace metals, oil decomposition products, and color bodies, such as chlorophyll, carotene, etc. from the oil.
3. The bleached oil is filtered. The filtered oil is called refined and bleached oil (RB oil).
4. The RB oil is deodorized at high temperature under very low pressure (high vacuum) using live steam stripping to remove the FFA, monoglycerides, oil decomposition products, etc. The oil at this stage is called RBD oil.
5. The RBD oil is cooled. Citric acid is added to the deodorized oil chelate metal ions to minimize oil oxidation and stored.
6. The oil is also winterized or hydrogenated as needed, before it is deodorized.

Antioxidants are added if needed by the end-users.
Fig. 5.1 shows the schematic diagram for the physical refining process.

5.3.1 Critical Control Points in the Physical Refining Process

Following are the critical control points in the physical refining process:

1. determination of the phosphorus (or phospholipids) content of the crude oil
2. amount of phosphoric acid added to the crude oil
3. acid addition temperature
4. degree of mixing between the oil and the phosphoric acid
5. contact time between the acid and the oil
6. type of clay

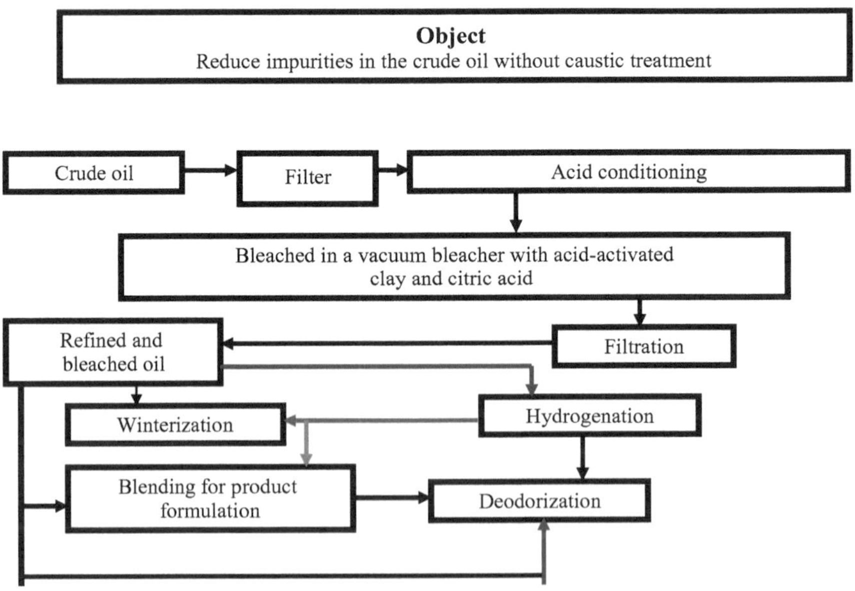

FIGURE 5.1 **Schematic flow chart for the physical refining process.**

7. amount of clay
8. degree of contact between the acid treated oil and the acid activated clay
9. contact time between the oil and the clay
10. bleacher temperature
11. absolute pressure in the bleacher
12. type of filter
13. oil temperature after bleaching
14. storage period between bleaching and further processing of the bleached oil
15. deodorization

Table 5.1 shows the limits for the critical control points for the physical refining process.

TABLE 5.1 Critical Process Control Points in Physical Refining

Critical control points	Limits/comments
Phospholipids (or Phosphorus) in the acid pretreated crude oil	Must be <750 ppm for most oils and <400 ppm for soybean oil
Amount of phosphoric acid to be used for pretreatment	For best results one should calculate it on the basis of Ca^{++} and Mg^{++} content of the crude oil as discussed under acid degumming in Chapter 3
Acid addition temperature	Discussed in Chapter 3
Degree of premixing with phosphoric acid	Discussed in Chapter 3
Holding time for the acid and the crude oil before bleaching	Discussed in Chapter 3
Type of clay	Acid activated
Amount of clay	As low as possible
Degree of mixing with the acid activated clay	Intimate mixing with a mechanical mixer
Contact time between oil and the clay	30–45 min
Bleacher temperature	110–120°C (230–248°F)
Absolute pressure in the bleaching vessel	<50 mm of mercury, preferable <36 mm of mercury
Type of filter	Totally enclosed
Bleached oil temperature before storage	<40°C, if the oil is *not* sent directly to the next processing step through close coupling
Holding time for bleached oil before further processing	As short as possible
Deodorization	Will be discussed in the Chapter on Deodorization

5.3.2 Bleached Oil Quality Parameters in the Physical Refining Process

Bleaching process and the quality criteria for the bleached oil will be discussed in great detail in Chapter 11. However, the oil quality parameters for the physically refined oils should be no different from the bleached oil from the chemical refining process. Commercially available physically refined oils tend to show somewhat higher phosphorus and trace metal contents as compared to those treated by the chemical refining process.

5.3.3 Troubleshooting Physical Refining Process

The typical process issues or the aberrations in the oil quality, probable contributing factors, and the recommended corrective measures are discussed in Table 5.2.

TABLE 5.2 Troubleshooting Physical Refining Process

Symptom	Contributing factors	Proposed solutions
Phosphorus >5 ppm	Nonhydratable phospholipids content may be high in the crude oil	Check crude oil for nonhydratable phospholipids content to make sure it is <750 ppm. Increase acid treatment
	Phosphoric acid dosage may be low	Check for calcium and magnesium content and adjust the amount of acid as discussed earlier
	Incomplete or improper dispersion of phosphoric acid into the oil	Make sure the high shear mixer is working properly
	Contact time between the oil and the acid might be short	Follow the procedure discussed in Chapter 4 on acid degumming
	Agitator in the contact residence tank is not working properly	Check and make sure the agitator is working
	Amount of clay used may not be adequate	Check the amount of clay and increase it
	Contact time between the oil and the clay may not be adequate due to:	
	1. short contact time 2. poor mixing condition	1. Increase contact time 2. Check the amperage on the mixer motor. If it is low, there may be problem with the impeller. Check and fix it
	Low bleaching temperature	Increase temperature

TABLE 5.2 Troubleshooting Physical Refining Process (*cont.*)

Symptom	Contributing factors	Proposed solutions
Bleached oil is high in: 1. Iron 2. Calcium 3. Magnesium	Same reason as for high phosphorus	Correcting the phosphorus issue generally resolves the high trace metal issue
In freshly bleached oil: PV > 0 as sent to storage tank	Poor vacuum	Increase vacuum
	Excessive contact between oil and air	Find the source of air coming in contact with the oil and stop it. Do not use atmospheric bleacher
	Bleached oil not cooled	Cool the oil to <40°C
pAV > 4.0	Poor vacuum	Increase vacuum
	Excessive contact between oil and air crude oil PV is high due to:	Find the source of air and eliminate it
	1. High PV in the incoming oil 2. Crude oil has been stored for long 3. Excessive aeration of the crude oil during unloading and storage	1. Be vigilant about the crude oil PV 2. Reduce storage time 3. Use bottom loading for the crude oil storage tanks
Chlorophyll is >30 ppb	Not enough bleaching clay	Increase the amount of bleaching clay
	Insufficient contact time between the oil and the clay	Increase the contact time
	Inadequate agitation	Check the agitator and take steps to rectify any issue with it
	Low bleaching temperature	Increase bleaching temperature by 5°C and recheck chlorophyll in the bleached oil. Increase it further if needed without going much >120°C (248°F)

5.4 CHEMICAL REFINING PROCESS

The chemical refining process is widely used in the United States and in other parts of the world for vegetable oils, especially the seed oils. In this process, the crude oil is pretreated with phosphoric acid and then neutralized with a caustic (sodium hydroxide) solution. The caustic reacts with the FFA in the crude oil and produces soap, which is then separated from the refined oil in a primary (refining) centrifuge; the remaining soap is mostly removed from the oil via water washing and separating the oil and water in a water washing centrifuge.

This method is particularly suitable for refining the following oils:

- crude soybean oil
- degummed soybean oil
- degummed canola oil
- crude, as well as degummed corn oil
- crude, as well as degummed cottonseed oil
- crude sunflower oil
- crude safflower oil

This method is not suitable for palm oil or coconut oil as mentioned earlier. Both palm oil and coconut oil are low in phospholipids and they do not absolutely require chemical refining. However, the chemical refining process is known to be more effective in reducing trace metals and phosphorus in these oils.

5.4.1 Batch Refining Process

This is the oldest method for chemically refining vegetable oils. The method can produce good quality oil but may not be low in trace impurities, such as phosphorus and trace metals when compared against the continuous chemical refining process.

Fig. 5.2 shows the schematic view of a batch reactor vessel. The vessel must be a pressure vessel if it is to be used for bleaching in addition to refining and water washing.

The step-by-step procedure for the batch refining process is outlined below:

1. Crude oil from the storage tank is analyzed for FFA, moisture and impurities, refining loss, and lab-bleach color. The method for the refining loss for

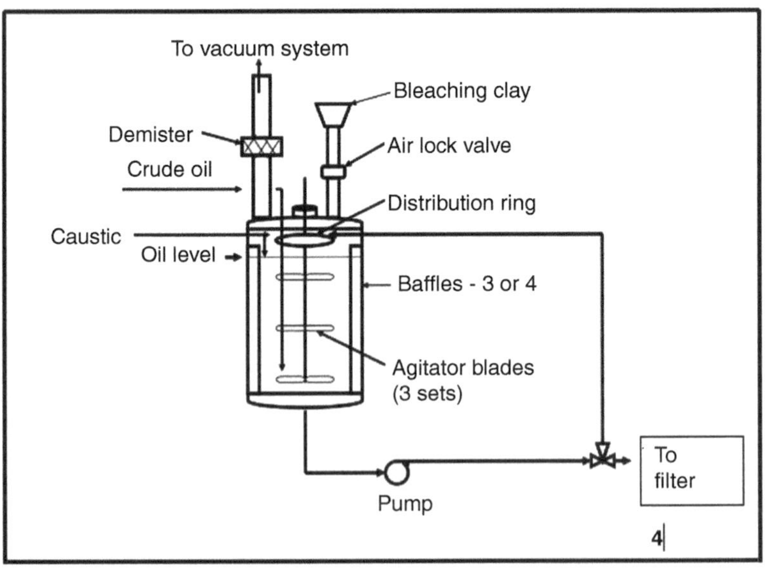

FIGURE 5.2 Combined batch chemical reactor and vacuum dryer.

the crude oil depends on the type of crude oil, as discussed in Chapter 2. Lab-bleach color is a predictor for the deodorized oil color. Generally, the plant processed oil turns out to be lighter than the lab-bleach color. Thus, darker than the lab-bleach color in the plant-produced oil indicates some process related issues that should be investigated and corrected.

2. Measured amount of crude oil is collected in a refining kettle. The kettle is equipped with top entering agitator with variable speed. Baffles are provided to prevent vortex formation.
3. Acid-pretreatment of the crude oil is very common in this process. Batch refining process is used by plants with low production requirement.
4. Caustic strength and the amount of caustic treat depends and are discussed later under Caustic Strength under Continuous Caustic Refining on page 96 of this chapter.
5. The required amount of caustic solution is added into the oil. Caustic solution and the crude oil are mixed at low agitator speed to avoid emulsification. The temperature of neutralization is 95–104°F (35–40°C), for seed oils. Palm oil or coconut oil has a tendency to form emulsion in this process.
6. Check for the FFA in the oil at the end of the reaction. It should be 0.01%–0.02%, expressed as oleic acid.
7. The oil should also be checked for phosphorus content. It should be <5 ppm.
8. The oil is heated to 185–195°F (85–90°C) and 10%–15% of hot deionized water is added into the refined oil. Hard water makes it difficult to remove soap from the refined oil.
9. The agitator speed is increased for obtaining intimate mixing between the refined oil and the water for improved soap removal.
10. The agitator is turned off and the soap in the oil is allowed to settle to the bottom of the tank.
11. A sample of oil is collected from the top and analyzed for soap.
12. If the soap content is >1000 ppm, the soap layer is drained and the oil is washed again.
13. The process of water washing and draining of the soap are continued until the soap content is <400 ppm. It will be seen later that the soap in water washed oil from the continuous process is typically <100 ppm. The amount of soap in the water washed oil in a batch process could be reduced further by repeated water washing but the oil loss would be extremely high.
14. The oil is then treated with 1%–3% of acid activated clay (for continuous caustic refining the typical clay usage is <1%) and bleached under vacuum at 230–248°F (110–120°C).
15. The agitator speed is set at maximum [roughly 120 revolutions per minute (rpm)].
16. The operating pressure for vacuum bleaching is 36–50 mm of mercury. However, most of the batch-refining systems around the world are old and they use atmospheric bleaching.
17. To achieve better deaeration, the oil should be circulated by a pump located at the bottom of the bleacher vessel and through the distribution ring at the top.

18. It is recommended to add 50–100 ppm of citric acid (or phosphoric acid) in addition to the bleaching clay. This helps chelate the trace metals and reduce the phosphorus content of the oil. Some of the acid reacts with the soap in the oil and hydrolyzes the soap. It slightly raises the FFA in the bleached oil.
19. The oil is analyzed for bleached oil color, phosphorus, soap, and chlorophyll contents. These analyses must be performed to confirm the quality of the bleached oil before it is filtered.
20. The bleached oil can be sent directly to the deodorizer. However, the oil must be dried in a vacuum dryer to moisture content of <0.1% if the oil is sent to the hydrogenation plant or is stored before deodorization.

5.4.2 Critical Control Points in Batch Refining

A number of critical control points for the batch refining process are similar to those in the continuous caustic refining process. The fundamental points of departure are shown.

5.4.2.1 *Agitator Speed*

The agitator must be operated at three speeds: low speed for neutralization, medium speed for water washing, and high speed for bleaching.

5.4.2.2 *Bleaching Clay*

The process requires much higher dosage of bleaching clay compared to the continuous caustic refining process. This is because the refined and water washed oil contains high levels of soap. Phospholipids deactivate some of the bleaching clay so not all of the clay added is effective in reducing the color bodies in the oil. This aspect will be further discussed later in this chapter in the section on continuous chemical refining.

5.4.2.3 *Refining Loss*

Refining loss in the batch process is very high as compared to that in the continuous chemical refining process.

5.5 CONTINUOUS CHEMICAL REFINING PROCESS

The continuous chemical refining process is also called the continuous chemical neutralization process. In the United States and other countries, this process is used to refine crude seed oils. This process effectively removes the trace impurities from the oil. There are two major continuous chemical neutralization processes, namely:

1. The Long Mix Process by Alfa Laval
2. The Short Mix Process by Westfalia

Fig. 5.3 shows the flow diagram for the modified Alfa Laval Long Mix Process. The process is referred to as modified because there are a couple of

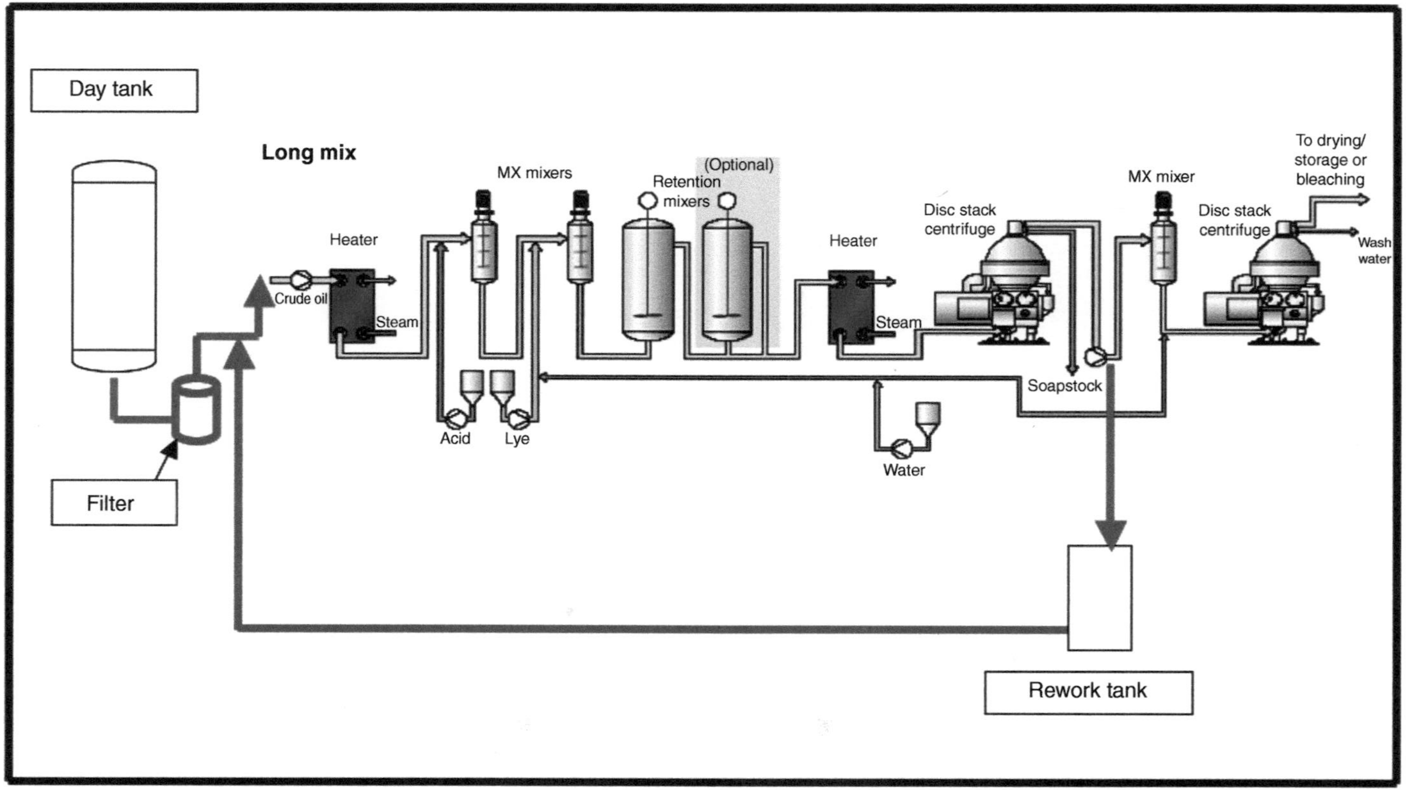

FIGURE 5.3 Alfa Laval Long Mix Process with modification.

additional units included here for obtaining better refining of crude oil according to the current author.

The step-by-step procedure for this process is described later:

1. At the oil refinery, the crude oil is pretreated with phosphoric acid. The oil is at ambient temperature and the acid is dispersed in the oil using a high shear mixer.
2. The acid pretreated oil is stored in a "day" tank. The tank holds up to 24 h of acid treated crude oil supply for refining. The day tank is equipped with a mechanical mixer, which enables one to maintain a uniform composition of oil feed to the refining process. The original Alfa Laval Long Mix Process does not include a day tank. The author strongly recommends that the process include a day tank.
3. The pretreated oil is pumped through a filter and a flow meter into a high shear mixer where metered amount of caustic solution is added. The two liquids are mixed in the high shear mixer (MX Mixer for Alfa Laval process or Centrifugal mixer for Westfalia process). Another high shear mixer, IKA, has been found effective in this application. The caustic solution is dispersed into microdroplets to provide large specific surface area for the reaction between the caustic and the FFA, as well as other impurities in the oil. The temperature of this mixing step is 90–104°F (32–40°C). The amount of caustic solution added is referred to as "caustic treat." The MX mixer has two inlets (Fig. 5.4). The oil and caustic can be mixed 100% in the dispersion (high shear) section or it can be split between the dispersion and the mixing section. This flexibility allows the refiner to prevent any emulsion formation, especially where poor quality crude is refined with an extra amount of caustic treat.
4. The caustic treated oil then passes through a set of retention mixers (Fig. 5.5). These are vertical vessels with top entering agitators with spacers and knife blades to maximize mixing of the two liquids under gentle condition. The retention time ranges from 4–15 min (Table 5.6), depending on the type of oil being refined. A certain amount of retention (contact) time is essential for the caustic to hydrate the nonhydratable phospholipids in the crude (or degummed) oil.

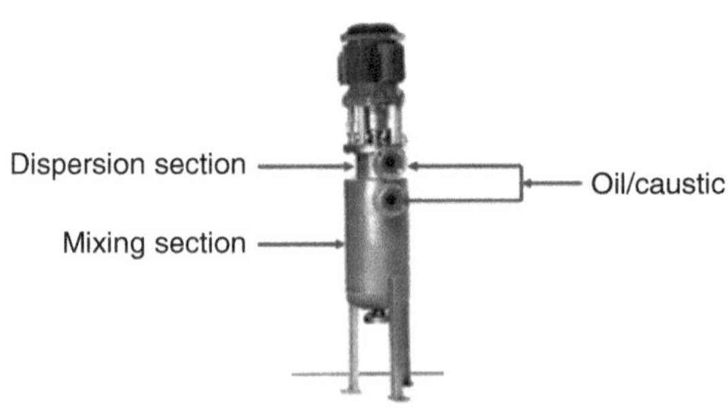

FIGURE 5.4 Alfa Laval MX mixer.

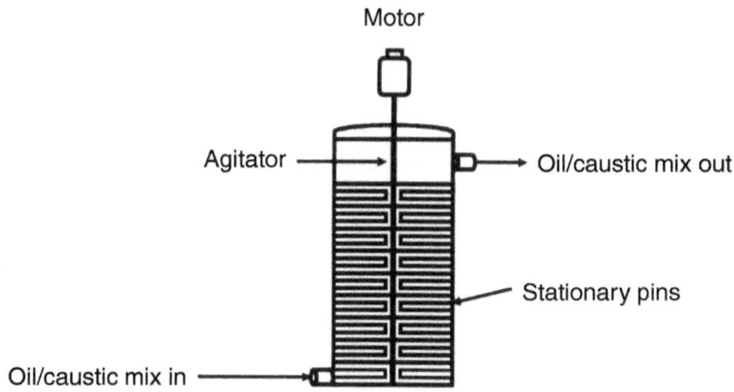

FIGURE 5.5 Vertical mixer.

5. The caustic treated oil from the retention tank is then heated to 150–164°F (65–73°C) in a steam heater (also called an emulsion heater), where the soap begins to separate from the oil.
6. The heated oil then passes through a primary centrifuge (also referred to as a primary separator). Under the centrifugal force of the bowl in the centrifuge, soap separates from the oil and goes toward the outer periphery of the bowl, leaving the oil toward the center of the bowl. There is an interface between the oil and the soap and water, which is formed toward the outer edge of the centrifuge bowl. It is critical to maintain an effective soap/oil interface in the centrifuge bowl in order to control the amount of residual soap in the oil and also the amount of neutral oil in the soap stream. Soap and oil leave the centrifuge through two separate ports. See Fig. 5.6 for the cut-away view of the primary centrifuge from Westfalia.

   Oil and caustic mixture from the emulsion heater enters at the top of the disc stacks of the centrifuge bowl. The centrifugal force of the bowl pushes

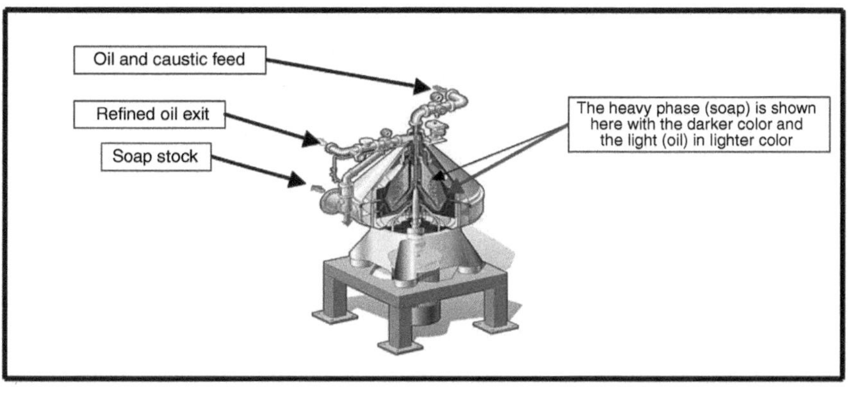

FIGURE 5.6 Cutaway view of primary separator (Westfalia).

the heavy phase (soap) toward the outer periphery of the disc stacks. The lighter phase (oil) goes toward the center and rises up vertically and then exits the centrifuge. Soap rises up the outer periphery of the bowl and then exits the centrifuge.

Along the vertical cross section of the centrifuge, the soap is on the outside, oil is inside and there is an interface between the two phases. The interface is a fine divider between the soap and the oil where the oil/soap emulsion is not totally separated. The key to proper separation of oil and soap is to maintain this interface in the right position in the centrifuge bowl and maintaining minimum oil content in the soap and minimum soap in oil. Therefore, it is necessary to maintain a certain amount of back pressure on the oil side (light phase) to control the amount of neutral oil in the soap. The object is to minimize the amount of soap in the neutralized oil. A lower back pressure allows more soap to remain in the refined oil (light phase). A higher back pressure on the oil discharge line increases the amount of neutral oil left in the soap (heavy phase). If a very high back- pressure is applied, the oil breaks over into the soap discharge and the phenomenon is called "break over." This results in heavy oil loss in refining. Therefore, a careful balance is maintained between the back pressure, neutral oil in the soap, and the amount of soap in the refined oil. The back pressure setting varies with the type of crude oil, as well as the quality of the crude oil. This is determined by the operator based on ppm soap in the oil and neutral oil content in the soap.

7. Oil leaving the primary separator at start up is diverted to a "work tank" until the refined oil quality is satisfactory. This is not included in the Alfa Laval Long Mix Process. This is the current author's recommendation. The start-up oil must not be sent back to the day tank, because it can upset the setting on the caustic meter, adjusted for some specific FFA and the estimated excess. Therefore, a work tank should be considered as an essential part of this process.
8. Continue to send the refined oil to the work tank until the FFA in the refined oil is 0.01%–0.02% and steady. The other analyses to be checked are phosphorus and lab-bleach color. Send the refined oil forward to water washing when the FFA in the refined oil is 0.01%–0.02% (not to exceed 0.03%), and other analyses are satisfactory. Generally, only FFA is checked before the refined oil is sent to water wash because checking for phosphorus and lab-bleach color is time consuming. Some modern refineries in the United States and also in some refineries around the world have installed an ICP unit, which is accessible to the operators to obtain the phosphorus analysis on the refined oil as well as bleached oil in minutes. This makes the control of the refined oil quality much easier.
9. If the FFA is >0.03%, continue to divert the refined oil to the work tank and check the following:
   a. If the high shear mixer and/or the vertical mixers are not working. If the mixer/mixers are not working, shut down the system and fix it. Restart

the process; wait for 15 min to recheck the FFA in the refined oil. Send the refined oil to water washing if the FFA is <0.03%.
   - **b.** If the mixers are working, increase the caustic treat and recheck the FFA in the oil after 15 min.
   - **c.** If the FFA is still high, recheck the FFA content in the crude oil. If the crude oil FFA is same as before, but the refined oil FFA continues to be high, check the strength of the caustic solution.
   - **d.** If the caustic strength is unchanged, increase the caustic treat. Recheck oil analysis after 15 min.
   - **e.** Continue the above process until the refined oil FFA is <0.03%, phosphorus is <3 ppm, and the steady state condition is reached.

10. If the FFA is satisfactory but the phosphorus did not reduce, check the high shear mixer and the vertical mixers to make sure they are working. If the mixers are working, reduce the crude oil flow rate to allow additional residence time in the vertical mixers for improved hydration of the nonhydratable phospholipids.
11. If the FFA and phosphorus are satisfactory but the lab bleach color is high, increase the strength of the caustic solution, keeping the total caustic treat unchanged.
12. Once the steady state conditions are reached in the primary separator, the refined oil is ready to be water washed.
13. At steady state the refined oil should have <0.01%–0.02% FFA, and never >0.03% FFA, <3 ppm of phosphorus, and <500 ppm of soap and never to exceed 1000 ppm.
14. Once the refining operation reaches the steady state, the oil from the work tank is slowly blended at 5%–10% into the main oil stream without disturbing the FFA of the feed.

The refined oil is mixed with 10%–15% of hot soft water in a high shear mixer and then centrifuged in a water wash centrifuge to remove the residual soap from the oil. The detailed step-by- step procedure for water washing is discussed in Section 5.6.

The water washed oil is sent to bleach. Bleaching is discussed in Chapter 6. The soap is sent to acidulation where it is split with sulfuric acid to recover the fatty acids. The bleached oil is sent to various processing steps, such as:

- winterization
- hydrogenation
- product formulation
- deodorization

The final step in oil processing is deodorization as illustrated in Fig. 5.7.

The oil leaving the refining centrifuge must be sampled for analysis at a certain frequency. Table 5.3 lists the sampling frequency for the continuous chemical refining process.

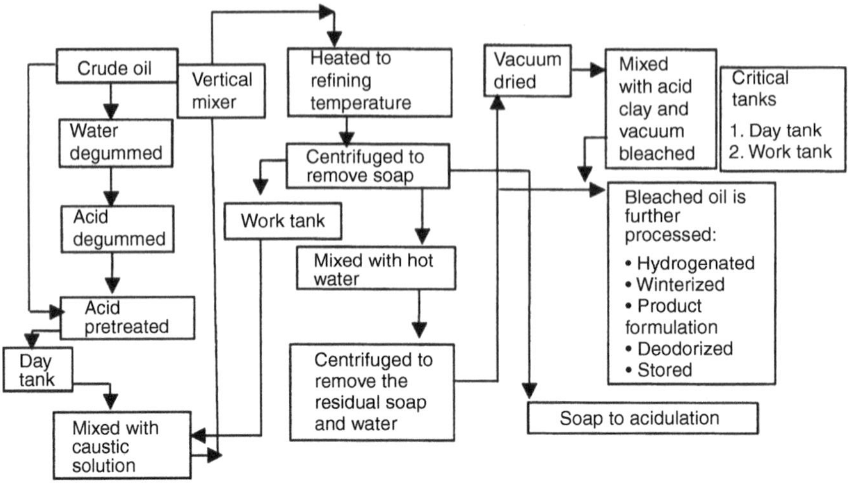

FIGURE 5.7 The schematic flow diagram for complete oil processing.

TABLE 5.3 Sampling Frequency in the Chemical Refining Process

Sample location	Frequency of analysis	FFA (%)	Soap (ppm)	Moisture (%)	Lab-bleach color	Phosphorus (ppm)
Crude oil (from day tank)	Once/Shift	X	—	X	—	X[a]
Exit primary separator at:						
Start	Every 15 min	X	X	—	—	X
Steady state	Every 2–4 h	X	X	—	—	X

[a]*Once per day in the tank.*

5.5.1 Critical Control Points in Continuous Chemical Refining Process

There are specific operating standards that one must follow in the refining process. Deviation from these recommended limits prevents the refiner from producing high quality refined oil. This can also cause certain quality or process issues in the later stages of oil processing that result in poor oil quality and an increase production cost. Following are the critical control points in the continuous refining process.

1. crude oil filtration
2. crude oil pretreatment (day tank)
3. uniformity of crude oil feed composition

4. uniformity of crude oil flow
5. uniformity of caustic flow
6. caustic strength
7. caustic treat
8. degree of mixing between the crude and the caustic
9. contact time between crude oil and caustic
10. caustic addition temperature
11. refining temperature (emulsion heater temperature)
12. segregation of the start-up oil (work tank)
13. condition of the primary separator
14. back pressure on the light phase of the primary separator

1. Crude oil filtration
   It is very important that the crude oil coming into the refining process be filtered. It is done either in basket strainers or in disc-stack filters. The solid impurities, such as the meal, the hull, dirt, etc., are removed in this process. Without this step, the centrifuge bowl of the primary separator gets dirty soon and requires more frequent cleaning. This is true even for the self-cleaning centrifuge.
2. Crude oil pretreatment
   Proper acid pretreatment is essential for all crude oils and especially for those that have been derived from poor quality seeds or damaged during storage. Without this step, refining of the crude oil is difficult. In some cases, it becomes impossible to refine the crude without this step. Acid degumming and acid pretreatment of crude oil have been discussed in detail in Chapter 4. Following are the important considerations for this step:
   Proper acid dosage must be determined on the basis of calcium and magnesium content of the crude oil plus some excess as discussed under degumming (Chapter 4).
   a. Low acid dosage cannot reduce the nonhydratable phospholipids in the crude oil and makes it difficult to refine.
   b. Higher acid dosage:
      - Can cause breakdown of the chlorophylls in the crude oil, making the oil more sensitive to photooxidation. This is discussed in detail in Chapters 6 and 12.
      - Will need a higher caustic treat. This can increase the amount of diglycerides formation in the refined oil. This can increase the refining loss.
3. Uniformity of crude oil feed composition
   It must be remembered that vegetable oil refining, water washing, and hydrogenation are chemical reaction processes. Therefore, one should consider each of these process units as chemical reactors, designed to perform certain functions under specific operating standards. This includes the feed to the refining centrifuge, which must be maintained at a uniform

composition, once the process has started and the reaction parameters have been set.
The caustic treat is set on the basis of the oil analysis and flow rate. A nonuniform feed oil quality can cause under- or overrefining of the oil.
Use of a day tank can alleviate the variations in the feed oil quality.

4. Uniformity of crude oil flow
A variation in oil flow can change the oil/caustic ratio causing:
   a. underrefining, or
   b. overrefining of the crude oil
The crude oil flow rate must be maintained at $< \pm 1\%$ of the set flow rate.
5. Uniformity of caustic flow
It is critical that the caustic flow is controlled within $\pm 1\%$ of the set caustic flow. A variable caustic flow results in over- and underrefining.
In older refineries, the oil flow, caustic flow etc. were controlled by the following means:
   a. manual adjustment
   b. controllers operated by the pneumatic control devices
   c. controllers operated by the relay system
The modern refineries use the programmable logic control (PLC) system for all flow, temperature, and pressure control. This system provides more uniform control of the process parameters and can be done from the central control room.
6. Caustic strength
The recommended caustic strengths for various types of oil are listed in Table 5.4. Generally, the oils with low phospholipids require dilute caustic solution. On the other hand, the nondegummed crude oils require stronger

TABLE 5.4 Recommended Caustic Strength for Various Crude Oils for Refining

Oil type	Recommended strength for the caustic solution, °Be
Crude soybean oil	18
Degummed soybean oil	14
Super degummed oil	14
Corn	18–22
Cottonseed	28–32
Sunflower	14
Safflower	14
Coconut oil (normally physically refined)	10–12
Peanut (groundnut)	18

caustic solution. One needs to select the right caustic strength in order to obtain the desired refining results and prevent undue oil loss because:

a. A higher than required caustic strength can cause excessive reaction between the caustic and the neutral oil. This increases the level of diglycerides in the oil and increases oil loss in refining.

b. A lower than recommended caustic strength may not be adequate to reduce the color of the oil. It also reduces the density of the soap, which may cause higher oil loss due to poor separation in the primary separator.

It becomes difficult at times to reduce red color in the refined cottonseed oil made from crude cottonseed oil. This occurs when the crude oil is old or produced from poor quality seeds or crushed under less desirable operating standards. In such a case, a higher caustic strength is used without changing the total caustic treat. This reduces the Lovibond red color of the refined oil.

7. Caustic treat

In addition to the strength of the caustic one must ensure the total caustic treat is adjusted for each oil feed going to the refining step. It has been explained earlier that the amount of caustic needed is higher than that required just to neutralize the FFA in the crude oil. This is because some additional amount of caustic is needed to hydrolyze the nonhydratable phospholipids in the oil. Table 5.5 shows the recommended percent excess caustic treat for various vegetable oils for refining.

Using higher or lower caustic treat will have similar effects on the refining process as the variable caustic feed discussed earlier.

TABLE 5.5 Recommended Percent Excess Treat for Various Crude Oils for Refining

Oil type	Recommended percent excess caustic
Crude soybean oil	0.05–0.15
Degummed soybean oil	0.01–0.05
Super degummed oil	0.05–0.1
Corn	0.1–0.4
Cottonseed	0.1–0.4
Sunflower	0.05–0.15
Safflower	0.05–0.15
Coconut oil (normally physically refined)	0.0–0.02
Peanut (groundnut)	0.05–0.15

Calculating caustic treat: It consists of two parts:

a. Theoretical amount of caustic needed to neutralize the FFA and form soap.

b. Additional amount of caustic needed to convert the nonhydratable phospholipids to a hydratable one. This is referred to as percent excess caustic treat.

Sample calculation:

$$\text{FFA in the crude oil} = 1.0\%$$

$$\text{Excess treat used} = 0.15\%$$

$$\text{Total caustic treat} = (0.142 \times 1 + 0.15) \quad = 0.292\ \text{lb} / 100\ \text{lb of oil}$$
$$\text{Or} = (0.292\ \text{kg} / 100\ \text{kg of oil})$$

$$\text{Caustic solution} = 16°\text{Be} = 11.09\%\ \text{NaOH}$$

$$\text{Caustic treat} = (0.292 / 11.09) \times 100 \quad = 2.63\ \text{lb} / 100\ \text{lb of oil}$$
$$\text{Or} = (2.63\ \text{kg} / 100\ \text{kg of oil})$$

$$\text{Refining rate} = 15,000\ \text{lb of crude oil} / \text{h}$$

$$16°\text{Baume caustic solution needed} \quad = (15,000 \times 2.63/100) = 394.95\ \text{lb/h}$$
$$= (15,000/2.2) \times (2.63 / 100) = 179.32\ \text{kg/h}$$

Table 5.5 lists the recommended percentage of excess caustic for various seed oils in chemical refining. One can calculate caustic treat using the sample calculation.

8. Degree of mixing between the crude oil and the caustic

Oil and caustic do not readily mix with one another because the caustic solution is in an aqueous medium. The caustic needs to be dispersed intimately into the oil so that an enormous specific area (area/unit weight or volume) for the caustic solution is created. This can be accomplished with the help of a high shear mixer. The Alfa Laval process uses the MX mixer while the Westfalia process uses a Centrifugal mixer. Fig. 5.8 shows the cutaway view of Westfalia Centrifugal mixer.

The mixer looks and also operates on the basic principles of centrifugal force and mixing action. The two streams of liquid enter the mixer where they are intimately mixed, and the centrifugal force pushes the mixture toward the periphery of the mixer, and then it leaves the mixer.

The MX mixer from Alfa Laval is different from the centrifugal mixer. The former (Fig. 5.4) has a mixing section and a high shear section that accomplishes the dispersion of the caustic solution in the oil.

One can experience low FFA in the refined oil but significantly higher FFA in the water washed oil if the caustic and crude oil are not properly mixed

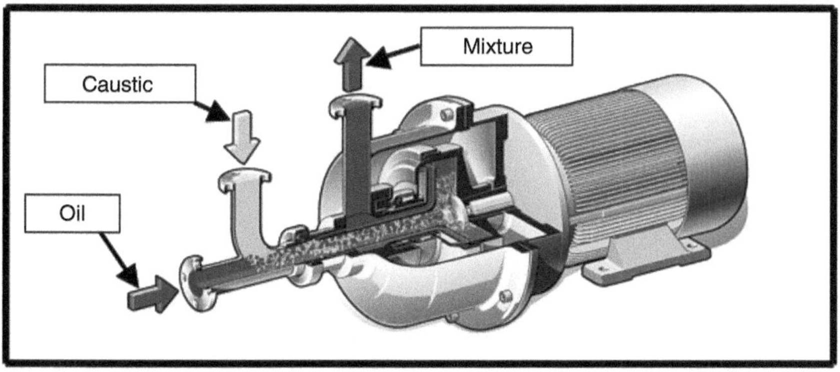

FIGURE 5.8 **Westfalia Centrifugal Mixer.**

together. This is because poor mixing of the oil and the caustic leaves excess amount of free alkali in the refined oil. This artificially reduces the amount of caustic solution needed to neutralize the FFA in the refined oil. The excess alkali gets removed from the refined oil in the water-washing step and the FFA in the water washed oil appears to be higher than expected. As stated earlier, one should expect to see 0.01%–0.02% FFA (not to exceed 0.03%) in the refined oil and no higher than 0.05% FFA in the water washed oil.

9. Contact time between crude oil and caustic
Alfa Laval Long Mix Process uses lower temperature for the caustic addition to oil but it allows several minutes of contact time for the caustic to react with the nonhydratable phospholipids in order to make it hydratable so the phospholipids content in the refined oil is low.
The suggested retention times for various types of oil are shown in Table 5.6. One can see that crude soybean and cottonseed oil take the longest retention time. Degummed soybean and canola oil rank second, sunflower, safflower oil contains very low levels of phospholipids (refer to Table 1.1). In reality, both sunflower and safflower oil, if water degummed, could be refined with 0–3 min of residence time or via physical refining.
The Short Mix Process, which will be discussed later does not use any contact time between caustic and crude oil after mixing.

10. Caustic addition temperature
Alfa Laval Long Mix Process uses an oil temperature of approximately 32–40°C (90–104°F) to add caustic to it before high shear mixing. The entire process is based on the longer mixing time and reduced temperature for the oil/caustic mixture. At temperatures lower than 32°C (90°F), the dispersion of caustic into the oil becomes less effective because of the higher viscosity of the oil. At temperatures higher than 40°C (104°F), the caustic may react with the neutral oil producing diglycerides and soap. This can increase the refining loss.

TABLE 5.6 Recommended Contact Time Between Various Crude Oils and Caustic in the Long Mix Process

Oil type	Retention time (min)
Soybean	
Crude Degummed	6–15 6–12
Canola	
Crude Super degummed	6–12 4–6
Cottonseed	9–15
Corn	4–6
Sunflower	4–6
Safflower	4–6
Peanut (groundnut)	4–6

11. Refining Temperature (Emulsion Heater)
In the Alfa Laval Long Mix Process, the oil from the retention tank is heated in an emulsion heater. Heating the caustic/oil mixture allows the soap to separate from the oil in the form of tiny globules.
Table 5.7 shows the recommended temperatures for oil/caustic mixture (or emulsion as it is commonly called).
The refining temperature is chosen on the basis of:
a. type of crude oil
b. difficulty of removal of the phospholipids
c. difficulty of color removal

TABLE 5.7 Refining Temperature

Oil type	Refining temperature	
	°F	°C
Coconut	130–140	55–60
Sunflower Safflower Palm	140–150	60–65
Soybean Canola Peanut (groundnut)	150–165	65–73
Cottonseed	140–150	71–79

Coconut oil, if ever chemically refined, must use the weakest caustic solution and the lowest refining temperature. On the other hand, crude cottonseed oil, which is typically the most difficult oil to refine, uses most concentrated caustic solution.

12. Segregation of refined oil at start up (use of a work tank)

Generally the FFA and other impurities take about 15 min to reach the desired level after start. The sample is taken after the primary separation. During this time the refined oil should be diverted into a tank called "work tank." This tank should have a capacity equivalent of 2–4 h of production. This acts as a buffer tank at the start because the oil/caustic ratio, the reaction between the caustic and the impurities are incomplete. There is always free caustic (also referred to as free alkali) in the oil/caustic mix at the start. This mixture, if sent to the day tank, can affect the uniformity of the oil quality in the feed. This can result in under- and overrefining of the oil. Therefore a work tank is highly recommended.

The oil from the work tank can be blended into the main feed stream at 5%–10% of the total oil flow rate until the work tank is empty.

13. Condition of the primary separator

The bowl of the separator must be clean in order to obtain a good separation between the oil and the soap. The older primary separators do not have the self-cleaning feature that the newer ones do. This makes it important to maintain a close watch on the refined oil quality and the neutral oil content in the soap. Without any other change in the process, an increase in the soap content in the refined oil is a good indicator that the bowl needs cleaning.

The disc stacks are taken out and caustic washed. The bowl is cleaned to remove all deposits. The self-cleaning type centrifuge bowl generally can be operated for several weeks before they need cleaning.

14. Back pressure on the light phase

It has been briefly discussed earlier that a certain amount of back pressure is maintained on the light phase (oil) exit from the primary separator in order to control the following:

a. amount of soap in the refined oil

b. amount of neutral oil in the soap

Maintaining an appropriate back pressure is very important because when the pressure is low, there is more soap in the oil. Conversely, when the back pressure is high, the amount of soap in the oil is less but the amount of neutral oil in the soap is high.

As discussed earlier, the amount of soap in the neutral oil must be less than 100 ppm (preferably <50 ppm). The soap produced by the Alfa Laval process contains 45%–50% moisture and the neutral oil content in the soap must not exceed 30% dry basis. In most well-managed operation it is less than 20%.

5.5.1.1 *Analyzing Percent Neutral Oil in the Soap*

Analyzing for neutral oil in the soap is rather tricky. This is because as the soap leaves the primary separator, it contains some amount of free alkali. This free alkali starts reacting with the neutral oil in the soap from the very moment the sample is collected.

Normally, the soap sample is collected and brought to the QC laboratory where it is analyzed for neutral oil content. It may take anywhere from 30 min to a few hours before the soap is mixed with the solvent for neutral oil analysis. A large portion of the neutral oil (in the soap) is converted to soap by the time the analysis begins.

In order to get an accurate reading on the neutral oil content in the soap, one must follow the steps outlined later:

1. In a 500 mL conical flask, measure out the correct volume of the solvent used for the extraction of the neutral from the soap [AOCS method: G 5–40 (09)].
2. Carry the flask to the oil refinery.
3. Weigh in the required amount of soap into the flask containing the solvent and mix the soap into the solvent with the help of a stainless steel spatula.
4. Take the flask to the QC laboratory and conduct the neutral oil test.

5.5.1.2 *Comments*

By dispersing the soap into the solvent the caustic is separated from the oil because caustic is insoluble in the solvent. Therefore, there is no further reaction between the neutral oil and the caustic, forming soap. To illustrate this point, the data from an actual refining test are shown in Table 5.8.

The neutral oil in the soap dropped significantly after 2 h of waiting before analysis. The neutral oil in the samples dispersed in the solvent did not change over 5 h of waiting before analysis. Many refiners do not have accurate information on the neutral oil loss (NOL) because they do not think about the excess caustic in the soap and its reaction with the neutral oil present in it.

TABLE 5.8 Impact of Storage Time on Neutral Oil in Soap

Storage time (h)	Neutral oil, % standard soap sample collection (dry basis)	Neutral Oil, % soap sample dispersed in solvent (dry basis)
0	26.8	27.6
2	14.7	27.6
5	11.5	27.6

5.6 WATER WASHING REFINED OIL

Once the primary separator reaches the steady state, the refined oil is sent to water washing. The object of water washing the refined oil is to reduce the soap to a lower level. The following steps take place in water washing.

1. The refined oil leaves the primary separator at 140–155°F (60–68°C) and is heated to 180–190°F (82–88°C).
2. The water-washing centrifuge is primed with soft deionized water and kept running for water washing of the refined oil.
3. Soft deionized water is heated to the same temperature as the oil: 180–190°F (82–88°C).
4. Approximately 10%–15% of hot soft water is added into the hot refined oil and mixed thoroughly in a high shear mixer (MX).
5. The water wash centrifuge reduces the soap content of the incoming refined oil by a factor of 10. The desired soap content in the water washed oil is <50 ppm and not to exceed 100 ppm. If the soap content in the refined oil is high, the water wash separator is unable to reduce the soap content to the desired low level. This happens when the refined oil contains more than 1000 ppm of soap.
6. If the soap content in the refined oil is approaching 1000 ppm, it is advisable to add some phosphoric acid in the oil/water mixture. The amount of acid added depends on the soap content of the refined oil. This offers the following benefits:
   a. The acid hydrolyzes the excess soap forming FFA and sodium phosphate. This reduces the net load of soap on the water-washing centrifuge for soap removal from the refined oil.
   b. This reduces the amount of acid activated clay required in the bleacher.
   c. No soap is left in the bleached oil even at the reduced usage of the acid activated clay.
   d. Reduction of soap to zero ppm prevents premature blinding of the filter screen. This improves process downtime to clean the filter press more frequently.
   e. It improves removal of the trace metals.

   However, the first priority of the refiner should be to determine and eliminate the cause of high soap in the refined oil.
7. The wash water is separated from the oil in a water wash centrifuge.
8. In an Alfa Laval B214C water wash centrifuge, the oil is pumped out of the centrifuge with the help of an integral component called a paring disc. This almost looks and functions like the impeller of a centrifugal pump.
9. A certain amount of back pressure is applied on the oil discharge line with the help of manual or an automatic control valve to accomplish the following:
   a. Control the amount of oil loss in the wash water discharge.
   b. Control the amount of soap in the water washed oil.

If the back pressure is too low, or the valve fails to maintain the back pressure, the oil flow through the centrifuge can become very high, overloading the vacuum dryer.

10. Other types of water washing centrifuges also use the back pressure on the oil discharge, even if they do not have a paring disc like the Alfa Laval Centrifuge, model B214C.
11. The oil is then dried in a vacuum dryer, maintained at 85–90°C (185–190°F), and operated under 36–50 mm of mercury absolute pressure.
12. The oil leaving the water wash separator typically contains 0.4%–0.5% moisture, 10–100 ppm of soap, and <3 ppm of phosphorus. This is ideal for the use of hydrogel silica like Trisyl in bleaching.
13. In a closed-coupled process, the water washed oil goes directly into the vacuum bleacher for bleaching. However, the water washed oil must be cooled to <40°C if the oil is not sent to bleaching immediately after water washing. The water washed oil must not be held without bleaching for more than a few hours, otherwise the FFA in the oil will begin to rise and the oil will begin to get oxidized.
14. Refined and water washed oil is analyzed. Table 5.9 lists the recommended frequency and the type of analysis for the water washed oil.
15. In some older plants water washing is conducted in a tank. The tank has a top entering agitator with high shear mixing capability and a set of three or four baffles along the vertical wall of the tank to prevent any vortex formation. The level of oil in the tank is maintained with the help of a level controller, a pump, and a three-way modulating valve (Fig. 5.9).

The level indicator controller (LIC) senses the oil level in the tank.

The pump pumps the oil out of the tank and sends it to the three-way modulating valve, which receives the signal from the LIC and allows a portion of the oil to move forward into the water wash centrifuge while maintaining the level in the tank by recirculating a portion of the oil back into the tank. At the end of the operation, the three-way modulating valve is put on manual mode to allow the oil from the tank to go to the water wash centrifuge.

TABLE 5.9 Oil Quality Standards in Continuous Chemically Refined and Water Washed Oil

Sample location	FFA (%)	Soap (ppm)	Phosphorus (ppm)
Exit			
Primary separator	0.01–0.03	< 500	< 3, Preferably <1
Water wash separator	0.02–0.04	< 50	< 3, Preferably <1

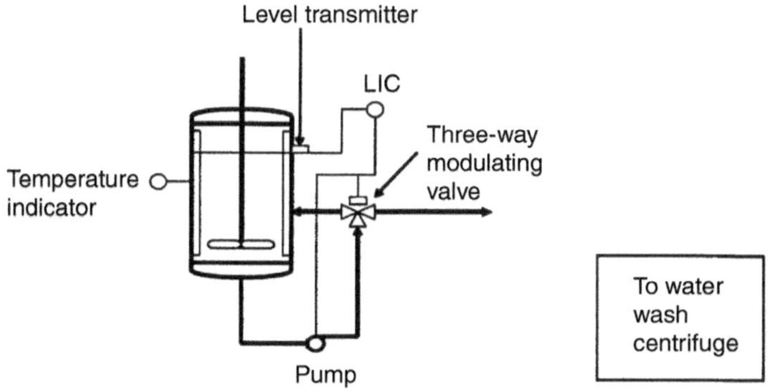

FIGURE 5.9 Water wash mixing tank.

5.6.1 Critical Control Points in Water Washing

Like the refining step, the water-washing step also requires careful control of the operating conditions. The critical process control points in water washing are listed later:

- ratio of wash water to the refined oil
- water wash temperature
- degree of mixing
- mixing time
- hardness of the water
- back pressure on the oil phase

1. Ratio of wash water to refined oil
   Refined oil to wash water ratio is critical. The refined oil must be mixed with 10%–15% deionized water. At a ratio of less than 10% water, the removal of soap is less efficient. If the water is much higher than 15%, the soap solution will have lower density and the separation by centrifugal force is less efficient. This can increase the oil loss in the water, as well as leave more soap in the oil.
2. Water wash temperature
   The temperature of refined oil and wash water mixture should be 180–190°F (82–88°C). At a lower temperature, the following process difficulties are experienced:
   a. Soap removal is inefficient because the water does not disperse well into the oil.
   b. The oil temperature is too low for proper moisture removal in the vacuum dryer.

   At a higher temperature, the mixture approaches the boiling point of water. This can cause cavitation in the pumps and the paring disc in the water wash separator. Cavitation causes physical damage to these process components.
3. Degree of mixing

Oil and water do not mix. The water must be dispersed into the oil phase like the caustic solution in the refining step. High shear mixing is needed to accomplish this type of mixing. Both MX mixer (Alfa Laval) and Centrifugal Mixer (Westfalia) are effective mixers for this process.

4. Mixing time
   The residence time in the MX or Centrifugal mixer is very low. On the other hand, where a mixing tank is used, the residence time for oil and water in the tank should be at least 6 min. At lower mixing time, there is less soap removal from the oil. This can also increase the oil loss in the wash water. The agitator impeller used in the tank must be of high shear type for efficient mixing of the refined oil and the wash water.
5. Hardness of water
   The wash water must be soft. The hardness of the water should be less than 20. At higher hardness, the water contains water-soluble calcium and magnesium salts. This interferes with the removal of the soap from the refined oil. In addition, some of the water-soluble calcium and magnesium salts can react with the FFA forming calcium and magnesium soap in the water washed oil. This can affect the refined oil quality in the long run.
6. Back pressure on the oil phase
   A back pressure of approximately 10 psig is applied to the oil discharge from the water wash centrifuge, Alfa Laval B-214C. This pressure is necessary in order to minimize the oil loss in the discharge water and at the same time have low soap in the water-washed oil.
   A lower back pressure will leave more soap and more water in the oil, which can reduce the efficiency of the vacuum dryer.

5.6.2 Importance of Oil Quality Parameters of the Refined and Water Washed Oil

It has been discussed earlier that the impurities in the crude oil are reduced in the refining process. The quality of the refined and water washed oil must meet the required standards before the oil can be sent to bleaching. Poor quality refined and water washed oil results in overall poor quality oil and also causes numerous difficulties in the bleaching, hydrogenation, winterization, and even in the deodorization process. The desirable quality standards for the refined and water washed oil are shown in Table 5.9. The quality of the deodorized oil depends greatly on the quality of the refined and water washed oil.

It is recommended that the refined and water washed oil should meet the following analytical standards:

FFA	0.03%–0.05%
Phosphorus	<3 ppm
Soap	10–100 ppm
Moisture	0.4%–0.5%
Neutral oil in water	<0.1%

5.6.3 Importance of Having Low FFA, Soap, and Phosphorus in the Refined and Water Washed Oil

FFA, soap, and phosphorus must be very low in the refined and water washed oil as shown in Table 5.9. The oil refiner can significantly improve the oil quality and reduce the processing cost if proper care is given in the refining step. The adverse impacts of the above impurities are discussed later.

5.6.3.1 FFA

The FFA in the refined oil, as well as the water washed oil, should be low, as indicated in Table 5.9.

A low FFA in the refined oil and significantly higher FFA in the corresponding water washed oil indicates the presence of free alkali in the refined oil due to:

- poor mixing of caustic and crude oil
- over treating the crude oil

In these instances, the excess alkali (caustic) left in the refined oil reduces the amount of titrating alkali giving the impression that the FFA in the refined oil is low. However, when the excess alkali is washed and removed in water washing, the FFA in the oil increases.

High FFA in the refined and water washed oil can increase the requirement for the bleaching clay because the soap poisons the acid activated clay.

High FFA in the oil can poison the hydrogenation catalyst.

The bleaching process generates some FFA in the oil because the acid activated clay hydrolyzes the soap in the water washed oil. This is why it is so important that the soap content in both the refined and water washed oil be as low as possible, as indicated earlier.

5.6.3.2 Soap

Soap in the refined oil must be as low as possible, preferably <500 ppm. Normally, the water wash centrifuge reduces the soap in the refined oil by a factor of ten (50 ppm or so). High soap in the refined oil makes it harder for the water wash centrifuge to remove it from the oil. Thus, high soap in the oil increases oil loss in the wash water from the centrifuge and also leaves high level of soap in the water washed oil.

High soap in the water washed oil has numerous detrimental effects on the oil quality, as well as in the subsequent stages of oil processing, such as bleaching, hydrogenation, and deodorization. The following is a list of detrimental effects of high soap in the oil:

5.6.3.3 Bleaching

It poisons the bleaching clay, requiring an increased dosage of the clay.

It blinds the filter and slows down the oil filtration rate, reducing productivity in bleaching.

The combination of higher amount of soap and clay produces higher FFA in the bleached oil due to the hydrolysis of the soap by the acid activated clay. The higher FFA in the oil poisons the hydrogenation catalyst and also increases NOL in the deodorizer distillate. High FFA in the oil can reduce the deodorizer throughput.

High use of bleaching clay can oxidize some of the natural antioxidants in the oil reducing the oil stability. This is commonly observed in the oil that has been bleached repeatedly for some reason.

If the soap is carried through all the way into the deodorizer, the finished oil may have darker color, poor flavor, and it may get darker in storage.

5.6.3.4 Hydrogenation

Soap, left in the bleached oil, poisons the hydrogenation catalyst; which shows the following detrimental effects:

- Increases catalyst usage.
- Increases reaction time, lowering productivity.
- Produces colloidal nickel that must be removed in the via additional bleaching step with acid activated clay.
- Slows the catalyst filtration rate. This lowers the productivity in hydrogenation.
- Changes the solids profile in the hydrogenated oil.
- Increases *trans* fat content in the hydrogenated oil.

5.6.3.5 Phosphorus

- High phosphorus in the refined and water washed oil impacts the bleaching and hydrogenation processes very much like high soap.
- High phospholipids in the bleached oil can prevent the reduction of FFA in deodorization.

This is because a certain amount of FFA is also formed in the deodorizer due to hydrolysis. However, an equilibrium condition is reached where the amount of FFA removed from the oil is higher than the amount of FFA formed. In the presence of a surfactant, such as phospholipids, the hydrolysis of the neutral oil is high and may exceed the rate of removal of the FFA from the oil by the stripping steam. In such cases, the FFA cannot be reduced below a certain limit. This is sometimes observed during deodorization of oil that still contains high phosphorus after bleaching. The author has experienced this phenomenon with corn oil, cottonseed oil, and palm oil (from improper physical refining of the oil) in the physical refining process, where the FFA does not come down below 0.06% when the phosphorus contents of these oils were 3–4 ppm.

5.6.4 Comments on Chemical Refining Process

Chemical refining is essential for most seed oils as discussed earlier. The chemical refining process is also very effective in removing the trace impurities from the crude oil, where the physical refining process might not do the same and proponents of physical refining dispute this. As mentioned earlier, some processors claim that they are successfully refining crude or degummed soybean oil via physical refining process. But in many instances, they do not check the oil for the phosphorus content either in the crude or in the refined oil. This is a contradiction! One must reduce the phosphorus in the crude oil to a very low level before the oil can be properly physically refined.

5.6.5 Troubleshooting Chemical Refining Process

The refined, as well as the water washed oil exhibit certain behavior if the critical process control standards are not properly maintained. Table 5.10 discusses the oil quality, aberrations, and provides guidelines to understand the underlying cause or causes behind the issue and offers solutions.

5.7 REFINING LOSS

One of the major objects of the oil processor is to produce oil with minimum losses at the various processing stages. Most of the oil loss occurs in refining and water washing. Therefore, the topic of refining loss needs some discussion.

Refining loss in the laboratory is determined either by the refining cup method [AOCS Method: Ca 9a-52 (11)] or by the chromatographic [AOCS Method: Ca 9a-52 (11)] method. The cup refining loss method can be used for all oils except the solvent extracted cottonseed oil from prepressed cake. The chromatographic method can be applied to all vegetable oils.

Alfa Laval Long Mix Process indicates that the actual refining loss at the plant must be no more than 20% of the chromatographic loss in the laboratory. The actual plant refining loss in comparison with the cup refining method should show at least a 25% gain over the cup loss figure.

Many oil refiners place emphasis on the refining loss and monitor the loss by an automatic monitoring system. This method measures the oil flow into the refinery (before the caustic addition) with the help of a mass flow meter, such as Micromotion. The oil flow out of the primary separator is also measured with another Micromotion mass flow meter. A correction factor is applied in order to adjust for the moisture in the refined oil. The flow data are fed into a PLC system that computes the percentage of refining loss. The operator can set up the refining loss monitor to give a report for any time span desired.

Frank Sullivan introduced the original oil loss monitor. This system worked well but went out of production a long-time ago. The next commercial oil loss monitor used in oil refineries was made by Elliott. This unit worked reasonably well, but it too is out of production. At present, loss monitors are designed and installed by individual engineers. Today, all engineering firms designing oil refineries design their own system using PLC for monitoring and data collection.

TABLE 5.10 Troubleshooting Chemical Refining Process

Symptom	Probable cause/causes	Recommended solutions
FFA in refined oil >0.03 %	Caustic treat may be low	Check the caustic strength and the treat and adjust, if necessary
	FFA in the incoming oil might have increased due to new lot of crude oil in the day tank	Check crude oil FFA and increase caustic treat if needed because of higher FFA in the crude oil
	Inadequate mixing of caustic and crude oil	Check the high shear mixer. Fix it if it is not working. Also check the retention mixers to make sure they are working. Shut down the line if either or both need to be repaired.
	Too many start ups and there is no work tank	Install a work tank to divert the oil from the primary separator during start up
Soap in the refined oil is >500 ppm	Back pressure on the oil exit at the primary separator is too low	Increase back pressure and check for ppm soap in the refined oil. Also check and make sure the neutral oil in the soap has not changed
	Caustic treat is too high	Check crude oil FFA, flow rate and the caustic treat. Reduce the caustic treat if it is high and recheck ppm soap in the refined oil after a few minutes
	Crude oil quality is poor and requires higher caustic treat	Reduce crude oil flow rate and adjust caustic treat. Recheck ppm soap
	Separator bowl is dirty	Shut down and clean the separator bowl
Lab-bleach color is high	Caustic strength may be weak	Increase caustic strength
	Crude oil quality may be poor	Conduct lab-bleach color to verify the best attainable oil color

Neutral oil in soap is high	Back pressure on the oil exit line on the separator is high	Reduce the back pressure and recheck ppm soap in the refined oil
	Crude oil quality is poor	Reduce crude oil flow rate, adjust caustic treat and recheck ppm soap
	Caustic treat is too high for the FFA in the crude oil	Check caustic treat and reduce it if it is high
	Separator bowl is dirty	Shut down and clean the separator bowl
Phosphorus is >3 ppm in the refined oil	Caustic treat is low	Verify and increase the caustic treat
	High shear mixer may not be working	Check and fix the high shear mixer if it is not working
	Low contact time in the vertical mixer	Increase dwell time by increasing the number of vertical mixers if not all of them are on line
	Vertical mixer agitator may not be working	Check the mixer and fix it. Shut down the line, fix the mixers and restart.
FFA in the water washed oil is >0.05% while FFA in the refined oil is 0.03% or less	High caustic treat	Verify if the high shear mixer and fix it
	Inadequate mixing	
Soap in the water washed oil is >50 ppm	Refined oil contains >500 ppm of soap	Check soap in the refined oil. Increase back pressure on the oil exit line and recheck ppm soap in the refined oil
	Insufficient wash water (<10%–15% as recommended)	Increase wash water without exceeding the 15% limit
	Low water wash temperature	Increase water wash temperature
	Inadequate mixing of wash water and refined oil	Check the oil water mixer and fix it if it is not working
	Wash water hardness is high	Check for the water hardness and properly treat the water
	Water wash centrifuge bowl is dirty	Shut down and clean the water wash centrifuge

(Continued)

TABLE 5.10 Troubleshooting Chemical Refining Process (*cont.*)

Symptom	Probable cause/causes	Recommended solutions
Neutral oil in the discharge water from water wash separator is >0.1%	Back pressure on the oil discharge line is high	Decrease back pressure on the oil exit from the water wash separator, Recheck neutral oil in the wash water
	High soap in the refined oil	Check soap in the refined oil and reduce it if found to be high
Oil at the exit of vacuum dryer is not dry (Moisture content is >0.1%)	Operating pressure of the vacuum dryer is >50 mm of Hg	Check vacuum ejector or vacuum pump to reduce the absolute pressure in the vacuum dryer to <50 mm of Hg
	Water washed oil temperature is <82°C. This must be 82–88°C	Increase water washed oil temperature
	Spray nozzles in the vacuum dryer may be dirty	Shut down and clean vacuum dryer nozzles
	Moisture in the water washed oil is >0.5%	The separator bowl is dirty. Shut down and clean the separator bowl

The author believes that it is always a good idea to have an oil loss monitor that measures the crude oil flow into the refiner and measures the vacuum dried oil going to the bleacher. The reasons for this preference are:

- Vacuum dried oil has very little moisture and low level of soap.
- There is some oil loss that takes place in water washing and vacuum drying.
- Oil loss monitoring downstream of the vacuum dryer is difficult on a short time frame. This is because the variations in the preprocess inventory cannot be accurately pinned down on an hourly basis. Therefore, a longer time is needed to obtain any meaningful information on oil loss. Usually a minimum of 8 h of operation is suggested if someone wants to monitor oil loss on short-term basis.

5.7.1 Manual Checks on the Oil Loss

Many plants check their oil loss manually. This involves the following steps:

The crude oil volume in the day tank is measured by a differential pressure (DP) cell situated at the bottom of the tank.

This indicates the pressure exerted by the oil, which is then converted to the volume of oil in the tank. The DP cell is precalibrated to the weight of oil per inch of height in the tank at the storage temperature of the oil.

Alternatively, the oil outage from the day tank and gain in the RB oil storage tank can be manually determined by a tape.

A temperature correction factor is applied for the actual oil temperature compared to the ambient temperature, which, unless otherwise stated, is considered to be 77°F (25°C).

The typical specific gravity numbers for common vegetable oils are listed here:

Oil	Specific gravity at 25°C	Source of information
Crude soybean	0.918–0.926	CRC Handbook, Boca Ratton
Crude soybean	0.915–0.928	Hodge and Lang
Crude sunflower	0.920	Handbook of Chemistry and Physics
Canola	0.920	
Cottonseed	0.925	
Coconut	0.924	
Palm	0.915	

The specific gravity of vegetable oil changes by a factor of 0.0004/°F or by 0.00072/°C.

Thus, the specific gravity of sunflower oil at 100°F would be:

$$0.92\ [1-0.0004\times(100-78)]=0.92-0.0088=0.9112$$

Thus, one can calibrate the tank using the appropriate oil volume or weight per inch (or foot) of oil and multiplying it by the specific gravity of the oil corrected for the oil temperature. For example:

Tank diameter	50 ft.
Cross sectional area	0.7854 × 502 = 1963.5 ft.^{2}
Oil volume per feet height	1963.5 cubic feet
Weight of oil per feet height	1963.5 × 0.9112 = 1789.1 lb at 100°F
Weight of oil per inch height	1789.1/12 = 149.1 lb at 100°F

The amount of crude oil pumped out of the day tank is determined over 24-h period.

A similar calculation is needed for the tank where the oil from the vacuum dryer is stored over the same period.

The oil loss is calculated.

The measured oil loss from earlier calculations is compared against that obtained from the oil loss monitor.

A discrepancy between the two numbers should be reconciled.

The loss of oil is not just confined in the refinery. There is oil loss in bleaching, winterization, hydrogenation, deodorization, product filling, and packaging. These will be discussed in Chapter 12. Many oil refiners look at the total oil loss at the end of each month and for cost and profitability purposes they assign all of the losses to the refining step because that is the least cost oil that there is in an oil refinery. This is a wrong approach. This does not allow the refiner to determine any major loss point in the operation. In reality, the loss can be quite substantial in the hydrogenation, deodorization, or winterization stages where the cost of the oil is much higher than that in the refining step.

However, an oil loss monitor should be considered an essential component in refining.

5.8 SHORT MIX PROCESS

The Short Mix Process uses a higher refining temperature and no retention time like the Alfa Laval long Mix process. Westfalia introduced this process. The refined oil is water washed and vacuum dried in the same manner as in the Alfa Laval Long Mix Process. Fig. 5.10 shows the schematic flow diagram for Westfalia Short Mix Process.

The Short Mix Process uses a stronger caustic solution as compared to the Alfa Laval Long Mix Process. For example, the process uses 18°Be for degummed soybean oil while the Alfa Laval Long Mix Process uses 14°Be caustic for the same oil. The step-by-step procedure for the Short Mix Process is listed here:

1. The degummed crude oil is heated to 195°F (90°C) and mixed with approximately 200–400 ppm of phosphoric acid in a centrifugal mixer.
2. The oil then passes through a holding tank and then mixed with 18°Be caustic solution in a second centrifugal mixer before the oil passes through the primary separator.
3. An optional retention tank is sometimes provided for hard to refine crude oils before the oil goes to the primary separator.

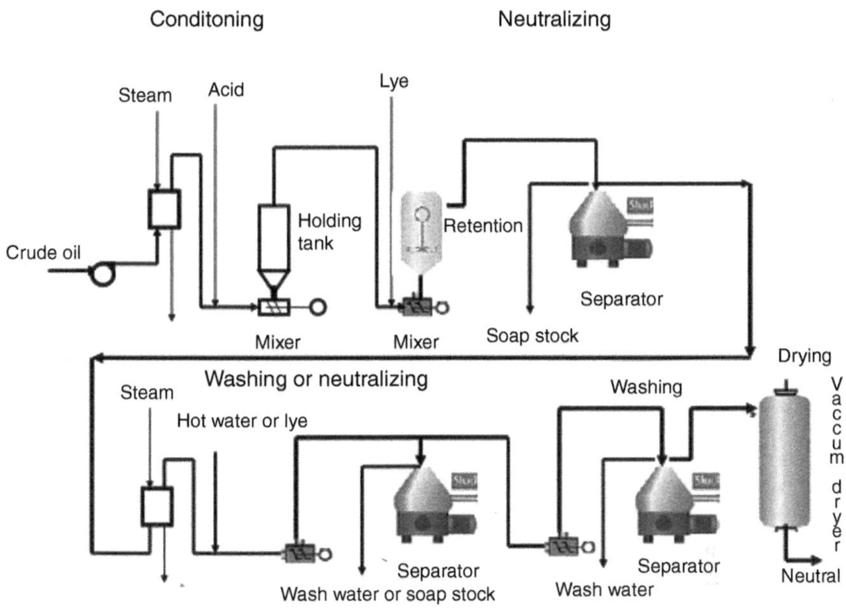

FIGURE 5.10 **Westfalia Short Mix Process flow diagram.**

4. Oil and soap are separated in the primary separator. A certain amount of back pressure is applied on the oil discharge line to control the neutral oil in the soap and the amount of soap in the refined oil as in the Alfa Laval process. This can be done manually or with an automatic control valve operated with a PLC system.
5. Oil from the primary separator is either heated back to 90°C and mixed with hot deionized water at the same temperature in another centrifugal mixer, or it is treated with caustic for the second time if the quality of the crude or crude degummed oil is poor.
6. The oil and water mixture is not heated further before it enters the water wash centrifuge.
7. The oil then passes through a water wash separator.
8. Oil leaving the water wash separator is approximately at 176–185°F (80–85°C).
9. The water washed oil is vacuum dried under a vacuum of 36–50 mm of mercury absolute pressure.
10. The vacuum dried oil goes to bleaching or to storage after it is cooled to <104°F (40°C).

There is a strong preference for the Alfa Laval Long Mix Process in the United States. The American oil processors believe that the longer contact time between the weaker caustic and the crude oil helps reduce the nonhydratable phospholipids without causing any measurable reaction between the caustic and the neutral oil in the crude.

European processors prefer the Westfalia Short Mix Process. They believe that the higher temperature facilitates the intimate mixing between the oil and the caustic. In addition, they believe that the stronger caustic solution is needed for achieving the reaction in the short time provided to the oil and caustic mixture.

5.8.1 Critical Control Points and Troubleshooting Short Mix Process

The fundamental critical control points and the trouble shooting procedure in the Westfalia Short Mix Process are very much the same as those for the Alfa Laval Long Mix Process.

5.9 VACUUM DRYING

Refined and water washed oil is vacuum dried if the oil is to be sent for hydrogenation or if it is to be stored for a few hours before it is bleached. Normally, the bleaching step follows immediately after vacuum drying. It will be discussed in Chapter 6 that it is essential to have the moisture content of the oil at 0.3%–0.5% for treatment with hydrated silica (e.g., Trisyl). Hydrated silica does not work on dry oil. The oil from the water washed centrifuge contains 0.4%–0.5% moisture. Therefore, it is ideal to send the water washed directly for hydrated silica treatment without vacuum drying.

A vacuum dryer is a pressure vessel because it has to be operated under full vacuum. It is also designed for the full discharge pressure of the crude oil pump. Fig. 5.11 shows the schematic view of a vacuum dryer. The design concept was developed by Walter E. Farr of the Farr Associates.

The step-by-step process of vacuum drying is described:

1. In some operations the water washed oil is dried to a moisture content of <0.1% in the vacuum dryer, cooled to <104°F (40°C), and stored before it is bleached.
2. A vacuum dryer is operated at 36–50 mm of mercury of absolute pressure.
3. In most cases, a two-stage vacuum steam ejector is used. Vacuum pumps are used in many plants instead of the steam ejector.
4. The oil leaves the water wash separator at 180–190°F (82–88°C). The oil is sprayed into the vessel through a spray header arrangement at the top of the vessel. The slightly higher temperature is needed to remove the dissolved moisture from the oil.
5. The water washed oil is heated to a temperature of 185–195°F (85–91°C) and sprayed inside the vacuum dryer through the distribution nozzles.
6. A minimum residence time of 15 min is provided to the oil by maintaining a certain oil level in the dryer. This is accomplished with the help of a centrifugal pump that continuously draws oil from the bottom of the vacuum dryer. The oil passes through a three-way modulating valve operated by a level controller as shown. Part of the oil goes back into the vacuum

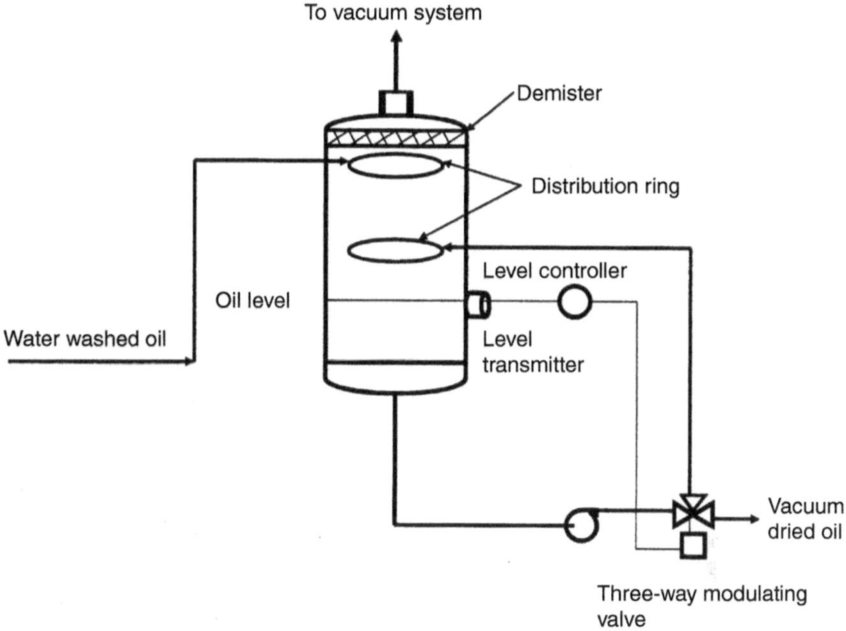

FIGURE 5.11 **Schematic diagram for vacuum dryer.**

dryer and is distributed through a second distribution ring. The recycled oil stream helps maintain the oil level in the tank. The rest of the oil goes to bleaching.

7. The vacuum dried oil leaves the vessel at a moisture-content of <0.1%.

5.9.1 Critical Process Control Points in Vacuum Drying

These are the critical process control points in vacuum drying listed in order:

1. oil temperature in the vacuum dryer
2. oil distribution in the dryer
3. residence time in the dryer
4. operating pressure in the vacuum dryer
5. condition of the nozzles

Vacuum drying is an essential step in oil refining. The oil must be deaerated and dried to a moisture content of <0.1%, otherwise several productivity and oil quality issues are experienced. For example, in bleaching:

- High moisture in the oil can reduce the effectiveness of the bleaching clay, requiring more clay to bleach the oil.
- Filtration of oil is slowed down, reducing production throughput.
- Poor air removal increases the oxidation in the oil in bleaching.

TABLE 5.11 Critical Control Points and Troubleshooting in Vacuum Drying Process

Control points	Recommended limits	Significance/consequence for noncompliance
Operating temperature	185–195°F (85–90°C)	Moisture removal is ineffective at temperature <176°F (80°C)
		Temperature >190°F (88°C) is not needed.
		Higher temperatures can cause cavitation in the water wash centrifuge and the pumps in the process
Operating pressure	<50 mm of mercury absolute	At higher than 50 mm of mercury absolute pressure the moisture and air removal efficiency is diminished
		The oil tends to get oxidized and also polymerize.
Residence time	Minimum 15 min	Lower residence time reduces the efficiency for moisture and air removal from the oil
		A longer residence time does not harm the oil unless the operating pressure is too high (Poor Vacuum) in the vacuum dryer can oxidize and polymerize the oil.
Oil distribution	A uniform dispersion in fine spray is required	Improves moisture and air removal from the oil

Table 5.11 lists the critical process control points in vacuum drying and the possible consequence of noncompliance to them.

5.10 SOAP SPLITTING FOR RECOVERING THE FATTY ACIDS (ACIDULATION OF SOAP STOCK)

This process is commonly referred to as acidulation. In this process, the soap is mixed with concentrated sulfuric acid and heated. The strong sulfuric acid reacts with the soap forming two main streams of liquid, namely fatty acid (also called acid oil) and the acid–water, which contains sodium sulfate and many other water-soluble compounds. The fatty acid is separated from the acid–water by allowing the reaction mass to separate into two phases in a settling tank where the fatty acids float to the top and the acid water settles at the bottom of the settling tank. In some operations the separation of the fatty acids and the acid water is done in a centrifuge. The acid oil is properly dewatered and then sold as industrial grade fatty acid, which is used in animal feed and in various industrial products. The acid water is neutralized with caustic solution before it can be discharged into the municipal waste water system.

The soap leaving the primary separator is maintained at a moisture content of 45%–50%. The moisture content should be maintained by controlling the amount of water used at the seals of the separator. At higher moisture content, the soap has lower viscosity and is easier to mix with the sulfuric acid. However, this increases the consumption of sulfuric acid for the acidulation step and that of the caustic for the neutralization of the acid water. This increases the cost of acidulation.

Acidulation is conducted via:

1. batch process
2. continuous process

5.11 BATCH ACIDULATION PROCESS

The schematic flow diagram for the batch acidulation process is shown in Fig. 5.12.The step by step procedure for the batch acidulation process is shown:

1. Soap stock from the primary centrifuge is collected in a tank (not shown in Fig. 5.12). It is recommended that the amount of water to the seal of the primary separator be kept to a minimum in order for the soap stock to meet the following analyses:

Total fatty acid (TFA)	35% minimum
Moisture	Target 45% and not to exceed 50%

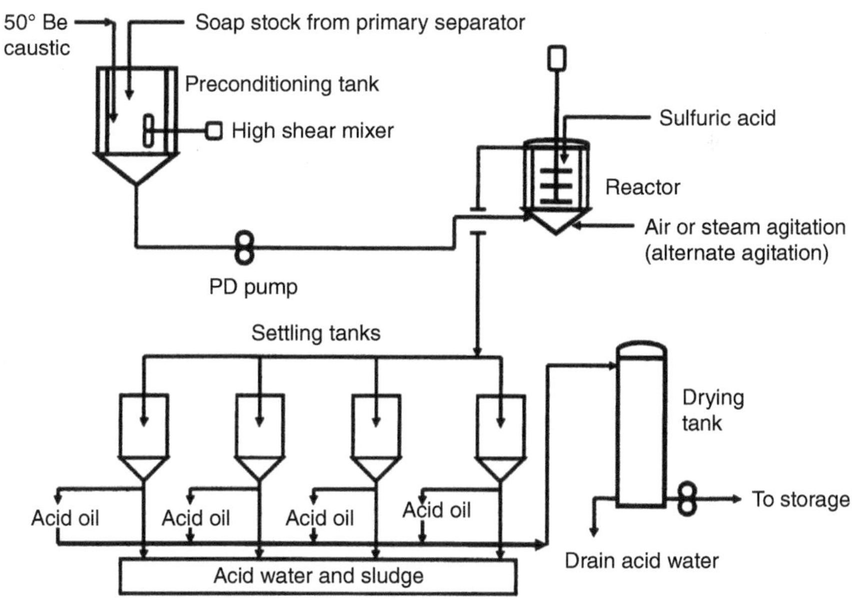

FIGURE 5.12 Schematic flow diagram for batch acidulation process.

2. The soap stock is then pumped into a steam-jacketed preconditioning tank with a high shear mixer. The mixer can be top entering or side mounted depending on the space available. A top entering agitator costs more but it also works better.
3. The tank must have three or four baffles along the straight vertical wall of the tank to prevent swirling and vortex.
4. The tank can also have steam coils to heat the soap stock.
5. A metered amount of 50°Be caustic solution is added into the soap stock in the tank. The purpose of this caustic addition is to saponify any residual neutral oil, as well as the monoglycerides and diglycerides in the soap stock. This is essential to prevent emulsion formation, which then reduces the formation of the middle phase in acidulation.
6. The soap stock with the caustic solution is heated to 185–195°F (85–91°C) and is agitated for several hours until the soap stock has become very fluid. The neutral oil in the treated soap stock should be 1% or less. The pH of the soap stock may be 10.5–11. Add extra caustic solution if the neutral oil in the soap stock is higher.
7. Using a positive displacement pump, the soap is then pumped into an atmospheric reactor.
8. The reactor has a top entering agitator with baffles. The material of construction for all wetted parts is Carpenter 20.
9. Instead of a mechanical mixer, some plants use compressed air or live steam for agitation. This method does not produce good agitation to mix the reactants.
10. Concentrated sulfuric acid is added into the reactor to split the soap.
11. The pH in the reactor is maintained between 1.5 and 2.5.
12. The reaction takes 4–8 h to complete (sometimes longer).
13. The pH is checked. Additional sulfuric acid is added to the reactor if the pH is higher than 2.5.
14. The reaction is allowed to continue with agitation and the pH is checked again to make sure that proper acidity is maintained.
15. The agitator is stopped and the reaction mixture is pumped into a series of settling tanks. These tanks must be either glass lined or epoxy coated carbon steel tanks or fiberglass tanks. The composition of the fiberglass tanks must be suitable to withstand a temperature of 260°F (127°C) or higher.
16. These tanks are maintained at 180–190°F (82–88°C) and the reaction mixture is allowed to stand for the separation of the acid oil and the acid water.
17. The acid water is drained from the bottom of the settling tanks and sent to the wastewater treatment plant where it is neutralized with caustic before it can be discharged as plant effluent. In many plants, this water goes through a primary water treatment process before it is discharged to the sewer.
18. The middle phase is at its minimum when the neutral oil in the soap stock is properly saponified in the caustic pretreatment step and it is maintained at 1% or less. Sometimes there can be more middle phase even if the neutral

oil is reduced in the preconditioning step for the soap stock. It is known that certain protein/lipid complexes have emulsifying property and can interfere with the separation of the acid oil from the acid water.

19. The middle phase is stored in separate tanks (not shown in Fig. 5.12) for further acid oil recovery. Sometimes the middle phase is reacidulated.
20. The acid oil is pumped into a set of drying tanks (only one is shown in Fig. 5.12). These tanks are of relatively narrow diameter and are tall in order to facilitate the separation of the acid water. The liquid inside these tanks is maintained at 180–190°F (82–88°C). The material of construction of the drying tanks is the same as that for the settling tanks.
21. The acid oil is ready to be transferred into the storage tank or shipped when the following analyses are met:

TFA	90% minimum
Moisture	<1%
Mineral Acid (MA)	<5%
Sediments	Balance

5.11.1 Critical Control Points in Batch Acidulation Process

Following are the critical control points in the batch acidulation process:

1. The soap stock from the primary separator should have 45%–50% moisture. Sulfuric acid and caustic usage increase significantly as the moisture content of the soap stock goes up.
2. The soap must be reacted with the caustic to reduce the neutral oil in the soap to 1% or less to minimize the middle phase formation.
3. Temperature of this reaction is maintained at 185–195°F (85–91°C). Acidulation of the soap proceeds very slowly and is less complete at lower temperatures.
4. The pH of the reaction mix is maintained between 1.5 and 2.5. At higher pH the reaction is incomplete. A lower pH increases the cost of chemicals.
5. The temperature of acidulation reaction is 185–195°F (85–91°C).

5.12 CONTINUOUS ACIDULATION PROCESS

The schematic flow diagram for the continuous acidulation process is shown in Fig. 5.13. The only difference between the continuous and the batch process is that the reaction is carried out in a continuous reactor at 240°F under pressure. The separation of the acid oil and the acid water is very similar to that in the batch process.

The continuous acidulation process is operated under the same principle of reducing the neutral oil in the soap by reacting with caustic in the preconditioning tank and then acidulating the mix with concentrated sulfuric acid.

Steps 1–7 under the batch acidulation process also apply to the continuous acidulation process. The rest of the steps follow.

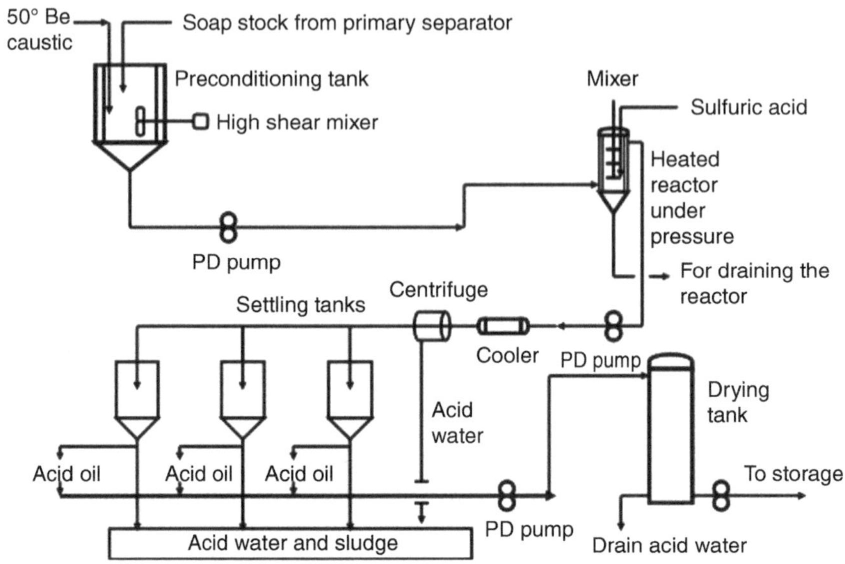

FIGURE 5.13 Schematic flow diagram for continuous acidulation process.

1. The heated mixture of soap stock and caustic is metered into a reactor continuously. Simultaneously concentrated sulfuric acid is metered into the reactor continuously.
2. The pH (1.5–2.5) of reaction is monitored with the help of a sensor and additional acid is added if the pH in the reactor goes above the set point.
3. The reactor can also be atmospheric, operating at 185–195°F (85–91°C) or it is designed to operate at 240°F (116°C). The reactor and all wetted parts are made of Carpenter 20 alloy.
4. The reactor must be a pressure vessel when it is operated at the elevated temperature 240°F (116°C).
5. The reaction mixture from the reactor is pumped into a series of separating tanks as shown in Fig. 5.12.
6. The reaction product is either separated in a centrifuge made of Carpenter 20 alloy, or it is sent directly to the separating tanks. The reaction mixture is cooled down to 190°F (88°C) before centrifuging in order to minimize cavitation. Use of a centrifuge is optional.
7. Separation of the acid oil from acid water is conducted in the same manner as described under the batch process.
8. As in the batch process, the acid oil goes to the settling tanks, where the majority of the acid water is drawn off. The acid oil is transferred into the drying tanks. The settling tanks, as well as the drying tank, are made of either glass lined or epoxy coated carbon steel tanks or made of fiberglass capable of withstanding up to 260°F (127°C).

5.13 TROUBLESHOOTING ACIDULATION PROCESS

The goal of acidulation is to:

1. Recover most of the fatty acids (FA) from the hydrolyzed soap stock.
2. Convert the neutral oil present in the soap stock into fatty acids.

In order to achieve the objectives, the soap stock from the primary separator must meet the standards listed in the previous section. The neutral oil in the soap stock from the primary centrifuge should be 30% or less dry basis, and the moisture content between 45% and 50%. The quality of the final product (i.e., the acid oil) depends on how well the process operating conditions are maintained. The quality aberrations, probable causes, and the recommended solutions for the acidulation process are listed in Table 5.12.

5.14 COLD CHEMICAL REFINING PROCESS FOR SUNFLOWER OIL

This is a very effective way to reduce the wax content in sunflower seed oil in addition to the other impurities. Although this process was originally introduced at ADM in 1975, it is not followed often in the United States but is followed in Europe and Latin America. This is a simple one-step process for refining and dewaxing sunflower oil or possibly other wax bearing oils.

This process is very similar in principle to the standard chemical refining process except:

1. The acid pretreated crude oil is heated to 131°F (55°C) to make sure that all the wax in the oil is melted.
2. Caustic solution is added to the oil and mixed in a high shear mixer.
3. The oil is centrifuged to reduce some of the soap with careful adjustment of the back pressure; the refined oil leaving the primary separator contains 1600–2000 ppm soap. The primary separator is operated at reduced throughput because the feed to the separator has higher viscosity due to the colder temperature than the standard Long Mix Process.
4. The oil is then slowly cooled down to 43–46°F (6–8°C) in an insulated tank, equipped with a specially designed top-entering agitator running at very slow speed (Approximately 5–10 rpm).
5. The rate of cooling is 5–7°F/h (3–4°C/h).
6. The oil is kept at this temperature for 12–24 h (preferably for 24 h) to allow complete crystallization of the wax.
7. The cold oil is then sent to a second primary separator where the soap and the crystallized wax are separate from the oil.
8. The oil is then heated for water washing as in the Alfa Laval Long Mix refining process and centrifuged to remove residual soap and the wash water.
9. The remainder of the process is the same as described under continuous chemical refining process.

TABLE 5.12 Troubleshooting Acidulation Process

Symptom	Probable cause/causes	Recommended solutions
Viscosity of the soap stock is too high	Low temperature in the soap stock storage tank	Raise the temperature in the soap stock storage tank
	The soap stock has been stored for days	Use heat and high shear mixing to reduce the viscosity
The pH of the mixture in the preconditioning tank is <10	Not enough caustic in the system	Add more caustic to bring the pH to 10.5–11
	Neutral oil in the soap stock is higher than normal and is consuming more caustic	
The neutral oil in the preconditioning tank is not dropping	Low caustic	Add more caustic and let the pH stabilize at 10.5–11
	Temperature <185°F	Raise the temperature
Acid oil and water are not splitting well after the reaction	Insufficient preconditioning with caustic leaving high neutral oil in the reaction mix	Make sure the neutral oil after preconditioning is 1% or less
	Reactor pH is high	Add more sulfuric acid to lower the pH to <2.5
	Insufficient mixing in the reactor	Check and make sure the mechanical agitation or air/steam agitation is satisfactory
	Low reaction temperature	Increase the temperature in the reactor
Heavy amount of middle phase is formed	Insufficient preconditioning with caustic leaving high neutral oil in the reaction mix	Make sure the neutral oil after preconditioning is 1% or less
	The crude oil is of very poor quality and had to be overrefined or rerefined	
Mineral acid in the acid oil is >5%	Incomplete separation of acid oil and acid water	Allow longer settling time and also make sure the heat is on. The temperature in the settling and the drying tank must be 180–190°F (82–88°C)
	Incomplete draining of acid water from the acid oil	
Acid oil has <90% TFA	Incomplete separation of acid oil and acid water	Allow longer settling time and also make sure the heat is on in both settling and drying tanks. The temperature in the settling and the drying tank must be 180–190°F (82–88°C)
	Incomplete draining of acid water	

Note

- The oil being colder, viscosity of the oil is high.
- The primary separator throughput is nearly 50% of that one gets under standard chemical refining process, Long as well as Short Mix methods.
- This process allows very effective separation of the wax and soap at the same time.
- The oil is also protected against the high temperature condition used in the other methods for refining the oil.

5.15 MODIFIED PHYSICAL REFINING PROCESS

MPR eliminates the primary centrifuge and soap stock effluent. As a result, refining NOL and effluents are completely eliminated. Economical and environmentally friendly, MPR used in conjunction with packed bed (discussed in Chapter 6) offers several benefits as discussed at the end of this section.

This process is based on the fact that hydrated silica (or silica hydrogel) removes phospholipids, trace metals, and oil decomposition products very effectively. In order for the silica to be effective, a certain amount of soap is required for the hydrated silica to adsorb phospholipids from the oil. Therefore, a certain amount of caustic is mixed into the oil before adding hydrated silica gel. This method is used effectively to treat palm oil but it is not suitable for seed oils in general. The semiphysical refining process is more suitable for seed oils. This is described later in this section.

The schematic flow diagram for the MPR process is shown in Fig. 5.14. In this process, the water degummed oil is analyzed for the ppm of phosphorus.

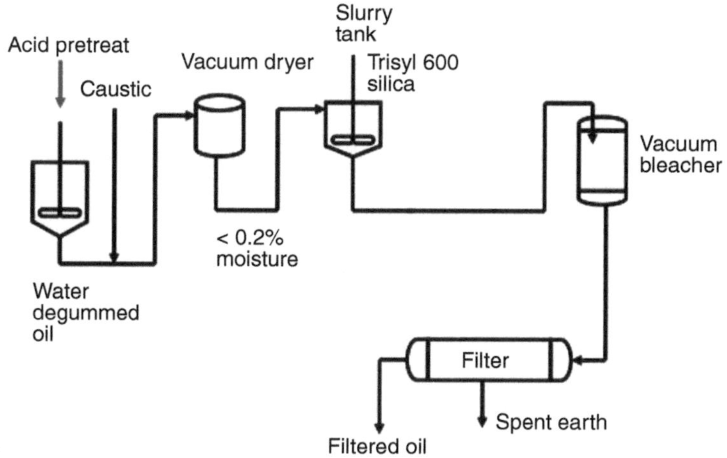

FIGURE 5.14 **Schematic flow diagram for the modified physical refining process.**

A calculated amount of caustic is then added to the oil to produce soap. The concentration of soap produced (ppm) must match the ppm of phospholipids in the oil. The following steps of procedure are followed:

1. The oil is vacuum dried to a moisture content of 0.2%–0.4% in a vacuum dryer.
2. The oil and the caustic solution are mixed in a high shear mixer.
3. Hydrated silica (Trisyl 600) is added to the oil in a slurry tank under atmospheric condition.
4. The oil is then bleached with acid activated clay under vacuum and at a temperature of 230–248°F (110–120°C) in a vacuum bleacher.
5. The bleached oil is filtered through a packed bed filter (discussed in Chapter 6).

The MPR Process offers the following benefits over the standard physical refining process:

- reduced NOL
- reduced adsorbent usage
- reduced cost of filter cake disposal
- longer filter cycle time
- improved oil quality because MPR removes more phosphorus and trace metals than the standard physical refining process
- reduced operating cost as long as the palm oil is low in phosphorus

5.15.1 Critical Control Points in Modified Physical Refining Process

The following are the critical control points in the MPR process:

1. moisture in the oil
2. amount of caustic added to the oil

5.15.1.1 Moisture in the Oil

Hydrated silica does not work with dry oil. The moisture content of the oil must be 0.2%–0.4% for the silica to function properly.

5.15.1.2 Amount of Caustic

Presence of soap in the oil facilitates the removal of the phospholipids. Therefore, a calculated amount of caustic is added as described earlier to produce some soap. The amount of soap produced must be equal to the ppm of phospholipids in the oil. Additional soap will increase the bleaching clay consumption and will also produce higher FFA in the bleached oil.

5.16 MODIFIED CAUSTIC REFINING PROCESS

The W. R. Grace Company in the United States introduced this method to eliminate the water-washing step in caustic refining. In this method, the refined oil from the primary centrifuge is treated through the following steps:

- The oil is heated to 90°C in a vacuum bleacher under atmospheric conditions.
- Trisyl (hydrated silica) is added into the oil and mixed for 15 min.
- Vacuum is applied to the vacuum-bleaching vessel.
- Acid activated bleaching clay is then added.
- The oil is heated to 110–120°C (230–248°F) under vacuum (<50 mm of mercury absolute pressure) and bleached for 30–45 min.
- The oil is filtered and handled as described in Chapter 6 under bleaching.

Fig. 5.15 shows the schematic flow diagram for the Modified Chemical Refining Process.

This process eliminates the use of water washing and also produces high quality oil. The key to achieving high quality refined oil is to have a supply of crude oil with very low levels of phospholipids and soap; otherwise the cost of hydrated silica can become prohibitive. This procedure is being used by some crusher refiners who can properly control the phosphorus content in the degummed crude oil. Crude oil from freshly crushed, good quality beans also contains low FFA and nonhydratable phospholipids. Therefore, once properly degummed and caustic refined, the oil contains low phosphorus and soap. Thus, silica refining becomes economical. This produces similar benefits discussed under MPR. However, the process becomes costly when the soap level in the refined oil exceeds 200 ppm.

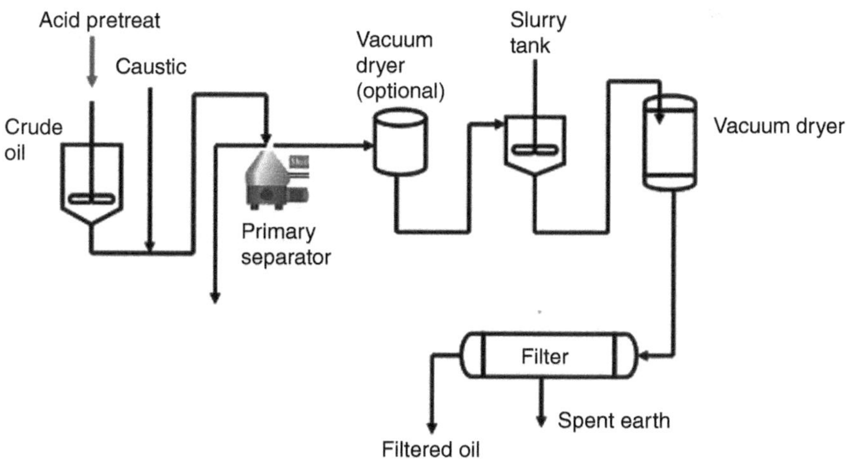

FIGURE 5.15 Schematic flow diagram for modified caustic refining process.

5.17 SEMIPHYSICAL REFINING PROCESS

It has been mentioned earlier that the physical refining process is not suitable for seed oils unless the nonhydratable phospholipid level is very low after degumming (phosphorus <10 ppm and preferably <5 ppm).

In this process, as introduced by Walter Farr, the water degummed crude oil is treated with sufficient amount of caustic using the ultrashear mixer. This reduces the nonhydratable phospholipids to a very low level. The oil then becomes suitable for physical refining.

This process may not work if the crude oil is of extremely poor quality. In that case the chemical refining process using the ultrashear-mixing step can significantly reduce the caustic usage, as well as the refining loss.

Chapter 6

Bleaching

6.1 INTRODUCTION

Bleaching follows the water washing and vacuum drying step discussed in Chapter 5. Unlike chemical refining, this is a physical process where the impurities in the oil, such as phospholipids, trace metals, color bodies, oil decomposition products, and so on, are removed with the help of an *adsorbent*. The impurities and the active sites on the adsorbent are attracted to each other, by the Van der Waal's force of attraction. The amount of attraction depends on several factors as follows:

1. The amount of electrostatic force on each of the impurities and the adsorbent.
2. The size of each of the components.
3. The distance between them or the degree of intimate mixing between the oil and the adsorbent.
4. Porosity of the adsorbent particles.
5. Specific surface of the adsorbent (surface area per unit weight).

In the earlier days, the term bleaching meant that some bentonite would be added to the oil, stirred for some time until the oil reached the desired lighter color, and then the bentonite would be filtered out of the oil.

The main concern in those days was the color of the oil. Later, the red color in soybean oil and cottonseed oil became a matter of esthetic concern and the oils were bleached to achieve the desired red color. Around this time, the oil processors started to pay attention to the green color in the refined oil, especially in soybean and in some instances in cottonseed and sunflower oil and later the green color became a major concern in canola oil. The green color in the oil is contributed by chlorophyll. There are two types of chlorophylls: chlorophyll A and chlorophyll B. The bleaching clay industry developed the acid-activated clay to reduce the amount of chlorophyll in the oil.

Effectiveness of the acid-activated clay has improved over the years. The vegetable oil industry also learned the fact that the red color in the refined oil, which is contributed by the carotenoids present in the oil, can be heat bleached in the final step of oil refining which is the deodorization process. However, the green color from the chlorophyll cannot be reduced in the deodorizer.

Practical Guide to Vegetable Oil Processing. http://dx.doi.org/10.1016/B978-1-63067-050-4.00006-4
Copyright © 2017 AOCS Press. Published by Elsevier Inc. All rights reserved.

Therefore, the oil refiners started to make sure the chlorophyll in the soybean oil was reduced to the desired level (this will be discussed later in the chapter) and allow the red color to remain higher in the refined oil for its removal in heat bleaching step in the deodorization process.

The oil refiners also discovered that using an extra amount of bleaching clay to reduce the red color was not only unnecessary but was harmful to the oil because it reduced its storage stability.

As analytical techniques became more sophisticated, the researchers in the companies in fields, such as bleaching clay manufacturing, oil processing, and USDA Research uncovered the real meaning of the phrase "bleaching vegetable oils." They learned that the bleaching of the oil reduces many impurities in the oil that are detrimental to the oil for its stability. Today's understanding of bleaching is that it not only makes the oil appear lighter in color but it performs the following beneficial functions:

1. Reduces the chlorophyll and some of the other color bodies present in the oil.
2. High chlorophyll in the oil is undesirable for the appearance, as well as for poor photooxidative stability of the oil. This will be discussed later in this chapter.
3. Reduces trace metals, such as calcium (Ca^{++}), magnesium (Mg^{++}), iron (Fe^{+++}), sodium (Na^{+}), etc. The impact of these trace metals will be discussed later.
4. Reduces the level of nonhydratable phospholipids in the refined oil.
5. Removes decomposition products, such as aldehydes, ketones, polymers, nontriglycerides produced from oil oxidation.

The researchers found that the oil loses some of its natural antioxidants in the bleaching step. This is particularly significant when the oil is bleached (1) improperly, (2) repeatedly, (3) with excessive amounts of bleaching clay, and (4) under atmospheric condition. Besides the loss of the antioxidants, the chlorophylls present in the oil undergo oxidation under acidic pH in presence of oxygen. The decomposition products of the chlorophylls make the oil 10 times more prone to photooxidation. This will be discussed further in Chapter 12. The oil processors need to take all the necessary steps to ensure proper bleaching of the refined oil.

6.2 GENERAL OPERATING STEPS IN BLEACHING

As discussed at the beginning of this chapter, bleaching is a physical process where the impurities in the oil are adsorbed by an adsorbent (bleaching clay, silica, activated carbon, etc.) through the physical force of attraction between the adsorbent and the adsorbate (impurities in oil). Effective adsorption and removal of the oil impurities depend on the following factors:

1. degree of intimate contact between the bleaching medium and the oil
2. oil temperature
3. time of contact

The significance of the above mentioned factors will be discussed in detail later in this chapter.

The water-washed oil is treated with acid-activated clay at elevated temperature, and preferably under vacuum, to protect the oil from oxidation. The oil and the clay are brought to intimate contact with the help of mechanical mixing. Some equipment manufacturers use steam agitation instead of mechanical mixing. In the author's judgment, mechanical mixing is a better option.

The oil and the bleaching clay are kept in contact for 30–45 min. In addition to the clay, a certain amount of citric or phosphoric acid is added into the oil, which improves the removal of trace metals from the oil. Many modern refineries are using silica hydrogel pretreatment followed by bleaching clay treatment. This procedure seems to be very effective in removing the trace metals and phospholipids from the oil.

The bleached oil is filtered through leaf filters with stainless steel mesh screens, cooled, and stored. Alternatively, the oil is sent directly to the next processing step, such as hydrogenation, winterization, or deodorization.

All modern plants are equipped with vacuum bleachers. These can be either batch or continuous systems. The end results from the batch or continuous bleacher are similar. Many older installations are still using an atmospheric bleaching process. Atmospheric bleaching is harmful to the oil. This aspect will be discussed later in this chapter and also in Chapter 12.

6.3 DRY BLEACHING VERSUS WET BLEACHING

The conventional bleaching process used water-washed and vacuum-dried oil as the feed to the vacuum bleacher where only bleaching clay was added. The bleaching clay could be acid activated or neutral. Acid-activated clay is more effective in removing the color bodies and trace metals in the oil as discussed earlier. The neutral clay is gentle to the oil and is used for treating animal fats, palm oil, and oils that are processed under the "organic" or "natural" label.

Fig. 6.1 shows the schematic diagram of the dry bleaching system. The step-by-step procedure for the dry bleaching process is outlined as follows:

1. The water-washed and vacuum-dried oil is pumped through a plate and frame heat exchanger where the heat from the outgoing bleached oil is recovered.
2. The oil then passes through a plate and frame heater where it is heated to a temperature of 100–110°C (212–230°F).
3. The oil is then introduced into the vacuum bleacher through a distributor with spray nozzles.
4. The absolute pressure in the vacuum bleacher is maintained at <50 mm of mercury (50 Torr).
5. The oil temperature is maintained at 100–110°C (212–230°F).
6. The top entering agitator operates at high speed (typically at 120 rpm).

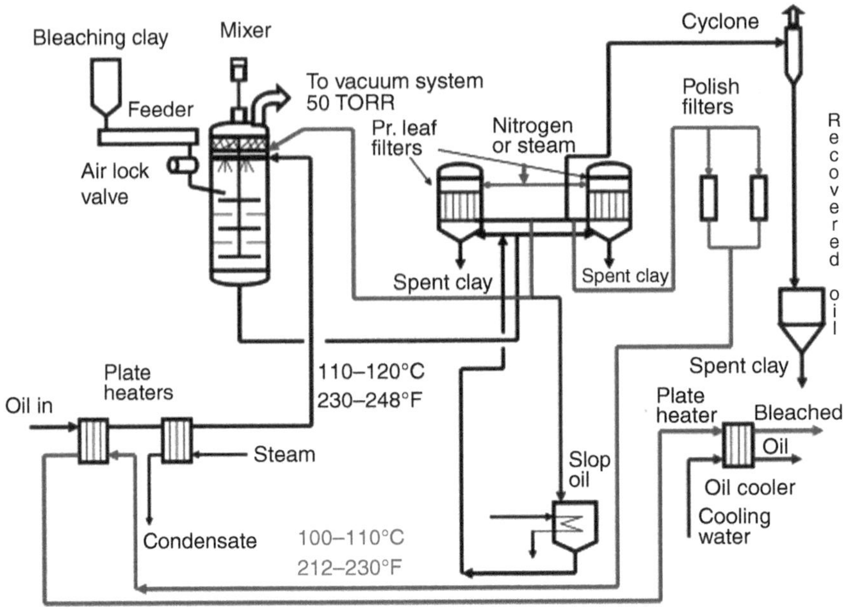

FIGURE 6.1 **Schematic diagram for Dry Bleaching System©.**

7. Acid-activated clay (typically 0.5% of the oil or less) is added to the bleacher. The clay usage can be higher for the oil that is produced from poor quality crude, naturally darker color oil, such as cottonseed oil or palm oil.
8. The oil and the clay are kept in contact for 30–45 min.
9. The oil is then filtered through a pressure leaf filter which is normally precoated with diatomaceous clay.
10. Precoating of the filter screens is done using diatomaceous earth through the filters. The procedure for precoating is discussed later in this chapter.
11. The filtered oil is recycled back into the bleacher until the oil is clear as determined by a filter test.
12. The oil is sent forward through a polish filter and then through a plate and frame heat exchanger to preheat the incoming oil.
13. The oil then passes through a cooler where it is cooled down to 40°C (104°F) by using cold water. In a closed coupled situation the oil goes to the next operation without cooling. However, it is recommended to cool the oil to <40°C (104°F) if it is to be stored before further processing.
14. Filtration is carried out for a predetermined time, which is calculated based on the dirt load capacity of the filter. The pressure drop across the screen at this stage is about 30–35 psig (2.18–2.54 kg/cm^2). These instrumentations are not shown in Fig. 6.1.
15. The filter is drained by blowing the oil through the screen by using nitrogen or dry steam.

16. After the oil from the filter is drained, dry steam is blown through the filter for 5–10 min to dry the cakes on the filter screens.
17. At the end of the drying cycle the cakes are dropped by vibrating the screens using the automatic vibrator.
18. The dry cake is collected from the bottom of the filter.
19. Fig. 6.1 shows two pressure leaf filters. One is on standby while the other is being used. The idle filter goes into service as the used filter is cleaned.
20. Sometimes there can be cake left on the screens after vibrating them. This happens more when the filter is not completely dry. It is recommended not to use any metal scrapers or plastic scrapers with rough edges. One must always use a plastic scraper with a smooth rounded tip and use it gently to dislodge the cake from the surface of the screen.
21. Nitrogen or steam blow creates some mist of oil at the end of oil drain step. The oil mist and droplets pass through a cyclone or other separation device to separate the oil from the gas. The oil is collected in the recovery oil tank. This oil is transferred into the slop oil tank and reprocessed in acidulation.

6.4 CRITICAL CONTROL POINTS IN DRY BLEACHING

The critical control points in the dry bleaching process are listed next.

1. the incoming water-washed oil quality
   a. soap (ppm)
   b. phosphorus (ppm)
   c. moisture (%)
2. type of bleaching clay
3. amount of bleaching clay
4. bleaching temperature
5. degree of mixing
6. contact time between oil and bleaching clay
7. use of phosphoric or citric acid as metal chelator
8. filter precoat
9. filtering area/oil flow rate through the filter
10. spacing between filter screens

1. Incoming oil quality
   The water-washed and vacuum-dried oil must meet the required oil quality standards to obtain good results in bleaching. As mentioned in Chapter 5, the quality requirements for the vacuum-dried oil are:

Phosphorus	<3 ppm, max.
Soap	<100 ppm, max.
Moisture	<0.1%, max.

Moisture, phosphorus, and soap are polar in nature and therefore become attached to the active sites of the bleaching clay. This results in fewer active sites left for adsorbing the impurities. Thus, an additional amount of bleaching clay is required to remove the impurities in the oil. This is observed when the bleached oil is high in phosphorus, soap, and moisture. As a result, the bleaching process suffers the following setbacks when the incoming oil does not meet the specified quality:

- An additional amount of bleaching clay is required for the process in order to compensate for the loss of the clay's bleaching efficiency.
- Soap, phosphorus, and moisture have a tendency to plug up the porosity of the filter bed, decreasing the filtration rate.
- Increased bleaching clay usage also builds up the cake faster in the filter, reducing the filtration cycle time.
- Higher soap in the oil produces higher free fatty acids (FFA) in the bleached oil.
- Higher amount of bleaching clay reduces the level of tocopherols and other natural antioxidants in the bleached oil.

Therefore, remember that the vacuum-dried oil must meet the specified quality standard before the oil is sent for bleaching.

2. Type of bleaching clay

There are two broad categories of bleaching clay used by the oil refiners. They are:

1. natural and
2. activated (acid activated).

The bleaching clay must have the following properties:

- High adsorption capacity to remove the impurities.
- Acidic pH (for the acid-activated clay).
- Numerous active sites for the adsorption of the impurities.
- Appropriate porosity.
- Good oil flow rate through the filter bed.

Neutral clay

In the early years, all naturally occurring activated clays having good bleaching properties were called fuller's earths. The name fuller's earth originated from the practice of "fulling" or cleaning the grease and stains from wool and cloth. Neutral clay was used by the oil processors for color removal from the oil. This type of clay is suitable when the oil is lighter in color. However, this is not suitable for bleaching darker oils, oils containing chlorophyll, or the oils from poor quality crude. Neutral clay is also used for bleaching oils that are grown and marketed under the "organic" or "natural" label.

Neutral clay does not reduce soap or phospholipids from the oil. Some refiners feel that this type of clay is adequate to remove the color from micella refined cottonseed oil because the oil is light in color compared to that obtained from the conventional refining process. However, the micella refined oil contains some soap and, if not removed with acid-activated clay, can exhibit many process and oil quality issues. The negative impacts of soap and phospholipids in the oil on the bleaching process have been discussed already in this chapter, as well as in Chapter 5. The impact of these impurities on hydrogenation will be discussed in Chapter 7.

Acid-activated clay

This refers to the surface-modified acid-activated bleaching clay. Traditionally, acid-activated bleaching earth is manufactured by digesting and leaching an optimum blend of attapulgite and montmorillonite. The process of blending, surface modification, acid activation with sulfuric acid, and calcination are quite complex. The clay manufacturers have determined that the mineral deposits located at all geographic locations do not produce the same effective bleaching clay. Their information and the process for the treatment of the minerals are their trade secrets.

Manufacturers of activated bleaching clay optimize their raw material and process them to make the activated clay that exhibits the following properties for the benefit of the oil refiners in oil bleaching:

- High selectivity for the removal of red color
- High selectivity for chlorophyll reduction
- High efficiency in removal of trace metals, phospholipids, etc.
- High filterability of oil through the press
- Ability to reduce the soap content in the feed oil to zero ppm

The acidic pH of the acid-activated clay hydrolyzes the soap forming FFA. This is the reason for increased FFA in the bleached oil compared to the water-washed and vacuum-dried oil coming into the bleacher. The FFA in the bleached oil normally seems to rise by 0.03%–0.04% over that in the vacuum-dried oil. However, this increase can be significantly higher if the soap in the vacuum-dried oil is much higher than 100 ppm as discussed earlier in this chapter.

3. Amount of bleaching clay

Bleaching clay concentration is typically <0.5% of the vacuum-dried oil or less, unless the oil contains high levels of soap, phospholipids, and/or moisture. The dosage can also be higher for oils derived from poor quality oilseeds, fruit palm or if the crude oil is abused before or during refining. The bleaching clay feeder must control the flow rate within ±0.03% of the set target.

The bleaching clay absorbs (not adsorbs) approximately 0.35 lb of oil/lb of clay. This is a substantial amount of oil loss in the process over the course of a year. Therefore, all efforts must be directed toward improving the quality of the vacuum-dried oil and the bleaching process to reduce the amount of bleaching clay used in the process. Proper drying of the spent clay in the filter is also important for reducing the amount of absorbed oil in the spent clay.

It has been mentioned earlier that the bleaching clay also removes some of the natural antioxidants from the oil, which can reduce the oxidative stability of the refined oil. Therefore, one must try to use the least amount of bleaching clay to get the desired quality bleached oil.

It has been pointed out earlier in this chapter that the amount of bleaching clay used for soybean or canola oil should be adjusted so that the chlorophyll content of the bleached oil is <30 ppb (parts per billion) in addition to the other impurities, such as phosphorus, soap, and trace metals that must meet the standard. Chlorophyll is a strong catalyst for photooxidation and is known as a photosensitizer. Therefore, it is advisable to target the green color (or the chlorophyll content) of the oil and it is not necessary to reduce the red color because the red color is removed in the deodorization process.

However, higher concentration of acid-activated bleaching clay is needed under the following situations:

- High amounts of impurities left in the vacuum-dried oil because of poor process control in refining and water washing.
- The vacuum-dried oil has been made from poor quality crude oil.
- If harvested under wet conditions, the seeds contain very high level of chlorophylls.
- Harvesting seeds that are *not* fully mature because they exhibit high chlorophylls in the crude oil.

There are two types of chlorophylls in soybean oil. They are:

1. chlorophyll A
2. chlorophyll B

In the presence of the acidic clay and oxygen, the chlorophylls are broken down to three types of compounds:

- pheophytins
- pheophorbides
- pyropheophorbides

These compounds are not detected on the spectrophotometer at the same wavelength for chlorophyll. This gives a false indication of the reduction of the chlorophyll while, in reality, a more unstable situation is created by the breakdown products of the chlorophylls. These breakdown products of chlorophylls are 10 times stronger photosensitizers than their respective parent compounds. Chlorophyll B and its breakdown products are stronger photosensitizers than their corresponding counterparts from chlorophyll A.

It is necessary to optimize the level of the acid-activated clay in bleaching by using the following target for the bleached oil quality parameters:

• Chlorophyll (soybean and canola oil)	30 ppb, max.
• Soap	0 ppm
• Phosphorus	<0.5 ppm; 1 ppm max.
• Iron	<0.2 ppm; 0.5 ppm max.
• Calcium	<0.2 ppm; 0.5 ppm max.
• Magnesium	<0.2 ppm; 0.5 ppm max.

Additionally, the freshly bleached oil must also meet the following quality standards:

• Peroxide value (PV)	0 MEq/kg
• Anisidine value	<2 AVU; 4 AVU units max.

4. Bleaching temperature
Temperature of the oil is critical in this process because it affects the efficiency of the process in two ways, as follows:
- The viscosity of the oil decreases as the oil is heated.
- This makes it easier to move the oil around via agitation. This increases the contact between the adsorbent and the adsorbates in the oil.

However, there is an optimum temperature for adsorption. Above the optimum temperature the desorption process begins when some of the impurities begin to get released by the adsorbent. This lowers the efficiency of bleaching.

Additionally, a higher than optimum temperature begins to damage the oil through oxidation and polymerization.

At lower temperatures, the viscosity of the oil is high. This reduces the dispersion efficiency of the mechanical mixer.

The bleaching process also produces a very small amount of *trans* fatty acid in the oil. Therefore, a higher than the optimum temperature is undesirable.

The recommended bleaching temperature for most oils is 100–120°C (212–230°F). Sometimes a higher temperature is required if the chlorophyll content of the oil is too high.

5. Degree of mixing
The oil and the adsorbent must be in intimate contact for better adsorption of the impurities by the adsorbent. This is why mechanical mixing becomes a very important factor in making the bleaching process effective.

Fig. 6.2 shows the schematic diagram for a vacuum bleacher. The bleacher is a pressure vessel with three to four baffles and a top entering agitator. The agitator has multiple sets of impellers. The top one is of axial design, which pushes the oil downward continuously. The middle and the bottom impellers have a blade design capable of shear action so the adsorbent and the oil are brought to intimate mixing continuously. The baffles prevent any vortex formation.

The mechanical mixer has a pumping capability of 4000–8000 gallons per minute (gpm). Therefore, in a bleacher with 30,000 lb of oil (4,000 gallons approx.), the entire oil in the vessel is turned over approximately twice per minute. Steam agitation does not provide the same degree of mixing. Besides, the use of steam increases the size of the vacuum ejector system.

6. Contact time between the oil and the bleaching clay
The impurities are not adsorbed instantly as the adsorbent and the oil are brought in contact. Therefore a certain amount of time is allowed for the oil

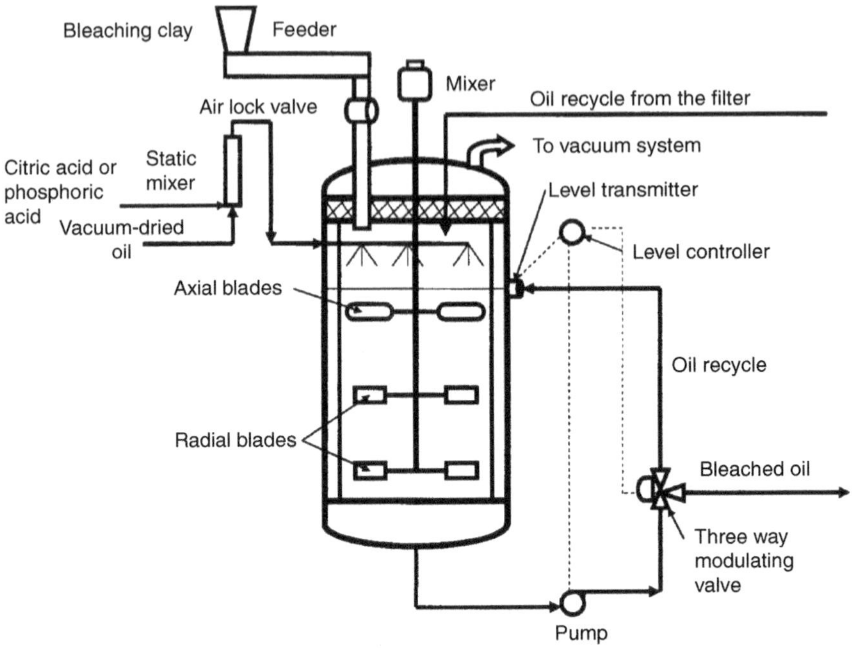

FIGURE 6.2 Schematic view of a Vacuum Bleacher©.

and the adsorbent to remain in the bleacher vessel. The typical time of contact is 30–45 min. In some instances, contact time as low as 20 min might be sufficient.

At short contact time, the adsorption of the impurities may not be complete. The process of adsorption is also compounded by the fact that there are different impurities in the oil that have different degrees of affinity toward the adsorbent. Therefore, more of the impurity with the greatest affinity for the adsorbent would be adsorbed initially and then some additional time is needed for the others to be adsorbed.

A longer contact time is not necessary. A very long contact time may have the following disadvantages:

- The acidic clay may react more with the oil and especially with the chlorophylls and cause their breakdown.
- There may be a higher loss of the natural antioxidants in the oil.
- There may be formation of dimers or polymers in the oil.

The oil level is maintained in the bleacher with the help of a recirculating loop. A level sensor senses the set oil level and the transmitter feeds the information to the level controller. The level controller modulates the three-way control valve to return part of the oil back to the bleacher and sends part of it to the filter. During start-up, the oil from the filter is also recycled back into the bleacher. Thus, the oil and the bleaching clay get the required contact time for bleaching.

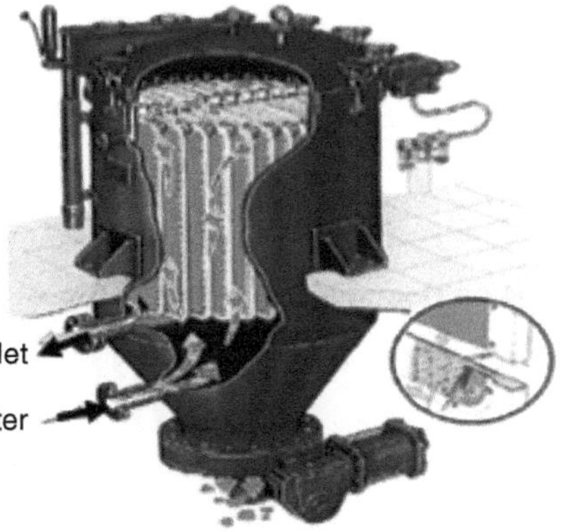

FIGURE 6.3 **Pressure leaf filter.**

7. Addition of phosphoric acid or citric acid
   A small amount of phosphoric or citric acid is added into the bleaching oil. The typical dose is 50–300 ppm, depending on the level of trace impurities in the oil. The purpose of acid addition is to hydrolyze part of the soap in the oil and improve the efficiency of scavenging the trace metals from the oil by the clay.
   The oil feed and the acid streams pass through a static mixer or a high shear mixer before entering the bleacher. The acid does not dissolve in the oil. However, the vigorous mixing by the mechanical mixer allows intimate mixing between the acid and the oil.
8. Filter precoat
   This is a very important step in filtering bleached oil, whether it is from dry bleaching or wet bleaching process. The suspended bleaching clay in the bleached oil tends to prematurely blind the filter screens and reduce oil flow through the filter. Therefore, filter precoat is necessary because the filter cycle time can be greatly reduced in the absence of proper precoat on the pressure leaf filter screens.
   Filter precoating is a simple operation. The overall bleaching system does not deliver the desired results if precoating is not done properly. It will be helpful to view the picture of the filter screen (Fig. 6.3) to explain the precoating operation.
   1. Oil with the bleaching clay enters the filter at the bottom.
   2. The oil moves up along the surface of the filter screens.
   3. The filtered oil flows across the filter screens and falls down to a manifold through which the filtered oil passes forward.

   Filter precoating system is shown in Fig. 6.4. The step-by-step procedure for filter precoating is outlined as follows:
   1. Filtered bleached oil at a temperature of 100–110°C (212–230°F) is pumped into the slurry tank. The volume of oil in the slurry tank must match the total volume of the filter, piping, and the accessories plus 25%–30% extra.

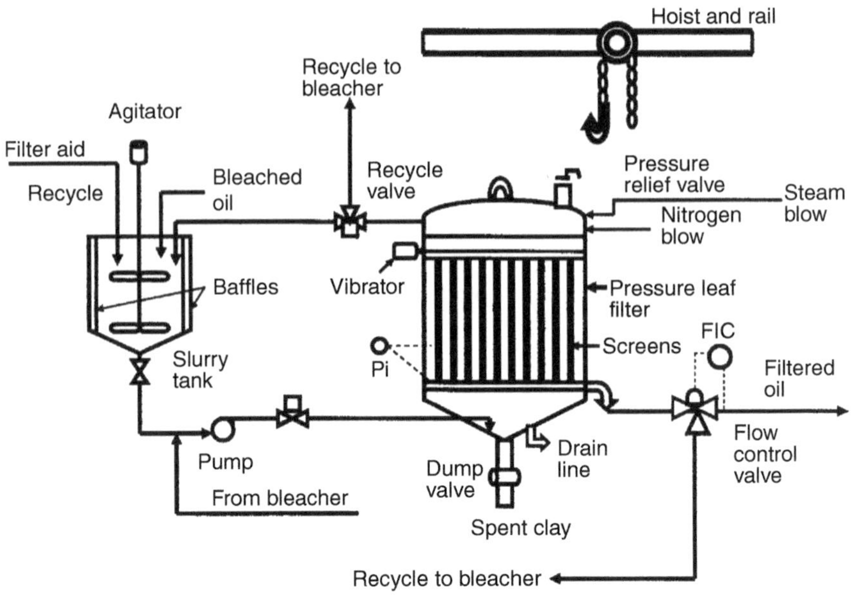

FIGURE 6.4 Schematic diagram for filter Precoat System©.

2. Diatomaceous earth (filter aid) is added into the oil in the slurry tank. Amount of the filter aid added is 0.1%–0.2% of the oil in the slurry tank.
3. The filter aid is dispersed into the oil with the help of a vertical mixer operating on medium speed (60 rpm). There are three to four baffles in the tank to avoid vortex formation. High speed is not recommended for the agitator to avoid physical breakdown of the filter aid.
4. Agitation in the slurry tank is continued for at least 10–15 min or longer to fully disperse the filter aid into the oil.
5. The slurry is pumped into the filter. The pump capacity should be twice that of the filtration rate of the bleaching. For example, for 30,000 lb/h production (67 gpm) of bleached oil, the slurry tank pump capacity should be 135 GMP.
6. The slurry enters the filter at the bottom and moves up in path parallel to the screen surface.
7. As soon as the pump starts, the recycle valve opens. Simultaneously, the oil outlet valve opens.
8. The flow indicator controller (FIC), located at the oil outlet line is set at the desired production rate for bleaching.
9. The oil coming out of the outlet line from the filter is recycled back into the slurry tank until the oil is clear.
10. The oil is also returned to the slurry tank through the recycle valve which is fully open.

11. The pressure drop across the filter is measured by the differential pressure sensor and it is displayed on the differential pressure indicator (PI).
12. At the start of precoat, there is no differential pressure across the filter screens as it is indicated by the PI.
13. After recirculating the oil through the filter for 1 min, the filter screens go into automatic vibrate mode for 1 min. The purpose of this step is to loosen up any accumulated material inside the screen.
14. The oil coming out of the filter through the oil outlet line is checked for clarity via filter test.
15. Precoating is complete when the oil is clear as determined by the filter test.

    Filtering bleached oil

    1. At this point the oil from the bleacher is pumped into the filter.
    2. Oil recycles to the bleacher through the recycle line, as well as through the oil outlet line and the recycle process is continued until the oil is found clear by filter test.
    3. The filter test can be done manually or the oil clarity can be determined with the help of a clarity meter (not shown in the diagram).
    4. The recycle from the oil outlet line is stopped if the oil is clear and the oil is sent forward.
    5. The three-way control valve then opens to send the oil forward to a cooler or to the next processing step.
    6. The filtration process stops automatically at a predetermined time to make sure the dirt load capacity in the filter is not exceeded. This is to avoid bridging and the consequent damage to the filter screen. At this point the following sequences of events occur:
        a. The filter blow cycle (using nitrogen) begins and the clean oil goes out through the oil outlet line.
        b. The recycle valve closes.
        c. The oil feed pump stops.
        d. The valve on the oil feed line closes.
        e. The flapper valve on the drain line opens and sends the oil slurry from the bottom of the filter to a scavenger filter.
        f. As soon as the oil flow stops, the flapper valve closes, steam blow cycle starts, and the nitrogen blow stops.
        g. Steam dries the cake. The steam blow comes out of the oil outlet line. The steam condensates and the residual oil from the filter screens go into a slop tank.
        h. The oil from the slop is recovered from the tank and is generally sent to acidulation.
        i. Steam blow stops after 5–10 min.
    7. Immediately after the steam blow, the vibrator is activated to discharge the cake from the screen surface.

8. The dump valve opens and allows the dry cake discharge to drop into a receptacle under the filter.
9. The spent earth is sent to one of the three following places:
   a. To a heat recovery system where the energy from burning the absorbed oil is recovered and used in the process.
   b. An oil recovery system where most of the oil is recovered via hot water and steam treatment of the bleaching clay along with some caustic. The absorbed oil from the clay is converted to soap. The soapy water is sent to acidulation. The water from the clay is drained.
   c. Sprayed with water and then sent directly to a landfill where there are no environmental regulations against it.
10. The screens need to be scraped if the cake discharge is not good.
11. The screen bundle is lifted and scraped gently using scrapers that do not have sharp surfaces because that would damage the screen.

9. Filtering area/oil flow rate
   It is critical that the appropriate filter size is chosen for a given production rate. Typically, the screen filters are designed for an oil flow rate of 0.1–0.2 US gallons/ft.^{2} of filter area/min. This means for a production rate of 150 US gpm, the filter surface should be at least 750 ft.^{2}.

Note

1. In the event that the nitrogen blow and/or the steam blow is either done inappropriately or interrupted, the cake on the screens would be wet and will not separate from the screen as cleanly. This will also have a tendency to blind the screens prematurely during the subsequent operation, as it will be indicated by the rapid rise in differential pressure across the screen before the normal cycle is completed.
2. Differential pressure at the beginning of bleaching oil flow should be 0–5 psi.
3. The differential pressure at the end of filter cycle should be 30–40 psi; most typically it is 35 psi.

10. Filter screen spacing
    For the bleaching process, the filter spacing is 6 in. for chemically refined oil. The recommended spacing is 5 in. for physical refining where the crude oil is bleached.

6.5 SAMPLING FREQUENCY IN BLEACHING PROCESS

As the bleached oil leaves the filter, the oil must be analyzed to ensure its quality before it is sent to the next processing step. Table 6.1 shows the sampling frequency and the required oil quality parameters for the bleached oil.

TABLE 6.1 Sampling Frequency and Oil Analysis in Bleaching

Sample	Filter test	Soap (ppm)	Phosphorus (ppm)	Chlorophyll (ppb)[a] (maximum)	Lovibond red color	Trace metals (ppm) (maximum)
At start-up and every 15 min until the oil is clear	Clean	—	—	—	—	—
At steady state					Varies with the type of oil	
Every 2–4 h	X	0 Max.	<0.5 not >1.0	30		Fe^{+++} < 0.2 not > 0.5 Ca^{++} < 0.2 not > 0.5 Mg^{++} < 0.2 not > 0.5

[a]*Parts per billion.*

6.6 TROUBLESHOOTING DRY BLEACHING PROCESS

Table 6.2 lists the typical oil quality symptoms observed in the dry bleaching process, their probable cause or causes, and the recommended solutions to correct them.

6.7 WET BLEACHING PROCESS

Wet bleaching is considered by many oil processors as a very effective way to bleach chemically refined and water-washed oil. This method differs from the dry bleaching method as outlined next:

1. The moisture oil in the dry bleaching process is 0.1% or less, while the moisture in the feed oil to the wet bleaching process is 0.4%–0.5%.
2. Water-washed oil is vacuum dried for the dry bleaching process. Vacuum drying of the water-washed oil can be skipped in the wet bleaching process. This is because the oil, leaving the water-washing separator, is suitable for the process because its moisture content is 0.2%–0.4%.
3. Since the wet slurry cannot be filtered because that would plug the filter screens prematurely (for the reason as mentioned in dry bleaching process), the oil must be dried in a vacuum bleacher to a moisture content of 0.1% or less, after it is treated with the clay.

The schematic diagram for the wet bleaching process is shown in Fig. 6.5. In this process, the oil from the water wash centrifuge is treated with acid-activated clay in a bleacher reactor as shown in Fig. 6.5.

TABLE 6.2 Troubleshooting Dry Bleaching Process

Symptom	Probable cause/causes	Recommended solution
Oil filtration rate is slow although the filter has been freshly cleaned	The vacuum-dried oil has the following: • Moisture level is >0.1% • Soap content is >0 ppm • Phosphatide (phosphorus) level is >1 ppm • The flow control through the filter is not adjusted properly allowing higher oil flow and blinding the screens	• Make sure the vacuum-dried oil has a moisture content of <0.1% • Check for the high soap in the primary or water wash centrifuge and correct it • Phosphorus level must be <1 ppm. Make sure that the oil is properly degummed and refined • Check and correct the flow rate
Filtration cycle time is too short	• The filter was not properly precoated causing premature blinding of the screens • The screens may have been blinded due to accumulation of material inside them • The oil flow may be too high • Oil temperature is low	• Check and follow proper precoating procedure • Clean the screens • Check and correct the oil flow • Increase oil temperature by 5°C without exceeding 239°F (115°C)
Filter grade is unacceptable	• Insufficient time allowed for oil recirculation • Improper precoating of the filter screens • One or more screens may have been torn • The "O" ring on the screen socket at the bottom of the screen may be worn out • The socket at the bottom of screen may *not* have been sitting properly on the manifold	• Make certain that sufficient time is allowed for oil recirculation to obtain clarity • Follow the required steps for precoating as discussed in Section 6.4 • Replace the damaged screens. It is *not* advisable to repair screens by the oil processor • Replace the "O" ring periodically to avoid this issue • Check every screen to make sure they are sitting properly on the manifold
Soap in the bleached oil is not zero	• High soap in the vacuum-dried oil	1. Check and correct the high soap in the water-washed oil 2. Increase the amount of bleaching clay 3. Add additional amount of phosphoric acid or citric acid in the bleaching process to hydrolyze the extra soap in the water-washed oil

TABLE 6.2 Troubleshooting Dry Bleaching Process (*cont.*)

Symptom	Probable cause/causes	Recommended solution

Bleached oil color is too dark	• Insufficient amount of bleaching clay
	• Inadequate contact between the oil and bleaching clay
		• Increase the amount of bleaching clay
		  • Check agitator and make sure all impellers are in place and the motor is drawing the rated amperage. Fix the agitator if needed
		  • Increase recirculation time through the filter

	• Bleaching temperature is low	• Increase oil temperature without exceeding 239°F (115°C)
	• The caustic strength may be low especially if the crude oil is of poor quality • Possibly crude oil is of very poor quality	• Increase the caustic strength without increasing the total caustic treat • Check the lab-bleach color of the refined oil. If the color is as dark as that produced at the plant, then the color of the oil is fixed and cannot be reduced.
Chlorophyll level is high in the bleached oil (>30 ppb for soybean or canola oil)	• The bleaching clay used may not have high selectivity for chlorophyll • Bleaching temperature is low	• Select the appropriate acid-activated clay for chlorophyll removal • Increase the oil temperature by 5°C, without exceeding 239°F (115°C) • Make sure the vacuum in the bleacher is <50 Torr
	• Insufficient contact time between the oil and the clay	• Follow the appropriate steps listed previously under "Bleached oil color is too dark"

	• The crude oil may have higher than normal level of chlorophyll because of:
	  • early harvest
	  • wet harvest condition
		• Do one or more of the following:
		  • Use most active acid-activated clay and at a higher dosage
		  • Blend with less green oil before or after bleaching
		  • Follow the procedure discussed under Very Green Oil.

In this process, the oil from the water wash centrifuge is treated with the bleaching clay in a bleacher reactor, which is operated at approximately 500 Torr. The objective is to maintain moisture content of 0.2%–0.4% in the oil. The oil temperature, bleaching clay dosage, mixing method, and mixing time are similar to those in the dry bleaching process. The oil processors find that the

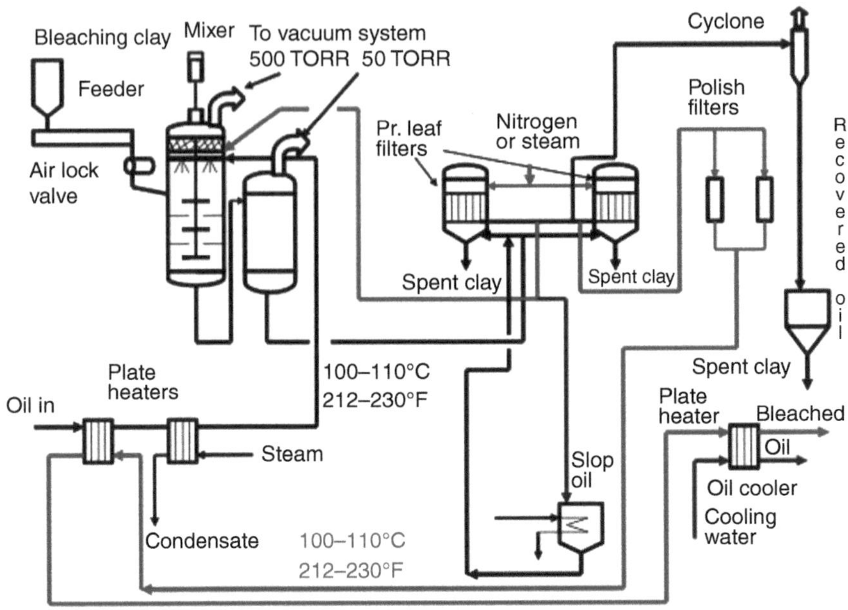

FIGURE 6.5 **Schematic diagram for Wet Bleaching Process©.**

efficiency of the clay is better in this method of bleaching, and the clay usage is lower. During the physical refining process in certain palm oil operations, a small amount of water is sprayed on the bleaching clay before it is used. This improves the performance of the bleaching clay.

After 20–30 min of contact time between the oil and the clay, the oil leaves the vacuum reactor and enters the vacuum dryer, which is maintained at 50 Torr. The moisture in the oil is reduced to <0.1%.

The oil is then filtered through a precoated screen filter as in the dry bleaching process. The filter precoating process, oil recycle, filtration, etc., should be carried out in the same manner as in the dry bleaching process.

6.8 CRITICAL CONTROL POINTS IN THE WET BLEACHING PROCESS

The critical process control points are similar to those discussed under the dry bleaching process with the following exceptions:

1. Moisture in the water-washed oil (as feed).
2. Absolute pressure in the bleacher reactor.
3. Absolute pressure in the vacuum dryer.

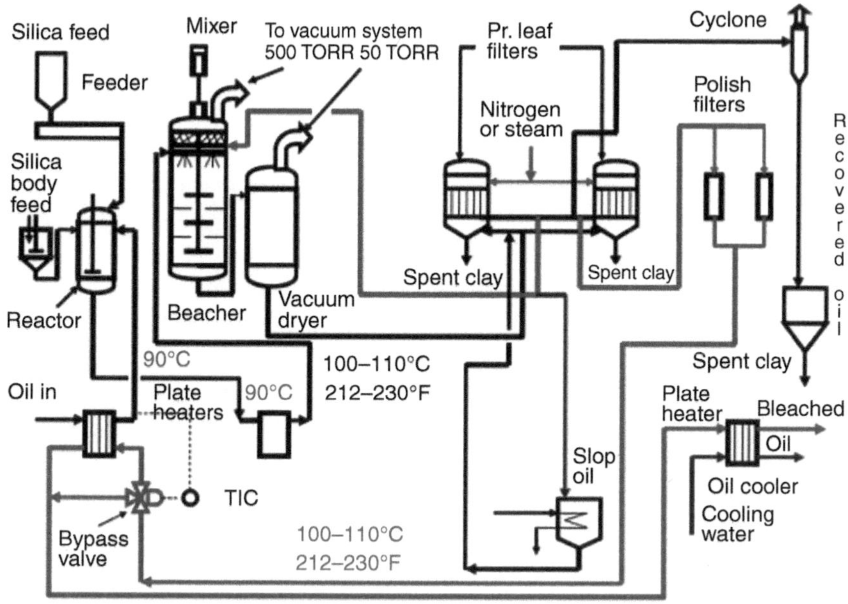

FIGURE 6.6 Two-Step Bleaching Process©.

1. Moisture in the feed oil
Moisture in the feed oil must be 0.2%–0.4%. The benefit of wet bleaching is not accomplished at moisture content <0.2%. On the other hand, a higher than 0.4% moisture increases the risk of having wet oil (moisture content >0.1%) leaving the vacuum dryer. This can cause premature blinding of the filter screens.
2. Absolute pressure in the bleacher reactor
The operating pressure in the bleacher reactor is 500 Torr. At higher pressure, the oil would have higher moisture content. At lower pressure, the oil might be too dry to derive the benefit of the wet bleaching process.
3. Operating pressure in the vacuum dryer
The maximum operating pressure in the vacuum dryer is 50 Torr. This allows the oil to be dried to moisture level of <0.1% before filtration.

6.9 TWO-STEP BLEACHING PROCESS (USE OF SILICA HYDROGEL)

Fig. 6.6 shows the schematic diagram for the two-stage bleaching process. In this process the oil from the water wash centrifuge is treated first with the silica hydrogel and then by the acid-activated clay to maximize the removal of soap, phosphorus, trace metals, and chlorophyll.

This is carried out in two steps:

Step-1

- Soap, phosphorus, and trace metals are removed from the oil by treating the oil with silica hydrogel at a temperature of 90°C (204°F) under atmospheric condition for 15 min.
- Sometimes a very small amount of silica hydrogel is added to the oil if the impurities in the bleached oil are still higher than the target levels. The body feed system is used and it feeds silica into the oil before the reactor.
- This system is used only when it becomes necessary to put an additional amount of silica because of the high impurities in the oil.

Step-2

- The silica-treated oil is then pumped into the bleacher reactor maintained at an operating pressure of 500 Torr. Under this condition the silica structure changes and traps soap, phosphorus, and the trace metals. Moisture content of the oil is 0.2%–0.4% as the oil enters the bleacher reactor.
- Activated clay is added and the oil is treated for 20–30 min.
- The oil then enters the vacuum dryer, which is maintained at an operating pressure of 50 Torr or less. The moisture in the oil is reduced to <0.1%.
- The oil is then filtered in the same manner as described under the dry bleaching process.

6.9.1 Benefits of Two-Step Bleaching Process (Use of Silica Hydrogel)

1. Hydrogel silica works effectively to remove the soap, phosphorus, and the trace metals so the activated clay can remove chlorophyll from the oil more efficiently.
2. This also reduces the amount of activated clay required to remove chlorophyll from the oil.
3. This increases the cycle time for the filter, requiring less frequent cleaning, thus increasing the productivity of the plant.
4. There is a net reduction in the cost of bleaching for the oil.
5. There is reduced oil loss through absorption into the bleaching clay because of the reduced clay usage.
6. The FFA rise in the bleached oil is measurably reduced because of the soap removal by the silica hydrogel and reduced clay usage.
7. The water-washing step can be eliminated via silica hydrogel treatment when high quality crude oil is refined and the soap content of the refined oil is 100 ppm or less.
8. The overall savings are not as attractive when the soap level in the oil is higher than 200 ppm.

6.10 CRITICAL CONTROL POINTS IN TWO-STEP BLEACHING PROCESS

Following are the critical control points for the two-step bleaching process:

1. Moisture in the feed oil.
2. Soap content in the feed oil.
3. Silica hydrogel dosage.
4. Oil temperature in silica hydrogel pretreatment.
5. Operating pressure in silica hydrogel reactor.
6. Contact time between silica hydrogel and the oil.
7. Acid-activated clay dosage.
8. Temperature in the bleacher reactor.
9. Operating pressure in the bleacher reactor.
10. Contact time in the bleacher reactor.
11. Operating pressure in the vacuum dryer.
12. Residence time in the vacuum dryer.
13. Moisture content of the oil to the filter.

1. Moisture in the feed oil
   Silica hydrogel works best when the oil contains some water. Generally, the target moisture in the oil is 0.2%–0.4%. Moisture content of the water-washed oil is normally 0.45% or less. Thus, the water-washed oil need not be predried before addition of the silica hydrogel.
2. Soap content in the feed oil
   Soap content in the water-washed oil should be 50–100 ppm under properly controlled refining conditions. The silica hydrogel dosage increases almost exponentially as the soap and phosphorus contents in the water-washed oil go up. Table 6.3 shows the relative amounts of TriSyl needed to reduce the soap content to zero ppm and the phosphorus content to <1 ppm. The Grace Davison Co. provided the data from their controlled laboratory bleaching study.
   The data in Table 6.3 illustrates the following points:
   - TriSyl 600 is a superior adsorbent for soap and phosphorus compared to TriSyl and TriSyl 300.
   - The dosage of the silica hydrogel goes up significantly with higher soap and phosphorus contents of the oil to be treated.
   - The cost of silica treatment also goes up in the same proportion.

   In reality, silica treatment tends to lose its cost advantage as the soap content in the water-washed oil goes above 200 ppm.
3. Silica hydrogel dosage
   In most situations, the hydrogel silica dosage is 0.1% of the oil or less when the soap content in the water-washed oil is 100 ppm or less. The oil processor must investigate the cause for high soap and phosphorus in the oil and take the necessary steps discussed on vegetable oil refining in Chapter 5. This will keep the silica hydrogel consumption under control.

TABLE 6.3 TriSyl Treatment Data From Grace Davison Co.

Soap and phosphorus in the feed oil	ppm	TriSyl dosage (% of oil)	TriSyl 300 dosage (% of oil)	TriSyl 600 dosage (% of oil)	Bleached oil analysis	ppm
Soap	40	0.09	0.063	0.032	Soap	0
Phosphorus	1.3				Phosphorus	<1
Soap	140	0.24	0.12	0.096	Soap	0
Phosphorus	5.2				Phosphorus	<1
Soap	290	0.30	0.135	0.135	Soap	0
Phosphorus	9.7				Phosphorus	<1

4. Oil temperature in silica hydrogel pretreatment
   The temperature of the oil is 90°C (±1°C). As the temperature drops below 85°C, the silica begins to lose the adsorption efficiency to remove soap and phosphorus. On the other hand, at temperatures above 95°C, the silica hydrogel can lose some of the moisture and become less efficient as adsorbent.
5. Operating pressure in the silica reactor
   The silica reactor must be operated under atmospheric conditions. The silica hydrogel loses all of the moisture under vacuum and cannot adsorb the impurities in the oil if it is subjected to vacuum without allowing the preadsorption opportunity.
6. Contact Time between silica hydrogel and the oil
   The recommended contact time between the silica hydrogel and the oil is 15 min. This allows the silica to come in contact with the impurities and adsorb them.
7. Acid-activated clay dosage
   The typical clay dosage in this process is around 0.1%–0.3% of the oil feed. This silica hydrogel dosage is much lower than that for either dry or wet bleaching process.

The remaining control points, 8–11 are the same as discussed under the wet bleaching process.

6.11 PACKED BED FILTRATION IN BLEACHING PROCESS

This is a very important concept of bleaching and is followed by many oil processors. This method uses both hydrogel silica and acid-activated clay. The bleached oil exhibits low impurities and the bleaching cost is lower.

This technique works best when high quality crude oil is refined and bleached in a vertically integrated crushing and refining operation. All oil processors,

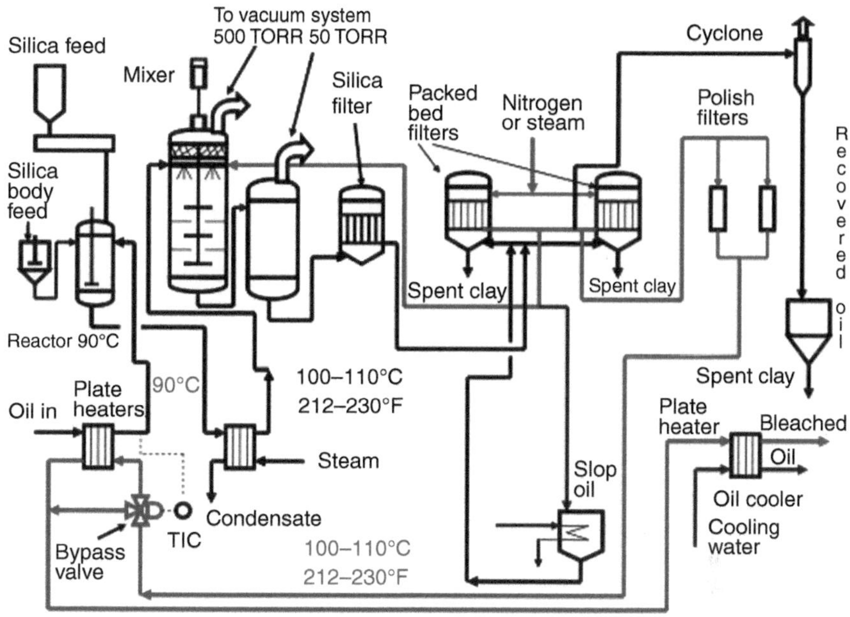

FIGURE 6.7 Schematic diagram for Packed Bed Bleaching Process©.

using this method in the United States, have vertically integrated crushing and refining operations, and they produce crude oil of high quality. Silica usage is high when refined oil obtained from poor quality crude oil is bleached by this method and makes the process uneconomical. Fig. 6.7 shows the schematic diagram of the packed bed bleaching system.

This process is similar to the two-step bleaching process in many aspects, but it has several distinctive features:

1. There are three filters: one is the silica filter and the other two are bleaching clay filters packed with fresh bleaching clay. This is the reason why this is called the packed bed bleaching process.
2. A precoat system is used initially to precoat the silica filter.
3. The packed bed filters are precoated with filter aid from a separate precoat tank and as soon as precoating is completed, bleaching clay is added to the precoat tank and the oil is recirculated until the filters are loaded with the designed dirt-load capacity of the bleaching clay.
4. The oil continues to recycle through all three filters until the system is ready.
5. The oil from the water wash centrifuge is treated with silica hydrogel in the silica reactor at 90°C, in the same manner as in the two-step bleaching process.
6. The oil leaves the silica reactor and is heated in the plate and frame heat exchanger to 100–110°C, like in the two-step bleaching process.

7. The oil enters the bleacher reactor which is maintained at an operating pressure of 500 Torr. As discussed in the two-step bleaching process, hydrogel silica is more effective under this condition when the oil leaving the bleacher reactor still contains >0.1% moisture.
8. *No bleaching clay is added into the bleacher reactor.*
9. The oil is allowed a residence time of 20–30 min in the bleacher reactor and then enters the vacuum dryer, which is maintained at an operating pressure of 50 Torr.
10. The hydrogel silica collapses under full vacuum and traps impurities like soap, phosphorus, trace metals (iron, calcium, magnesium), but not the chlorophyll.
11. The moisture in the oil is reduced to <0.1%.
12. After the designated residence time in the vacuum dryer, which is roughly 15 min, the oil enters the silica filter.
13. The pump on the silica precoat tank stops as soon as the oil from the vacuum dryer is pumped into the silica filter.
14. The oil comes out of the silica filter and enters the packed bed filters in series.
15. Like in the silica system, the precoat tank in the packed bed filter system stops as soon as the oil from the vacuum dryer is pumped into the system.
16. The oil leaving the packed bed filters is recycled back into the bleacher reactor in the beginning. The oil picks up moisture from the bleaching clay in the packed bed because the clay has not been exposed to any vacuum like in the other processes.

6.11.1 Oil Quality Checks

1. The initial oil leaving the silica filter has 0 ppm soap, 0 ppm phosphorus, and essentially 0 ppm of the trace metals.
2. The oil leaving the silica filter is monitored at a regular frequency.
3. The oil leaving the packed bed filters is checked for chlorophyll, using an automatic instrumentation.
4. The instrument is set to sound an alarm when the chlorophyll in the oil exceeds the preset upper limit.
5. The initial oil leaving the packed bed filters is very low in trace metals (<0.1 ppm), as well as in chlorophyll (0–10 ppm).
6. The level of chlorophyll in the bleached oil leaving the first packed bed filters may begin to exceed the preset upper limit after 12–24 h of operation.
7. At this point, the first filter is taken off the line, and it is taken through the automatic blow dry and cake discharge cycle, the second packed bed remains in operation.
8. The filter is precoated, reloaded with the bleaching clay. This filter then becomes the second filter in line.
9. In the meantime, the oil from the second filter (from the initial sequence) continues to move forward and is collected in the bleached oil storage tank.

10. As soon as the second filter (originally the first filter) is placed in service, the bleached oil is recirculated back into the bleacher reactor until the moisture in the oil reaches <0.1%.
11. The plant needs to develop the experience to determine how long the second filter can work alone without producing off-quality oil.
12. It is always recommended that the plant should maintain two or three bleached oil storage tanks so the oils from these tanks can be blended to meet the bleached oil quality standard.
13. The silica press generally maintains the oil quality for 24–36 h, depending on the incoming oil quality.
14. The silica filter is put on the blow and dry cycle when the trace metal content reaches a level between 1–2 ppm.
15. The filter then goes through the blow, dry, and discharge cycle.
16. The filter is precoated with fresh silica and restarted.
17. The packed bed filters remain on recycle mode during the time until the silica filter is ready.
18. After the silica filter is back in operation, packed bed filters are put back into filter forward mode.

6.12 CRITICAL CONTROL POINTS IN PACKED BED BLEACHING

Following are the critical control points in packed bed bleaching process:

1. Moisture in the oil in the silica reactor.
2. Operating temperature in the silica reactor.
3. Operating pressure in the silica reactor.
4. Residence time in the silica reactor.
5. Body feed.
6. Operating temperature in the bleacher reactor.
7. Operating pressure in the bleacher reactor.
8. Residence time in the bleacher reactor.
9. Operating temperature in the vacuum dryer.
10. Operating pressure in the vacuum dryer.
11. Residence time in the vacuum dryer.
12. Precoating of the silica as well as the packed bed filters.
13. Initial oil quality at the exit of the silica filter.
14. Termination point for the silica filter.
15. Recycle of the oil at the start of the packed bed filter operation.
16. Initial oil quality at the exit of the packed bed filter.
17. Termination point for the packed bed filter.

1–4: These control conditions are identical to those for the two-step bleaching process.

5: Body feed—A small amount of silica hydrogel is added into the oil as continuous body food.
6–11: These are identical to those discussed under the two-step bleaching process.
12: Precoating filters—This is the same as discussed in the dry bleaching process.
13: Initial oil quality at the exit of the silica filter—The oil leaving the silica filter has 0 ppm soap, 0 ppm phosphorus, and essentially 0 ppm of the trace metals in the beginning.
14: Termination point for the silica filter—The trace metal concentration in the oil is 1 ppm and not to exceed 2 ppm.
15: Oil recycle at the start of the packed bed filters—Bleaching clay contains a certain amount of moisture which is removed from it when the clay is added into the bleacher reactor. In the packed bed process, the filters are packed with the clay containing its original moisture.

The FFA in the oil reacts with the metals, such as calcium and magnesium in the bleaching clay which forms soap. This soap is solubilized by the moisture present in the bleaching clay. The soap acts as an emulsifying agent in the presence of the moisture and hydrolyzes some of the neutral forming more FFA. The soap gets hydrolyzed by the acidic clay to form FFA. Thus, there is a significant increase in the FFA content of the bleached oil if the initial oil leaving the packed bed filters is not recycled through the bleacher reactor to remove the moisture.

16: Initial oil quality at the exit of the packed bed filters—The initial oil leaving the packed bed filters is very low in trace metals (<0.1 ppm) as well as in chlorophyll (0–10 ppm).
17: Termination point for the packed bed filters—The first packed filter is taken off the line when the chlorophyll in the bleached oil approaches the preset upper limit. This limit is established by the plant through experience. A blending provision is needed at the plant as discussed earlier. The trace metal concentration in the bleached oil should not exceed 0.5% at the time when the chlorophyll level is allowed to rise.

6.13 FILTERS FOR FILTERING BLEACHED OIL

The oil industry uses three basic types of filters to remove the adsorbents from the bleached oil. They are:

1. plate and frame filters;
2. horizontal pressure leaf filters; and
3. vertical pressure leaf filters.

6.13.1 Plate and Frame Filters

Fig. 6.8A–B show the typical filter press used in the oil processing industry.

(A)

(B)

FIGURE 6.8 (A) Plate and (B) frame filters.

The filter press has been in use in the vegetable oil industry for many years. These are rugged and dependable filters but the older versions have certain drawbacks in terms of oil quality. This is described in item 2.

1. There are frames and plates stacked alternately with an end plate. The frame provides the cavity between each of the two plates on the filter. The oil slurry enters the cavity and the oil passes through a filter media (special cloth) and enters the oil discharge manifold through a spout connected to each plate.
2. In the older designs, the filtered oil drops into a trough. The oil then collects in a surge vessel and is pumped through a cooler and finally into the storage tank. This design causes excessive oxygen exposure to the oil causing oxidation. Modern plate and frame filters handle the oil in a totally enclosed environment, minimizing oil oxidation.
3. If the oil is to be hydrogenated or deodorized immediately, the oil goes to the next processing step without cooling.
4. Filtration is continued for the estimated time to fill the cavities in the filter (referred to as dirt load capacity of the filter).
5. The filter is blown at first with nitrogen to remove the oil and then with steam to dry the cake.
6. The filter is opened and then the cake is scraped off manually as shown in Fig. 6.8B.
7. The discharged cake is collected in a trough under the filter or in some cases, it is carried by a common auger (where there are several filters) to a common dump.
8. The bleaching clay contains as much as 35% of its weight in absorbed oil. Straight disposal of the spent bleaching clay is a fire hazard. The oil in the clay can start smoldering due to spontaneous combustion. Therefore, the disposal method for the spent clay is critical. The bleaching clay is disposed in one or more of the following manners:
   a. In some operations, the spent clay is burned in a furnace to eliminate the oil and in many instances the heat of combustion is captured and used at the plant.

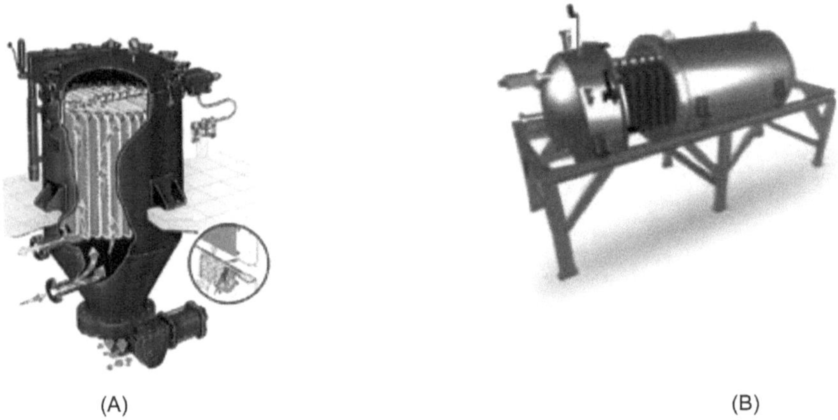

FIGURE 6.9 **Pressure leaf filters.** (A) Vertical leaf screen filter, (B) vertical leaf horizontal screen filter.

   b. The spent clay is treated with caustic solution. The oil in the clay is released and some of it is converted to soap. This is sent to the soap stock acidulation process for recovery of the acid oil.
   c. The spent clay is sprayed with water and transported to the facilities where they recover the oil from it.
9. Fully deoiled spent clay is no longer a fire hazard.
10. Filter presses do not require any precoat with filter aid.

The filter presses are manually operated in most of the older plants. However, the modern design of the filters enables one to operate them with automation. Initially the oil is recirculated back into the filter through a three-way automatic valve. Oil clarity is monitored with an instrument. The three-way control valve opens in the forward direction only when the oil meets the preset clarity criteria. This allows the oil to flow forward toward the next processing step. Clarity of the oil is also cross-checked via filter test in the QC laboratory.

6.13.2 Pressure Leaf Filters (Horizontal and Vertical Tanks)

Fig. 6.9 shows the pressure leaf filters with horizontal as well as vertical tanks. The leaves are all in the vertical position, regardless of the type of the tank.

Fig. 6.9A shows the vertical tank vertical pressure leaf filter where the screens have to be lifted for inspecting or complete cleaning of the screens.

Fig. 6.9B shows the horizontal tank vertical pressure leaf filter. Some filter manufacturers have the tank in the fixed position and the leaves are moved out by a hydraulic mechanism for cleaning and inspection of the screens. On the other hand, some manufacturers have the filter screens remain stationary and the tank slides out. The end result is the same in either case.

There are advantages and disadvantages in both designs (Vertical and Horizontal leaf filters). These are summarized in following sections.

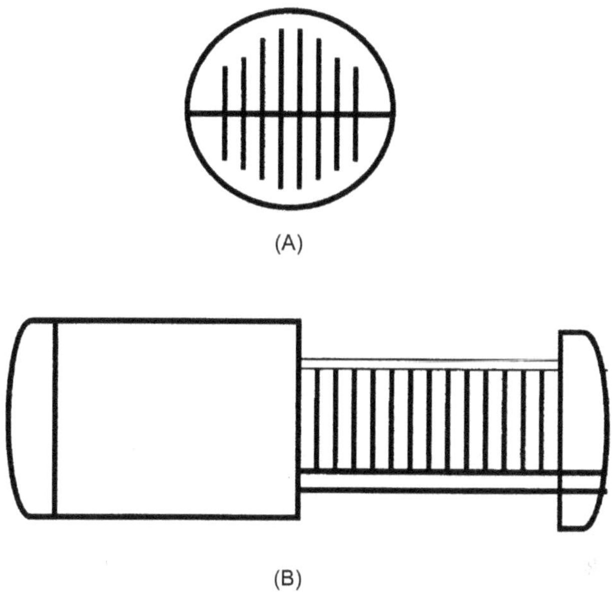

FIGURE 6.10 **Filter screen arrangements in Vertical and Horizontal Tank Pressure Leaf Filters©.** (A) Vertical tank vertical pressure leaf filter arrangement and (B) horizontal tank vertical leaf pressure leaf filter.

6.13.2.1 Vertical Tank Vertical Pressure Leaf Filter

6.13.2.1.1 Advantages

1. Lower capital cost to purchase and install.
2. Less floor space required.
3. Closed cake discharge: wet cake cannot be seen without lifting the filter screen assembly.
4. Fewer automation components are required.

6.13.2.1.2 Disadvantages

1. The filter screens are of varying sizes (Fig. 6.10).
2. The appearance of the cake on the screen cannot be observed unless the top of the filter is opened and the filter screens lifted.
3. The cake adhering to the screens can and will reduce the filter cycle time if the screens are not inspected and cleaned. This is a time-consuming and labor-intensive operation.
4. Cake adhering to the filter screen can cause spontaneous combustion under suitable conditions.
5. The filter dome and the vibrator must be disconnected to hoist the leaves.
6. Needs a hoist and requires the extra ceiling height to lift the screen bundle.

FIGURE 6.11 **Pear-shaped cake on the filter screen©.**

7. The very tall screens may exhibit pear-shaped cake formation as shown in Fig. 6.11. This can cause bridging at the lower parts of the screens, which can damage the filter screens.
8. The dump valve at the bottom of the filter may stick, causing difficulty in removing the spent clay. The oil flow control and oil bypass, nitrogen and steam flow described in the dry bleaching process can alleviate the wet cake issue, as well as the pear-shaped cake formation on the screens.

6.13.2.2 Horizontal Tank Vertical Pressure Leaf Filter

6.13.2.2.1 Advantages

1. All screens are of the same size.
2. The filter head is opened by a hydraulic system with the help of a push button switch.
3. Every filter screen can be inspected for cake discharge every time the filter is opened at the end of the filter cycle.
4. Opening, inspecting, and cleaning of the screens can be done at minimum time and with less manual labor.
5. The size of the filter can be larger without making it more labor intensive to clean it.
6. An individual screen can be removed easily by disconnecting it from the vibrator.
7. Extra ceiling height is not necessary.

6.13.2.2.2 Disadvantages

1. Requires more floor space because either the filter shell or the filter screens has to slide to open the filter screen carriage.
2. Higher installation cost because of more piping and automation components.
3. Higher capital cost.

TABLE 6.4 Bleaching Clay Treatment Data From Oil-Dri Corporation

Type of bleaching clay	pH	Delta FFA in the bleached oil (%)
Neutral	6.8	No change
Acid treated	3.0	+ 0.03
Acid activated-1	3.7	+ 0.06
Acid activated-2	3.4	+ 0.10

6.14 BLEACHING AGENTS

It has been discussed in the beginning of this chapter that the process of bleaching is to remove the trace metals, oil decomposition products, phosphorus, and the color bodies (primarily chlorophyll) from vegetable oils.

There are three types of adsorbents that are used for bleaching oil, they are:

1. bleaching clay
2. hydrogel silica
3. silicates
4. activated carbon

1. Bleaching clay

   There are two groups of bleaching clay, namely (1) neutral clay and (2) acid-activated clay. As the names imply, the neutral clay does not have any acidity and has a neutral pH. The acid-activated clay on the other hand is treated with mineral acid to modify the surface property and has an acidic pH.

   Bleaching earth, by far is the largest volume of adsorbent that is used in bleaching vegetable oil. As discussed earlier, the process of removal of the trace metals, phosphorus, and chlorophyll is a physical phenomenon known as adsorption. The bleaching clay also reacts chemically with the soap in the oil and hydrolyzes it to FFA and forms sodium salt of the mineral acid in the acid-activated clay. Table 6.4 shows the FFA data (presented by the Oil-Dri Corporation at the Latin American Oil Chemists Meeting in 2001) on the same oil bleached with the neutral and acid-activated clay.

   As one can see the neutral clay did not increase the FFA of the oil in the bleaching process. The acid-treated clay increased the FFA the least. The two acid-activated clays produced different amounts of FFA in the oil. This is attributable to the degree of acid activation of the bleaching clay.

   The neutral clay is good for reducing the red color from the oil but it is not as effective in removing chlorophyll, soap, and trace impurities from the oil.

There are various grades of both neutral and acid-activated bleaching earth available. Each one is specifically designed by the manufacturer to perform satisfactorily in some specific application.

Performance of bleaching clay is influenced by several factors. Some of these are inherently related to the properties of the bleaching clay while the others result from the bleaching process.

Factors inherently related to the bleaching clay are:

- Type of clay—neutral or acid-activated clay.
- Selectivity of the clay for the adsorption of the impurities in the oil.
- Pore size desirable for specific impurity.
- Good physical integrity, that is, does not break down in size during oil treatment, since this can reduce oil flow through the filter.

Factors related to the process that influence the performance of the bleaching clay are listed as follows:

- Quality of the refined and water-washed oil coming to the bleaching system.
- Bleaching clay dosage.
- Bleaching temperature.
- Degree of mixing in the bleacher.
- Contact time.
- Operating pressure of the bleacher.

Sources of bleaching clay

There are numerous suppliers of bleaching clay around the world. The clay supplied by these suppliers may perform adequately for most applications. It is important to understand that the manufacturing technology of bleaching clay has evolved over many years through the research work of some of these companies. According to the most prominent bleaching clay suppliers, such as Engelhard—based in Germany (BASF Catalyst LLC in USA), Oil-Dri Corporation (based in USA), and Sud-Chemie (based in Germany), the key to their success can be attributed to the following factors:

1. Identifying the right type of clay deposit worldwide.
2. Treatment of the raw clay material through their proprietary processes to obtain specific properties of the clay that determines its suitability in bleaching certain types of oil.

Each of these companies produces several grades of bleaching clay that will be mentioned in this book. As mentioned earlier in this chapter, there are neutral and activated (acid activated) clays that are used in oil bleaching. The neutral clay is good for the removal of trace metals, carotenes, some soap, and chlorophyll. For poor quality crude oils and oils containing high levels of chlorophyll, phospholipids, and soap, activated clay is recommended.

Some examples of the bleaching clays from each of the three suppliers are listed in Tables 6.5–6.7; with their recommended applications.

TABLE 6.5 Selected Bleaching Clays From Engelhard (BASF Catalyst in USA) for Treating Edible Oils

Clay	Activated/ neutral	Suggested applications
Grade F-160	Activated	Recommended for: • Oils hard to bleach due to high phospholipids, chlorophyll, soap, carotenes, and oil decomposition products • Technical grade vegetable and marine oils for extra light color • For extra light soybean, sunflower, rapeseed, and coconut oil • For physical refining
Grade F 115-FF	Activated	Recommended for: • Oils containing high phospholipids, chlorophyll, soap, carotenes and oil decomposition products • For operations with limited filter capacity because of the fast filtration rate of the clay
Nevergreen	Activated	Recommended for: • Oils with naturally high chlorophyll content, such as, soybean, canola, sunflower, and rapeseed • Oils containing high carotenoids, such as palm, palm stearin, and palm olein
Grade F-1	Activated	Recommended for: • General color reduction in all types of oils • Not as effective as some of the other activated clays mentioned earlier for removing the soluble impurities from the oil
Grade F-100	Neutral	Recommended for: • Color removal from oils by gentle treatment
Grade F-110	Activated	Recommended for: • Moderate-to-easy-to-bleach oils • Effective for removal of oil soluble impurities, such as phospholipids, oil oxidation products, color compounds
Grade F-118 FF	Activated	Recommended for: • Vegetable oils, marine oils, and animal fats for operations with limited filter capacity requiring fast filtration • Removes soluble impurities from the oil
Grade F-4	Activated	Recommended for: • Bleaching soybean, corn, palm, cottonseed, olive, peanut (groundnut), sesame seed oils, tallow and lard. • Nonedible products, such as bees wax

TABLE 6.6 Selected Bleaching Clays From Oil-Dri Corporation

Clay	Activated/ neutral	Suggested applications
Pure-Flo B80	Neutral	Recommended for: • Physical refining of palm oil • Color removal from cottonseed, tallow, coconut, and other oils • Chlorophyll removal from oils with low-to-moderate chlorophyll contents • Compatible with food grade acids for enhanced chlorophyll removal
Pure-Flo B81	Activated	Recommended for: • Physical refining of palm oil • Multioil processing with a single bleaching product that offers balance of activity and filtration rate • Color removal from cottonseed, corn, tallow, and other oils
Pure-Flo Supreme Pro-Active,	Activated	Recommended for: • Removal of chlorophyll from oils like canola, soybean, and sunflower • Removal of color compounds other than chlorophyll from oils like cottonseed, corn, canola, soybean • Removal of soap and phospholipids, trace impurities, such as iron, phosphorus, magnesium, nickel, and so on • Offers long filtration cycle between filter clean-ups
Perform 4000, 5000, 6000	Activated	• For high chlorophyll oils and difficult to bleach oils, including soybean, canola, and olive oils. • Removal of soap and phospholipids, trace impurities, such as iron, phosphorus, magnesium, nickel, and so on • Offers long filtration cycle between filter clean-ups

2. Silica hydrogel

Silica hydrogel has been used in treating vegetable oils to remove trace impurities for over 25 years. These compounds are especially effective in removing phospholipids, trace metals, and soap in the refined oils.

Like bleaching clay, many companies produce silica hydrogel products that have been used in vegetable oil refining. Owing to the high ability to

TABLE 6.7 Selected Bleaching Clays From Sud-Chemie

Clay	Activated/ neutral	Suggested applications
TONSIL SUPREME 126 FF	Neutral	Recommended for: • Difficult-to-bleach oils, especially canola, rapeseed, and soybean oil with high chlorophyll • Oils with normal levels of peroxide value and anisidine value • Oils with some oxidative problem • Mildly damaged oil
TONSIL SUPREME 1202 FF	Activated	Recommended for: • Difficult-to-bleach oils, especially canola, rapeseed, and soybean oil with high chlorophyll • Oils with higher than normal levels of peroxide value and anisidine value • Oils with heavy oxidative problem • Highly damaged oil
TONSIL SUPREME 1200 FF	Activated	Recommended for: • Effective bleaching clay for oils like avocado, canola, palm, linseed, and castor oil
TONSIL SUPREME 1211	Activated	Recommended for: • Similar applications as 1200 FF but is not as fast in filtration rate
Actisil 220 FF	Activated	Recommended for: • Bleaching oils with lower chlorophyll content like cottonseed, coconut, safflower, corn, palm, and soybean • Adequate for oils like rice bran, peanut (ground nut), and linseed oil • Can be used for canola oil containing low–moderate chlorophyll content
Actisil 221	Activated	Recommended for: • Similar application as Actisil 220 FF but is not as fast in filtration rate
Optimum 320 FF	Very lightly activated	Recommended for: • Bleaching sunflower, palm oil, and peanut (groundnut) oil • Adequate for bleaching oils like cottonseed, safflower, coconut, corn, palm, and soy

remove trace impurities from the oil, silica hydrogel is used in the following applications:

1. Pretreatment of the oil prior to bleaching
2. Elimination of water washing in the chemical refining process (modified chemical refining)
3. Physical refining

TABLE 6.8 Silica Hydrogel Products From Grace Davison

Product name	Recommended applications
TriSyl	• Pretreatment of chemically refined vegetable oils before bleaching • Physical refining of oils • Modified chemical refining to eliminate water-washing step • Modified physical refining to improve removal of trace impurities
TriSyl 300	All of the previously mentioned treatments. TriSyl 300 is more effective than TriSyl
TriSyl 600	All of the previously mentioned treatments. TriSyl 600 is more effective than TriSyl 300 and is most effective of all three

4. Modified physical refining

Item #1 has been described in this chapter under the two-step bleaching process. The rest have been described in Chapter 5 on vegetable oil refining.

As in the case of bleaching clays, there are several companies that offer silica hydrogel products for edible oil treatment. Some of the leaders in this area are:

1. Grace Davison (formerly W.R. Grace & Co.) of USA

2. INEOS Silica (formerly Crosfield) of United Kingdom

In this section a few selected silica hydrogel products from these leading producers are listed. Table 6.8 shows the list of the silica products from Grace Davison Co.

The relative effectiveness of the TriSyl products in reducing soap and phosphorus (P) are shown in Table 6.9 (data provided by Grace Davison).

TABLE 6.9 Relative Effectiveness of TriSyl Products in Reducing Soap and Phosphorus in Bleaching Soybean Oil (Final Soap = 0 ppm and Final P = <1 ppm)

Analysis of refined oil	Silica added (% of oil)	Residual TriSyl (%)	Residual TriSyl 300 (%)	Residual TriSyl 600 (%)
Soap = 40 ppm P = 1.3 ppm	0.09	0.00	36	65
Soap = 140 ppm P = 5.2 ppm	0.24	0.00	50	61
Soap = 290 ppm P = 9.7 ppm	0.30	0.00	51	59

TABLE 6.10 Silica Hydrogel Products From INEOS Silica

Sorbsil products	Recommended applications
SORBSIL R40	• Pretreatment of chemically refined oils before bleaching • Physical refining of vegetable oils • Elimination of water washing in chemical refining
SORBSIL R92	• Pretreatment of chemically refined oils before bleaching • Physical refining of vegetable oils • Elimination of water washing in chemical refining

The results shown in Table 6.9 indicates that with the various feed stock quality and the silica dosage, TriSyl was used up completely while there were significant amounts of TriSyl 300 and TriSyl 600 remaining in the oil even after the removal of soap and phosphorus to the same levels. This demonstrates the relative efficiency of the three TriSyl products in removing soap and phosphorus from the oil.

INEOS Silica, formerly Crosfield offers silica hydrogel adsorbents under the name, Sorbsil. These products are recommended for similar applications as TriSyl products from Grace Davison or Select products from Oil-Dri Corporation. Table 6.10 lists the Sorbsil products from INEOS Silica.

3. Silicates

Table 6.11 shows some of the silicate products from Oil-Dri Corporation that are recommended for vegetable oil processing. These products are different from the silica hydrogel, but are recommended for specific applications in bleaching vegetable oils for the removal of trace impurities.

Relative efficiencies of Select products from Oil-Dri Corporation are shown in Table 6.12.

The previously mentioned results indicate the capability of the Select products for reduction of soap and phosphorus form refined oil.

Table 6.13 shows the results provided by INEOS Silica on removal of soap and trace impurities in physical refining of palm oil.

TABLE 6.11 Silicate Products From Oil-Dri Corporation

Product name	Recommended applications
Select 350	• Pretreatment of chemically refined vegetable oils before bleaching • Physical refining of oils • Elimination of water wash in chemical refining process
Select 450	• All of the previously mentioned treatments. Select 450 is faster filtering than Select 350

TABLE 6.12 Removal of Phosphorus and Iron From Palm Oil in Physical Refining Process

Treatment	Phosphorus (ppm)	Iron (ppm)
Crude palm oil	13.5	5.6
Crude palm oil + 1.3% bleaching clay	<0.4	0.04
Crude palm oil + 0.9% bleaching clay + 0.15% SORBSIL R40	<0.4	0.03

One can see that in the addition of SORBSIL R40 allowed a 31% reduction in the bleaching clay usage to obtain the same trace impurities in the treated oil.

Table 6.13 shows the effect of SORBSIL R92 in treating chemically refined sunflower oil. Addition of 0.05% SORBSIL R92 in this case significantly reduced the trace impurities in the refined oil.

4. Activated carbon

Activated carbon is an effective adsorbent for aromatic hydrocarbons. Crude vegetable oils do not normally contain any appreciable amount of aromatic hydrocarbon unless the crude oil is seriously oxidized.

Coconut oil can contain noticeable amount of aromatic hydrocarbons when the copra is dried in dryers on wood flame. The color of light oil, such as coconut oil can be darkened by these impurities. Activated carbon can remove these contaminants from the oil.

The author has mixed experience with the use of activated carbon. With good quality crude oil, it is not necessary to use any activated carbon. However, in certain parts of the world it becomes necessary to use it for bleaching the oil.

The activated carbon used in oil treatment must not be derived from lignite source.

TABLE 6.13 Removal of Phosphorus and Iron From Palm Oil in Chemical Refining Process

Treatment	Soap (ppm)	Phosphorus (ppm)	Calcium (ppm)	Magnesium (ppm)
Caustic neutralized	278	4.7	0.24	0.9
Caustic neutralized + 0.75% bleaching clay	0	0.8	<0.02	0.06
Caustic neutralized + 0.9% bleaching clay + 0.05% SORBSIL R92	0	<0.4	<0.02	<0.02

6.15 BLEACHING VERY GREEN CANOLA OIL

Almost every year the oil processors are confronted with canola oil with very green color. The green color comes from the chlorophylls present in the oil and they have to be removed in the bleaching step. The bleaching agents from almost every supplier is designed to reduce the chlorophyll level in green canola oil but when the green color is too deep, the standard bleaching uses very high level of bleaching clay. This has some negative effects on the process, such as:

1. Cost of clay used for the process increases.
2. The oil loss is greater due to the additional oil absorption in the extra clay used.
3. There is more loss of the natural antioxidants in the oil due to higher bleaching clay usage.

Chlorophyll content of canola oil is normally 15–30 ppm. Over the years crude canola oil containing a chlorophyll content of 40–50 ppm has been reported by some processors. The fact is these very green oils are extremely difficult to bleach in order to reduce the green color to an acceptable level via the usual bleaching process. Canola oil containing as much as 50 ppm chlorophyll can be bleached by following a process patented in 1994 by Christopher R. Behari of USA and his coworkers, Levente L. Diosady, and Leon J. Rubin of Canada. In this process, water-degummed oil is used as the feed. The following steps are followed:

1. Water-degummed oil is mixed with 0.25%–0.3% phosphoric acid (dry basis) in a static mixer (or a high shear mixer, if available).
2. The mixture is heated to 194–203°F (90–95°C) and sprayed into a vacuum dryer, maintained at an operating pressure of 10 mm of mercury or less.
3. The oil is gently agitated with nitrogen sparging at the bottom of the dryer. A top entering turbine mixer could be used at low impeller tip speed (60 rpm). The chlorophyll separates in a crystalline form.
4. The mixture is allowed to remain in the vacuum dryer for about 15–20 min to allow the chlorophyll to precipitate. Under this condition the moisture in the oil drops to $<0.1\%$ (preferably $<0.07\%$).
5. The oil leaves the vacuum dryer at a temperature of 185–194°F (85–90°C) and goes to neutralization step.
6. The oil, after the caustic addition and mixing in a high shear mixer is allowed a retention time of 10–15 min.
7. The oil is then centrifuged to separate the soap and the precipitated chlorophyll.
8. The oil from the primary separator is now sent to bleaching.

6.15.1 Critical Control Points

The critical control points for the process are as follows:

1. Acid dosage
2. Temperature of reaction with the acid (also temperature of reaction with caustic in neutralization

3. Vacuum in the vacuum dryer (moisture of the oil in the vacuum dryer)
4. Agitation in the vacuum dryer
5. Time of contact between the acid and the oil
6. Caustic treat and excess caustic in neutralization

1. Acid dosage
   Acid dosage is higher than what is normally used in acid conditioning. It requires an acid dosage of roughly 0.3% (dry basis).
2. Temperature of reaction with the acid
   It appears that a reaction desirable temperature of 80–100°C is the range for obtaining good results and the range of 90–95°C is the desirable range. Below 80°C the precipitated chlorophyll begins to redissolve in the oil and higher green color oil is obtained after caustic refining. At temperature above 100°C the deodorized oil color begins to get darker.
3. Vacuum/moisture
   The vacuum must be <10 mm of mercury to reduce the moisture in the oil to <0.1% and preferably to <0.07% in order to achieve improved crystallization of the chlorophyll so as to achieve better chlorophyll reduction in the oil.
4. Agitation in the vacuum dryer
   The agitation must be as gentle as possible in order to retain the structure of the crystallized chlorophyll.
5. Time of contact between acid and oil
   It is important to have a minimum contact time of 15 min at the reaction temperature of 194–203°F (90–95°C). At lower temperature a longer contact time would be necessary.
6. Caustic treat and excess caustic in neutralization
   Due to the high phosphoric acid dosage in the reaction step a higher caustic treat as well as excess caustic is needed.

6.15.2 Bleaching of the Treated Oil

The patent shows that water-degummed oil with 6.7 ppm chlorophyll content needed 2.4% bleaching clay to reduce the chlorophyll content of 0.01 ppm (10 ppb) of chlorophyll, while the same oil, treated with 0.23% of phosphoric acid (dry basis) in the manner according to the patent, produced the same level of chlorophyll (0.01 ppm), using only 0.8% bleaching clay.

READING REFERENCES

Anderson, D.A., 1996. Primer on Oil Processing Technology. Bailey's Industrial Oils and Fat Products, fifth ed., vol. 4. Wiley Interscience, USA.

Brooks, D., Azzarello, S., 2001. Bleaching Options for Refining Palm Oil. Latin American Congress of American Oil Chemists Society, Cancun, Mexico.

Brooks, D., Unpublished. The Bleaching Process: The Common Problems and Solutions.
Bogdanor, J.M., Price, A.L., 1994. Novel synthetic silica adsorbents for refining of edible oils. Oil Mill Gazetter 99 (12), 32–34.
Chapman, D.M., 1994. Benefits and limitations of a novel chlorophyll adsorbent. JAOCS 71, 4.
Erickson, D.R. (Ed.), 1995. Practical Handbook of Soybean Processing and Utilization. AOCS Press, Champaign, IL, USA.
Guler, C., Tunc, F., 1989. Chlorophyll adsorption on acid-activated clay. JAOCS 66, 3.
Parker, P.M., Welsh, W.A., 1988. Method for refining glycoside oils using amorphous silica specifically phospholipids and associated metal ions, from glycoside oils. US Patent 4,734,226, March 29, 1988.
Patterson, H.B.W., 1992. Bleaching and Purifying Fats and Oils, Theory and Practices. AOCS Press, Champaign, IL, USA.
Pryor, J.N., Bogdanor, J.M., Welsh, W.A, In press a. Dual phase adsorption and treatment process for the removal of impurities from triglyceride oil. US Patent # A 940,809.
Pryor, J.N., Bogdanor, J.M., Welsh, W.A., In press b. Process for removal of chlorophyll, color bodies and phospholipids from glyceride oils using acid-treated silica adsorbent. US Patent # 4,781,864.
Taylor, D.R., Jenkins, D.B., Ungerman, C.B., 1989. Bleaching with alternative layered minerals: a comparison with acid-activated montmorrilonite for bleaching soybean oil. JAOCS 66, 3.
Welsh, W.A., Bogdanor, J.M., In press. Method for treating caustic refined glyceride oils for removal of soaps and phospholipids. EP 247411.
Welsh, W.A., Parent, Y.O., 1986. Method for refining glycoside oils using amorphous silica useful in processes for the removal of trace contaminants, specifically phospholipids and associated metal ions, from glycoside oils. US Patent 4, 629,588, December 16, 1986.

Chapter 7

Hydrogenation

7.1 INTRODUCTION

Hydrogenation of vegetable oil has been practiced for over a century. The process was originally introduced to convert some of the unsaturated fatty acids in vegetable oils, as well as marine or animal fats to make them more stable to oxidation.

In this process, the unsaturated double bonds (see Chapter 2) in the fatty acids of the oil molecules react with hydrogen atoms in the presence of a catalyst. Nickel catalyst is used in commercial hydrogenation of edible oils. Other catalysts, such as platinum, palladium, copper, etc., have also been applied in hydrogenation applications. These are not used in commercial hydrogenation of edible oils.

Hydrogenation has been used for a long time to improve oxidative stability of vegetable oils for improved shelf life and to modify the solids content and melting characteristics of the oil to formulate shortening and margarine products with the desired physical properties. Other methods, such as chill fractionation and interesterification processes, were developed later for making shortening and margarine. These processes also allowed the oil technologists to modify the melting characteristics in the final shortening or margarine product made from certain types of vegetable oils, such as palm oil, palm kernel oil, etc., to meet the required physical properties.

7.2 HISTORICAL BACKGROUND OF HYDROGENATION

The first patent on hydrogenation of vegetable oil was issued to W. Normann in 1903 in the United Kingdom. Although the credit goes to Normann for his pioneering work on the hydrogenation of fatty material, the principle of hydrogen reaction between unsaturated compounds and hydrogen gas was demonstrated in the gaseous phase by Paul. Sabatier prior to 1903. Paul Sabatier was awarded the Nobel Prize along with Victor Ginard in 1912 in Chemistry for their original work on hydrogenation of organic compounds using metal catalysts. The patent right belonged to Crossfield and Sons.

Practical Guide to Vegetable Oil Processing. http://dx.doi.org/10.1016/B978-1-63067-050-4.00007-6
Copyright © 2017 AOCS Press. Published by Elsevier Inc. All rights reserved.

Procter and Gamble Co. acquired the patent right in 1909. The first commercial application entered the market as Crisco shortening, which was made from hydrogenated cottonseed oil. The product was extremely successful.

J.J. Burchenal received a US Patent in 1915. Procter and Gamble was manufacturing the shortening under the broad claims of this patent. This created some strong opposition from the competitors and the patent was nullified. This opened the door for all shortening manufacturers using hydrogenation.

The process of hydrogenation itself went through evolutionary studies during the next 60 years. Scientists from the United States Department of Agriculture (USDA) contributed a great deal in understanding and developing the field of hydrogenation. Corporations, such as Procter Gamble, Lever (Unilever), Lurgi, and many others developed their own techniques that enriched the knowledge of hydrogenation. Simultaneously, the catalyst manufacturers in Europe and the United States devoted their studies toward the advancement of hydrogenation catalysts. During the last three decades of the 20th century, the catalyst manufacturers and the oil processors in Europe and the United States developed an in-depth understanding of the behavior of the hydrogenation catalysts and their interactions with the impurities in the oil, hydrogen gas, and process conditions. This has benefited oil processors throughout the world.

7.3 UNDERSTANDING THE PROCESS OF HYDROGENATION

Hydrogenation is a heterogeneous reaction process and complex in nature where the reaction occurs between the hydrogen (gaseous phase) and the unsaturated fatty acids (liquid phase), converting some or all of the unsaturated fatty acids into saturated fatty acid (stearic acid). This involves the following physical and chemical reactions:

- Mixing and dispersion of the catalyst by a mechanical mixer.
- Diffusion of hydrogen gas through the mass of oil to the catalyst surface.
- Adsorption of the reactants (hydrogen and unsaturated fatty acids) on the catalyst surface.
- Partial or complete saturation of the unsaturated fatty acids on the catalyst surface.
- Desorption of the products of reaction and the unreacted fatty acids and oil molecules from the catalyst surface.
- Release of heat due to the exothermic nature of the hydrogenation reaction.

Hydrogen gas and the unsaturated fatty acids diffuse through the bulk of the oil and reach the catalyst surface. Migration of these reactants through the bulk of the oil is facilitated by mechanical agitation.

The two reactants, hydrogen and the unsaturated fatty acids, diffuse through a stagnant microfilm of oil on the surface of the catalyst to reach the active sites on the catalyst.

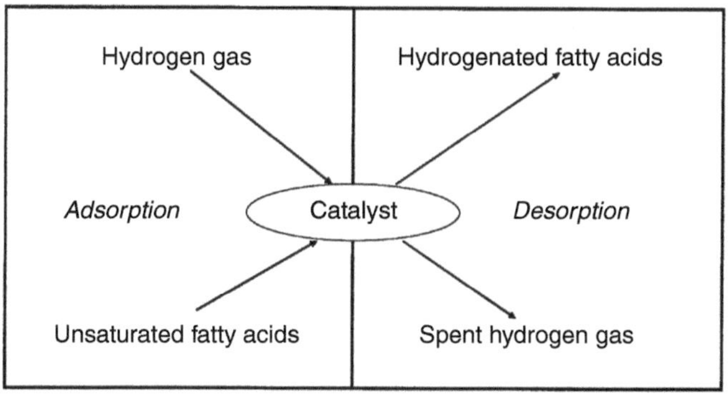

FIGURE 7.1 Pictorial representation of reaction mechanism in hydrogenation.

The reaction between the unsaturated fatty acids and hydrogen gas occurs on the catalyst surface. This is sometimes referred to as the chemisorption process, implying a chemical reaction aided by the adsorption process.

The products of reaction, which are typically saturated fatty acids, unsaturated fatty acids (*cis* and *trans* isomers), and triglyceride molecules, leave the catalyst surface via the desorption process. More fresh reactants are adsorbed on the surface of the catalyst and the hydrogenation reaction continues. Fig. 7.1 is a pictorial concept of the hydrogenation reaction.

Heat is generated in the reaction. The average heat of reaction in hydrogenation is 1.65 Btu/lb of oil for a drop in the iodine value (IV) by one unit. This is a valuable source of process heat and is recovered in many plants via a heat-recovery system.

7.3.1 Effects of Hydrogenation

The following events occur in the hydrogenation process:

- The unsaturated fatty acids become more saturated.
- The IV of the oil decreases.
- The melting point of the oil increases.
- The oxidative stability of the oil improves.
- The solid content of the oil increases.
- A multitude of side reactions take place, including the formation of certain alcohols and acids.
- Isomerization of some of the polyunsaturated fatty acids takes place.

Fig. 7.2 is a typical example of a hydrogenation reaction and the changes in the unsaturated fatty acid content of soybean oil. One can see in Fig. 7.2 that the linoleic and the linolenic acids decrease with the reduction of the IV and become essentially zero. However, the oleic acid reaches a maximum and then begins to decline as it gets converted to stearic acid. Palmitic acid content does not change with hydrogenation because it is already a saturated fatty acid and

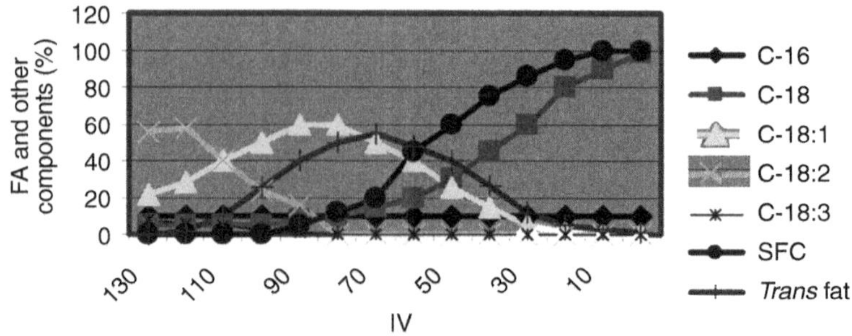

FIGURE 7.2 **Typical hydrogenation curve for soybean oil.** *FA*, Fatty acids; *IV*, iodine value; *SFC*, solid fat content.

not affected by the hydrogenation process. Similar to the oleic acid, the *trans* fatty acid concentration starts at zero, attains a maximum, and declines as the IV in the oil decreases further.

The unsaturated fatty acid isomers can be:

- geometric isomers or
- positional isomers.

It was stated in Chapter 2 that the hydrogen atoms on the fatty acids in the natural vegetable oil molecules are located on the same side of the carbon atoms in the fatty acid chain. This is known as the *cis*-form. Geometric isomers are formed during hydrogenation when one of the hydrogen atoms attached to the double bond moves to the other side of the fatty acid chain. This is the *trans*-form.

Positional isomers are formed when a double bond on the unsaturated fatty acid chain physically shifts. The resultant fatty acids have very different physical and chemical properties.

Isomerization and other side reactions result from several factors that will be discussed in detail later in this chapter. The model for isomerization during hydrogenation is shown in Fig. 7.3.

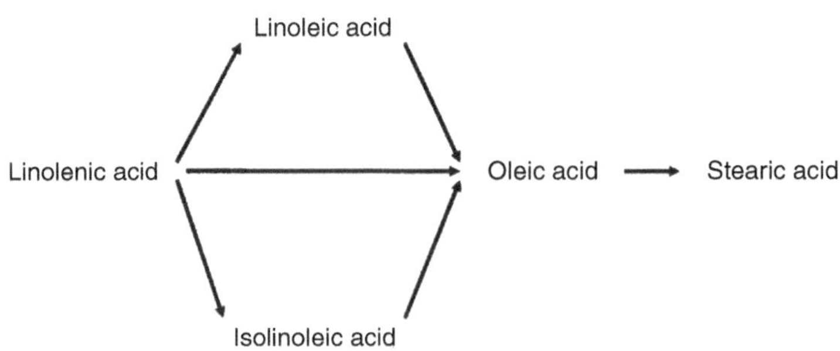

FIGURE 7.3 **Isomerization.**

Depending on the condition of the hydrogenation process, there may be isomers of both linoleic and isolinoleic acids present in the hydrogenated oil.

7.4 HYDROGENATION PROCESS

The commercial processors of the world use hydrated nickel catalyst to manufacture all edible hydrogenated products. Two types of reactors are used for the process. They are:

1. batch reactors
   a. deadend type
   b. recirculating type
2. continuous reactors

7.4.1 Batch Hydrogenation Reactor

Batch reactors are used worldwide to hydrogenate oils for various applications. Fig. 7.4 shows the schematic diagram of a batch hydrogenation reactor. These reactors are referred to as "converters" by many oil processors.

The basic design features and appearance of the reactors are very similar regardless of their capacity, which can range from a few tons up to 30 tons or larger.

The batch hydrogenation reactor (converter) is equipped with the following accessories:

- mechanical agitator,
- heating and cooling coils,
- hydrogen gas sparger,
- temperature controller,
- pressure controller,
- vent valve for the spent gas,
- safety rupture disc,

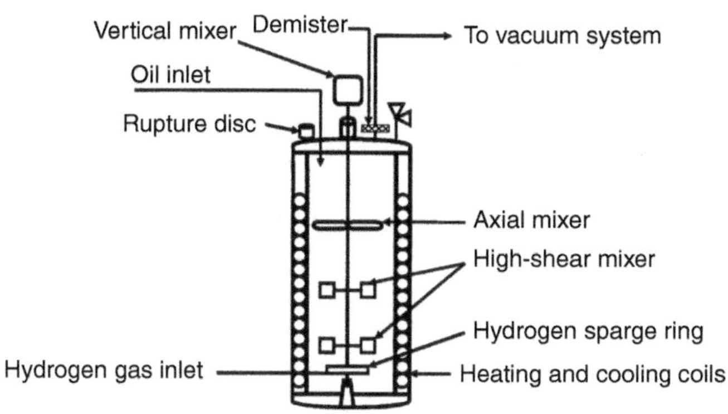

FIGURE 7.4 **Batch hydrogenation reactor.**

- agitator seal protector (lubricant),
- rupture disc to protect the reactor against overpressurizing,
- hydrogen gas flow meter with flow indicator,
- gas flow totalizer, and
- vacuum ejector or vacuum pump.

7.4.2 Operation of a Batch Hydrogenation Reactor

Fig. 7.5 shows the schematic diagram for a batch hydrogenation system with heat recovery.

1. The oil enters the reactor under vacuum.
2. The oil is heated to 185–195°F (85–95°C) in the reactor to deaerate it This removes the dissolved oxygen and moisture from the oil.
3. Typically, it requires 15–20 min to complete the deaeration process.
4. Reactors can have oil preheaters that heat the reactor feed to 185–195°F (85–95°C) as the oil enters the reactor under vacuum. This deaerates the feed oil as the reactor is being filled.
5. Fresh catalyst is added, using about 0.01%–0.02% of the oil (nickel basis) into the reactor when the oil temperature reaches 315–325°F (157–162°C) with most nickel catalysts. The nickel concentration may be as much as 0.04% of the oil, when reused catalyst is used in the slurry form or under some special conditions to be described later.
6. Presently, most oil processors use only fresh catalyst for every batch of oil hydrogenated.
7. With most nickel catalysts, the hydrogenation reaction starts when the oil temperature reaches approximately 315–325°F (157–162°C).

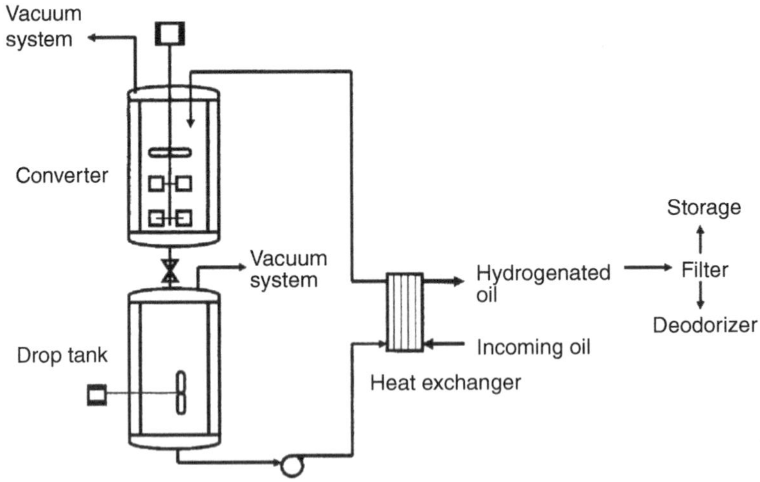

FIGURE 7.5 Schematic diagram for hydrogenation system.

8. The hydrogenation reaction then proceeds. Some catalysts have lower initiation temperatures.
9. The heat generated from the reaction raises the oil temperature in the reactor.
10. Steam heating is no longer required.
11. The reaction in a batch reactor is carried out either adiabatically or isothermally.

7.4.3 Adiabatic Reaction Process

In the adiabatic process:

1. The temperature of the oil in the reactor is allowed to increase, as the heat of reaction increases the oil temperature with the progression of the reaction.
2. Heat loss occurs only through the normal radiation process.
3. Flow of hydrogen gas is stopped when the predetermined volume of the hydrogen gas has been introduced into the reactor.
4. The refractive index of the hydrogenated oil is checked to verify if the desired end point for the reaction has been reached.

The reaction end point is also confirmed through laboratory analysis of the IV, solid content (SFC or SFI), and melting point of the hydrogenated oil.

7.4.4 Isothermal Process

In the isothermal reaction process:

1. The temperature of the oil in the reactor is allowed to reach the predetermined set point and then it is maintained within ±2°F (1°C) or less by passing cooling water through the cooling coils in the reactor.
2. The hydrogenation proceeds rapidly at the beginning. The IV drops at the rate of 1–1.5 IV units/min until it reaches around 70 (for soybean oil). The IV can vary with the type of oil being hydrogenated.
3. The reaction rate slows down as the IV of the oil begins to drop further.
4. The reaction is monitored with the help of a refractometer. The refractive index of the oil drops linearly as the oil gets hydrogenated (Fig. 7.6).
5. The reaction end point is determined by monitoring the refractive index in the oil in the reactor. The refractive index of the oil decreases with decreasing IV during hydrogenation. The refractive index drops in a linear fashion with hydrogenation. Therefore, the progress of the reaction can be monitored with the help of a Butyro refractometer. The prism chamber of the refractometer is maintained at 60°C with the help of hot water circulated from a temperature-controlled hot-water bath.
6. The reaction is terminated when the target refractive index is reached.
7. Hydrogen gas flow is stopped either automatically or manually depending on the system design as the desired refractive index is reached.

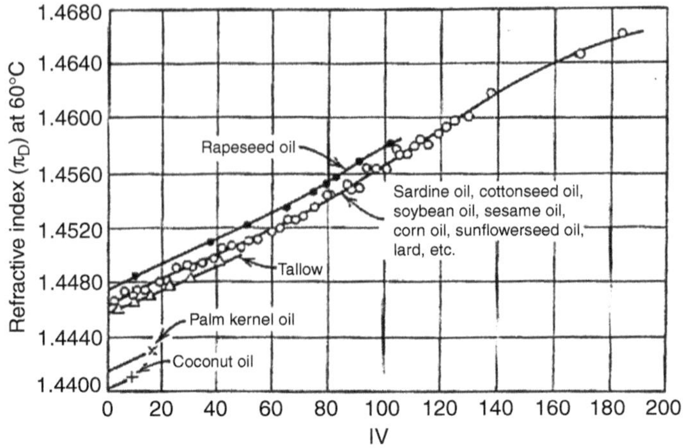

FIGURE 7.6 Refractive index at 60°C versus IV in various oils.

8. The reaction continues as long as there is dissolved hydrogen gas in the oil. The reaction stops soon after the dissolved gas in the oil has completely reacted.
9. The reaction stops as the oil is cooled after the reaction.
10. The oil quality is checked in the laboratory for confirmation of the reaction end point.
11. The gas accumulated at the top of the reactor is vented into the atmosphere to maintain proper gas purity in the head space. This minimizes side reactions that can adversely affect the quality of the hydrogenated oil. The venting process can be either automatic or manual.
12. The oil is cooled down to approximately 250–288°F (121–138°C) either in the reactor or in a drop tank built below the reactor.
13. In some plants there is a heat-recovery system that extracts most of the heat from the hydrogenated oil to preheat the oil entering the reactor.
14. The cooled oil is recirculated through the filter until the desired clarity is reached and then it is filtered.
15. The oil is also analyzed for IV, SFC or SFI, melting point, fatty acid composition (if needed), and soluble nickel and particulate matter.
16. The oil is either sent directly to the deodorizer or to storage tanks for blending and product formulation.

7.4.5 Deadend-Type Hydrogenation Reactor

The majority of the batch reactors around the world are the deadend type. The basic design of both deadend and recirculating reactors go back to early 20th century. There are a few recirculating-type reactors seen at the older installations around the world.

In the deadend type of reactor:

1. The oil is agitated with the help of a top-entering agitator.
2. There are typically three sets of impellers.
3. The one at the top is of axial flow design and it is located 15–24 in. below the surface of the liquid, depending on the diameter of the reactor. This height can vary with the size of the reactor. The axial flow impeller pushes down the oil from the top and recirculates the oil axially for achieving intimate mixing between the oil, hydrogen gas, and the catalyst. The impeller should not be too close to the liquid surface to prevent vortex formation, which can pull the less pure hydrogen from the head space of the reactor.
4. The two lower impellers are of high-shear design, which creates a very intimate mixing of the oil, catalyst, and the hydrogen gas.
5. The vent at the top of the reactor can be set to open automatically to release some of the spent gas from the top.
6. Hydrogen gas is distributed through a distribution ring. The diameter of the distribution ring is no larger than 80% of the impeller diameter at the bottom of the reactor to prevent channeling of the hydrogen gas up the reactor.
7. The distribution ring should preferably be made of sintered metal. However, if a tube is used, it is recommended that the holes on the ring be as small as possible and the total cross-sectional area of these holes does not exceed 65% of the cross-sectional area of the distribution ring.
8. The reactor is typically operated at pressure of 5–60 psig (34–408 kPa), depending on the type of product to be made.

7.4.6 Recirculating-Type Hydrogenation Reactor

In the recirculating type of reactor:

1. Agitation in the reactor is accomplished by bubbling hydrogen gas through the oil in the reactor.
2. The gas leaving the reactor is recompressed, water is drained, and the gas is recirculated through the reactor.
3. Part of the gas is vented to maintain its purity.

7.4.7 Comparison Between the Deadend and the Recirculating Types of Reactors

There are several points of differences between the two types of batch reactors. Table 7.1 shows the comparison between the two.

7.4.8 Continuous Hydrogenation Reactor

The schematic diagram for the continuous hydrogenation system is shown in Fig. 7.7 and is described:

TABLE 7.1 Comparisons Between the Deadend- and Recirculating-Type Reactors

Areas of comparison	Deadend-type reactor	Recirculating-type reactor
Construction:		
Agitation	Mechanical agitation is used	Hydrogen gas circulation is used to for agitation; no mechanical agitation is used
Vacuum system	Equipped with vacuum ejector or vacuum pump for deaeration of the oil as the reactor is filled	Vacuum system is not used; the hydrogen gas strips the dissolved air and moisture from the oil
Gas vent	Has manual or automatic gas vent at the top of the reactor to release the spent gas to maintain high-gas quality in the head space	Uses a different type of gas vent system: • A gas purity meter measures the gas quality and vents part of the recirculating gas automatically • In some installations, a manual valve is opened to release part of the spent gas • The accumulated moisture from the gas is removed by cooling the gas leaving the reactor and draining it • The gas is compressed before it is returned to the reactor • An intercondenser cools the compressed hydrogen to drain the moisture from the gas
Operating pressure	Depending on the type of product being made, the reaction is maintained at 5–60 psig (34–408 kPa)	The operating pressure can vary from 8 in. (3.15 cm) of water column to 60 psig (408 kPa)
Heating the incoming oil	The oil is heated either with the internal coils or externally via a heat exchanger	The oil is heated the same way as in the deadend reactor
Reaction time	More rapid because of mechanical agitation	Generally slower because the agitation is achieved through bubbling of the gas
Oil quality	The oil exhibits the following quality traits: • Lower oxidation • Lower hydrolysis	The oil exhibits the following quality traits: • Higher oxidation • Higher hydrolysis This is because of less complete deaeration and dehydration of the oil in the reactor

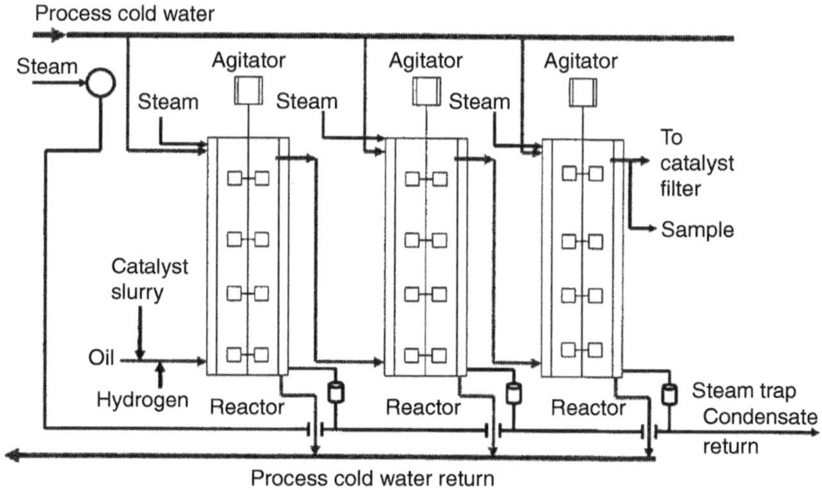

FIGURE 7.7 **Schematic diagram for continuous hydrogenation process.**

1. It is made of a series of vertical reactors where the refined and bleached (RB) oil enters at the bottom of the first reactor.
2. The reactors are jacketed so it is possible to either heat or cool the oil in the reactors.
3. The oil from a feed tank (not shown) is heated to the initiation temperature by steam before it enters the first reactor.
4. Catalyst slurry and hydrogen gas are added to the oil entering the system.
5. The reaction temperature is carefully controlled with the help of a temperature controller on each reactor (not shown).
6. The oil passes through the reactors in "plug flow," which means there is no back mixing. One can notice that there are no axial flow impellers in the reactors.
7. A sample is collected at the sample port for checking the refractive index.
8. Step #7 is repeated until the reaction reaches the steady state.
9. During this period the oil from the filter is returned to the reactor feed tank.
10. Once the steady state is reached, the oil is analyzed for IV, filter grade, soluble nickel, SFC, SFI, etc.

The oil is cooled in an external cooler (not shown), filtered (not shown), and stored in a tank for further processing.

7.4.9 Applicability of a Continuous Hydrogenation Reactor

A continuous reactor is suitable for the following applications:

1. Producing lightly hydrogenated oils with low-melt points.
2. Producing the same stock every day or for a long production time before any stock change.

7.5 CRITICAL CONTROL POINTS IN THE HYDROGENATION PROCESS

It has been mentioned earlier that the process of hydrogenation is complex. The readers will appreciate the complexity of the process from the discussion in this section. The critical control points for the hydrogenation process are listed:

1. catalyst activity,
2. catalyst selectivity,
3. catalyst concentration (nickel loading),
4. RB oil quality,
5. hydrogen gas quality,
6. hydrogen gas dispersion,
7. hydrogen gas venting from the reactor,
8. hydrogen gas supply,
9. reaction pressure,
10. reaction temperature, and
11. agitation.

7.5.1 Catalyst Activity

The hydrogenation reaction takes place on the active sites present on the catalyst surface. These active sites are numerous in the fresh catalyst and get reduced due to several factors. Some of them are listed:

1. poor activity in the fresh catalyst to start with,
2. repeated use of the catalyst,
3. oxidation due to exposure to air,
4. contact with moisture,
5. impurities in the RB oil, and
6. impurities in the hydrogen gas.

7.5.2 Manifestations of a Poor-Activity Catalyst

Normal manifestations of a poor-activity catalyst are:

1. slow reaction,
2. requires higher amount of catalyst to start and complete the reaction,
3. produces higher level of colloidal nickel in the hydrogenated oil,
4. alters the solids curve at the same IV of the oil altering the melt point for the hydrogenated oil,
5. increases the *trans* fatty acid content,
6. higher free fatty acid (FFA) in the hydrogenated oil, and
7. reduced filtration rate resulting in low productivity.

Note:

A one-time use of fresh catalyst is highly recommended because the reaction is fast, predictable, and it produces less quality variation in the hydrogenated oil.

7.5.3 Catalyst Selectivity

Selectivity of the catalyst refers to the propagation of hydrogenation of the unsaturated fatty acids as shown:

Linolenic acid → Linoleic acid → Oleic acid → Stearic acid

Lack of selectivity of a catalyst produces stearic acid even before the polyunsaturated fatty acids are depleted.

Selectivity ratio for hydrogenation is defined as follows:

Selectivity ratio (SR) = Conversion of linoleic to oleic acid/Conversion of oleic to stearic acid

SR versus fatty acid conversion is shown in Table 7.2

It should be clear from these discussions that in a low-selective hydrogenation a higher amount of stearic acid is formed, while the converse is true at higher selectivity.

7.5.3.1 Reaction Conditions That Promote High Selectivity

The following reaction conditions are known to produce high selectivity in the hydrogenation reaction:

- high-reaction temperature,
- low-reaction pressure,
- poor quality of the RB oil (containing high phosphorus, moisture, and positive in soap),
- generally low agitation,

TABLE 7.2 Selectivity Ratio Versus Fatty Acid Conversion

SR	Corresponding hydrogenation results
0	All unsaturated fatty acids are converted to *stearic acid without* going through the stages: Linolenic→Linoleic→Oleic→Stearic
1	Equal conversion of linoleic to oleic and oleic to stearic acid
2	Linoleic acid reacts 2 times faster than the oleic acid
50	Linoleic acid reacts 50 times faster than the oleic acid

SR, Selectivity ratio.

TABLE 7.3 Selective Versus Nonselective Hydrogenation

	Selective condition	Nonselective condition
Reaction temperature (°F)	392	284
Reaction pressure (bar)	0.7	3.0
Catalyst type	Nysosel 222	Nysosel 222
Nickel loading	0.01% (100 ppm)	0.01% (100 ppm)
IV	70	70
Reaction time (min)	80	**240**
Trans fatty acids (%)	**47.6**	27.7
Stearic acid (%)	11.4	**23.2**
Oleic acid (%)	**72.7**	53.8
Linoleic acid (%)	3.9	**10.6**
Linolenic acid (%)	0	0.5
Mettler drop point (°F)	99.5	**120.6**

The bold values indicate significant difference from the other conditions in comparison
Source: Courtesy of BASF Catalysts LLC, GA, USA.

- higher-catalyst concentration,
- poisoned catalyst,
- poor-hydrogen gas quality, and
- low-hydrogen gas supply.

Table 7.3 illustrates the example where soybean oil is hydrogenated by using Nysosel 222 catalyst from BASF Catalysts LLC, at equal nickel loading. Selective (high temperature and low pressure) and nonselective (low temperature and high pressure) hydrogenation conditions were used. The selective condition produced the following results as compared to the nonselective condition:

- the reaction time was significantly shorter due to a higher reaction temperature,
- *trans* fatty acid content was much higher,
- linoleic acid was much lower,
- linolenic acid was absent,
- oleic acid content was much higher,
- stearic acid content was much lower, and
- Mettler drop point was much lower.

7.5.3.2 Significance of Selectivity

It is important to note that selective hydrogenation produces oils with very distinct characteristics for the same endpoint of IV, compared to that attained via the nonselective hydrogenation method, such as:

- higher-oleic acid,
- higher–*trans* fatty acids,
- lower-stearic acid,
- lower-linoleic acid
- lower-melt point, and
- sharper-melt curve.

To produce margarine or shortening with high spreadability and a sharp-melting curve, oil processors have used catalyst with higher selectivity and/or the selective hydrogenation condition. Higher selectivity can be created with the same catalyst by manipulating operating conditions, such as pressure, catalyst concentration, agitation, reaction temperature, etc.

7.5.3.3 Catalyst Concentration

The reaction proceeds rapidly when the catalyst concentration (nickel loading) is increased. Table 7.4 shows the results of soybean oil hydrogenation carried out at two different levels of nickel loading, keeping the other variables, such as the reaction temperature and pressure constant. The following results were obtained at the higher nickel loading:

- The reaction time was greatly reduced.
- The oil contained a significantly higher level of *trans* fatty acids.
- The stearic acid level was significantly lower.
- The oleic acid level was significantly higher.
- The Mettler drop point was significantly lower.

TABLE 7.4 Effect of Catalyst (Nickel) Loading on Hydrogenation

	Catalyst Nysosel 222 Nonselective condition	Catalyst Nysosel 222 Nonselective condition
Reaction temperature (°F)	284	284
Reaction pressure (bar)	0.01	0.10
Catalyst type	3	3
Nickel loading	70	70
Reaction time (min)	**240**	**22**
Oleic acid (%)	**53.8**	**64.2**
Linoleic acid (%)	10.7	7.7
Linolenic acid (%)	0.5	0.0
***Trans* fatty acids (%)**	**27.7**	**37.4**
Stearic acid (%)	**23.2**	**16.1**
Mettler drop point (°F)	**120.6**	**107.2**

The bold values indicate significant difference from the other conditions in comparison
Source: Courtesy of BASF Catalysts LLC, GA, USA.

7.5.3.4 Refined and Bleached Oil Quality

The quality of the refined oil is extremely critical to the hydrogenation process. Trace impurities, such as phosphorus (phospholipids), soap, and moisture can reduce the catalyst activity by poisoning it. This significantly impacts the reaction, as well as the quality of the hydrogenated oil.

Therefore, it is important that the RB oil meet stringent quality standards before it is sent to hydrogenation. The most common harmful impurities in the RB oil are soap, phospholipids (phosphorus), moisture, FFA, aldehydes, ketones, and dissolved oxygen.

To obtain a good hydrogenation reaction, it is important to maintain the following quality standards in the RB oil:

Phosphorus	<1 ppm; preferably <0.5 ppm
Moisture	<0.1%; preferably <0.05%
Soap	0 ppm, max.
PV	<4.0; preferably <2.0
FFA	<0.15%; preferably <0.1%

Phospholipids are measured and expressed in the oil as parts per million (ppm) of phosphorus. These compounds are surfactants. They can poison (deactivate) the active sites of the catalyst. They are also highly polar in nature, which means they have free electrons that can be donated. These free electrons can poison the catalyst, reducing its active sites. The same applies for soap and moisture. Therefore, the catalyst gets poisoned if the level of any of these impurities (phosphorus, soap, or moisture) is higher than the levels recommended as in the above table. The impacts of these impurities are noticed in the longer induction period and slow reaction. This means the reaction does not start and is slow to take off when some of these impurities are present in the refined oil at higher, than the recommended, levels. In addition, this results in higher than normal catalyst consumption and inconsistent quality of the hydrogenated oil.

7.5.3.5 Impact of Catalyst Poisoning

The impacts of catalyst poisoning from the impurities in the oil were recognized by the early researchers. The consequence of catalyst poisoning can be quite serious in hydrogenation process in terms of overall productivity of the plant as discussed:

- The hydrogenation rate is greatly reduced. This requires a higher amount of catalyst to start and achieve the reaction within a reasonable time.
- A higher amount of catalyst slows down the oil flow through the filter at the end of the reaction, causing reduced productivity in the process.
- Poisoning of the catalyst can increase the colloidal nickel concentration in the hydrogenated oil. The colloidal nickel in the hydrogenated oil is not removed in the normal filtration step after hydrogenation. Colloidal nickel,

if not removed, can create a reddish or orange color in the fresh shortening. This effect can be more pronounced when the shortening is stored in the warehouse. It can also adversely affect the flavor stability of the shortening and any product formulated with it.
- Colloidal nickel plugs the filter, resulting in frequent filter cleaning and lower productivity.
- Soap in the RB oil feed to the hydrogenation process impacts the process and quality in the same manner as phospholipids.
- Part of the high moisture is lost during hydrogenation. However, the initial contact between the catalyst and the excess moisture in the oil can cause enough deactivation of the catalyst. The outcome of this is similar to those discussed for phospholipids and soap in the oil.
- High peroxide in the oil can break down to form free radicals that can attack the active sites on the nickel catalyst and reduce the activity.
- High FFA in the oil can also impact the activity of the catalyst. This is especially true when the RB oil is derived from poor-quality crude oil or the oil is poorly handled in the process. Some of this deactivation is due to the formation of nickel soap with the FFA.

7.5.3.6 Hydrogen Gas Quality

Like phosphorus, soap, and moisture in the RB oils, impurities present in the hydrogen gas can poison the catalyst and produce similar results. These impurities can cause catalyst poisoning even when they are present in trace amounts in the hydrogen gas.

The maximum levels for the commonly known impurities, which ate also catalyst poisons, in the hydrogen gas are listed:

Sulfur compounds	Trace
Carbon monoxide	Trace
Carbon dioxide	Trace
Nitric or nitrous oxides	Trace
Moisture	Trace

The steam-reformed hydrogen gas made from steam reforming of natural gas, before purification, may contain certain impurities that are listed:

Impurity	Level
Carbon monoxide, CO	0.2%–0.5%
Carbon dioxide, CO_2	0.5%–1.0%
Hydrogen sulfide, H_2S	0.05%–0.15%
Organic sulfur (mercaptans, etc.)	0.1–0.5 grains/100 $ft.^3$
Oxygen	0.0%–0.1%
Nitrogen	0.3%–1.0%
Hydrogen purity	98%–99%

The purified hydrogen gas typically shows the following analysis:

Impurity	Level
Carbon monoxide (CO)	0.001%
Carbon dioxide (CO_2)	0.001%
Methane (CH_4)	0.018%
Organic sulfur (mercaptans, etc.)	0.1–0.5 grains/100 $ft.^3$
Oxygen	0.005%
Nitrogen	0.007%
Hydrogen purity	99.968%

Cryogenic hydrogen gas has very high-purity. It has been the experience among the oil processors that the cryogenic hydrogen delivers the best reaction because of the high purity (99.995%). The reaction rate is high and the quality of the hydrogenated oil is more consistent.

Most oil refineries generate hydrogen gas by catalytic steam reforming of natural gas. The product of reaction is very carefully purified through a pressure swing adsorption (PSA) system. The impurities in the hydrogen gas can be high if:

- The PSA system is not properly tuned.
- The adsorption catalysts are old and have become less functional.
- The natural gas contains high levels of sulfur compounds and they are not removed at the beginning of the process. Sulfur compounds are harmful to the PSA catalyst, as well as the hydrogenation catalyst.

Sulfur compounds in the natural gas (used for generating hydrogen gas) are one of the worst poisons for the nickel catalyst. The nickel catalyst rapidly adsorbs these compounds and the damage caused by these poisons is irreversible. Fresh catalyst gets progressively deactivated as it adsorbs more sulfur compounds during reaction and can become ineffective when the sulfur in the catalyst is about 5.5%–9.0% of the nickel. A step-wise deactivation of the catalyst by successive addition of a sulfur compound to the catalyst can be seen and measured by the residual level of the active nickel in the catalyst. The author reported the results of an experiment, which are summarized in Table 7.5.

TABLE 7.5 Sulfur Content Versus Activity of a Poisoned Catalyst

Sulfur compound	Calculated percentage of active nickel combined with sulfur	Residual active nickel (%)	Hydrogenation time (min) for (Iodine value 107 down to 25)
Hydrogen sulfide	0.9	5.6	30
Carbon disulfide	2.9	3.6	48
Sulfur dioxide	6.0	0.5	280

Approximate active nickel content at start, 6.5%.

Three different sulfur compounds—hydrogen sulfide, carbon disulfide, and sulfur dioxide—were used for the poisoning of the catalyst to specific degrees of nickel deactivation. The time required to hydrogenate cottonseed oil from an IV of 107 down to 25 was recorded in every case. It was noticed that hydrogenation time increased with the increase of catalyst deactivation.

Adsorption of carbon monoxide is slower than for the sulfur compounds. Carbon monoxide from the hydrogen gas can be removed at high temperature and under good vacuum. However, most of the reaction is carried out under pressure. Therefore, the unadsorbed carbon monoxide accumulates in the reactor head space in a deadend reactor. The concentration of carbon monoxide increases and begins to impact the hydrogenation reaction if the head space is not purged properly. It is also observed that the impact of carbon monoxide becomes worse with reduced reactor temperature. The other gases, if present at higher levels, can act as diluents and retard the reaction.

Many oil processors in Latin America and in some other countries produce hydrogen gas via an electrolytic process. This gas has a high-moisture content. This moisture in the hydrogen must be removed completely. In many instances, the absorption tower does not completely dry the gas. Water also separates from the gas after it is compressed for use. The author observed a number of plants where the intercondenser was not adequately designed to remove the moisture from the compressed gas. The moisture in the gas was poisoning the catalyst, slowing down the reaction. This always caused a higher than normal use of catalyst and other issues as described earlier and not to mention the off-quality hydrogenated oil.

The extra moisture from the gas accumulates at the top of the oil in the reactor, which reduces the partial pressure of hydrogen gas and slows down the reaction.

Additionally, instead of high-pressure gas cylinders, many of these plants use gas holders that are designed with a floating top with water seals. The hydrogen gas becomes saturated with water vapor. The moisture then condenses when the gas is compressed. The compressed gas must be adequately cooled to remove the moisture separated during compression and then preferably passed through a set of dryers to eliminate the residual moisture in the gas before it is used.

The summary of all of the aforementioned factors that impact the hydrogenation reaction and the oil quality are listed in Table 7.6.

7.5.4 Hydrogen Gas Dispersion

Dispersion of hydrogen in the reactor must reach a submicron size for proper adsorption on the active sites of the catalyst. This is accomplished with the help of a hydrogen gas sparger made of sintered metal and a high-shear impeller at the lower part of the reactor. A proper ratio between the diameter of the hydrogen sparger and that of the impeller of the high-shear mixer is important to obtain proper gas dispersion. The recommended diameter of the hydrogen gas sparger

TABLE 7.6 Summary of Factors Producing Poor Hydrogenation and Poor-Oil Quality

Factors responsible	Source	Slow reaction	High FFA	High-soluble nickel	Melting point different from target	Poisons catalyst
Poor-oil quality: • High soap • High phosphorus • High moisture • High FFA • High peroxide	• Poor refining and bleaching • Abused or stored for a long time	X	X	X	X	X
Poor-hydrogen quality: • High CO • High sulfur compounds • High moisture	Poor reaction and purification	X	X	X	X	X
Insufficient hydrogen supply	Gas plant is working below capacity or low-storage capacity	X	X	X	X	X
Poor-quality catalyst	Poor quality of fresh catalyst or poor storage and handling	X	X	X	X	X

FFA, Free fatty acid.

is 60%–80% of the impeller pitch diameter. A sintered metal sparger is preferred because of its ability to produce micron and submicron size gas bubbles.

7.5.5 Hydrogen Gas Venting From the Reactor

As the hydrogenation reaction progresses, some volatile impurities from the oil are released, which accumulate above the oil surface in the reactor. The purity of the gas in the head space is low. If left in the reactor, the gas from the head space can be absorbed by the oil and produce some undesirable side reactions that can affect the flavor stability of the oil (after deodorization).

Therefore, it is necessary to vent this gas from the reactor head space from time to time. In some instances, an instrument checks the purity of the gas in the head space and the gas is then vented periodically to maintain the desired gas quality above the oil surface in the reactor. In the absence of an automated system, one must manually vent the gas from the reactor head space periodically to maintain the gas quality in the reactor. A small amount of carbon monoxide is also formed during hydrogenation. This is believed to be due to oil decomposition. Accumulation of carbon monoxide is undesirable as discussed earlier. Therefore, venting of the gas from the reactor head space is critical to reduce catalyst poisoning by carbon monoxide.

7.5.6 Hydrogen Gas Supply

Once the reaction is started, there should be ample gas supply to complete the hydrogenation of the batch of oil. The author has observed processing plants where the reaction had to be suspended for lack of gas. This was caused by an inadequate size of hydrogen generator and a lack of gas storage. The plant experienced high variability in the hydrogenated oil quality in terms of solids and melt points because of the interruptions in the hydrogenation process.

The hydrogenation reaction consumes 14.15 $ft.^3$ of hydrogen gas/1000 lb of oil/IV drop. The corresponding number in the metric system is 883.3 L of hydrogen gas/1000 kg of oil/IV drop. One can estimate the theoretical amount of hydrogen needed for a given batch of oil to attain a certain IV drop. In actual practice one should have 5%–10% excess gas supply for peak production at any given time. The loss of hydrogen gas is generally 3–5% of the theoretical amount.

It is suggested that the hydrogen gas have the storage capability where hydrogen gas is stored at 200 psi and can supply the plant for 4–6 h to prevent any short supply of hydrogen gas at the plant. Fig. 7.8 shows the suggested schematic diagram for hydrogen gas storage and supply.

7.5.7 Reaction Pressure

Lower-reaction pressure increases selectivity and higher reaction pressure reduces the selectivity. It can be seen in Table 7.7 that the reaction carried out at 3.0 bar pressure versus that conducted at 25.0 bar pressure produced significantly

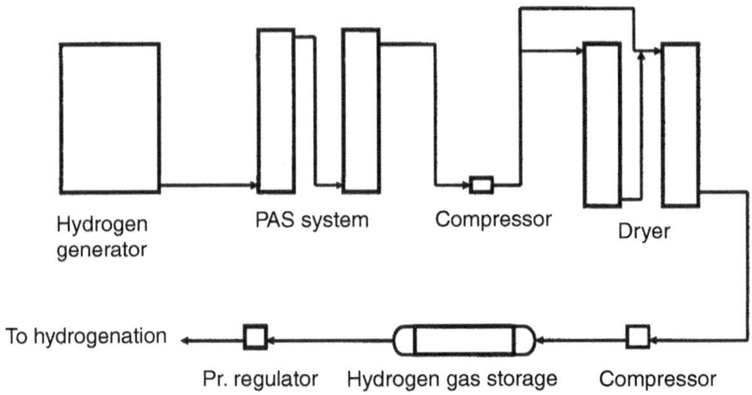

FIGURE 7.8 **Schematic diagram for recommended hydrogen gas system.** *PAS*, Pressure swing adsorption.

TABLE 7.7 Effect of Reaction Pressure

	Selective condition	Nonselective condition
Catalyst type	Nysosel 222	Nysosel 222
Reaction pressure (bar)	**3.0**	**25.0**
Reaction temperature (°F)	284	284
Nickel loading (%)	0.1 (1000 ppm)	0.1 (1000 ppm)
IV	70	70
Reaction time (min)	240	**<10**
***Trans* fatty acids (%)**	**37.4**	**32.5**
Stearic acid (%)	**16.1**	**24.2**

The bold values indicate significant difference from the other conditions in comparison
Source: Courtesy of BASF Catalysts LLC, GA, USA.

different results even though the nickel loading and the reaction temperatures were the same. The results showed that at the reaction pressure of 25 bar:

- the reaction time was reduced by a factor of nearly 25,
- produced lower *trans* fatty acids, and
- produced higher stearic acid.

7.5.8 Reaction Temperature

Higher-reaction temperature increases the hydrogenation reaction rate. The reaction end point is reached faster. With every other condition remaining the same, a higher reaction temperature produces the following results:

- the selectivity is increased,
- reaction time is reduced,
- lesser amount of stearic acid is produced,
- higher amount of *trans* fat is produced, and
- lower-melt point is obtained in the hydrogenated oil.

Higher-reaction temperature significantly reduces the viscosity of the oil. This improves the transport of the two reactants, namely hydrogen gas and unsaturated fatty acids (triglycerides with unsaturated fatty acids), to the surface of the catalyst to facilitate the reaction.

At higher temperature, the solubility of hydrogen gas into the oil becomes a significant factor. Solubility of hydrogen in oil is proportional to the pressure but it increases significantly with the temperature of the oil. The solubility data for hydrogen gas from this reference is shown:

Oil temperature (°F/°C)	Hydrogen gas solubility (mole/m^3/bar)
248/120	2.2
320/160	2.9
392/200	4.0

Additionally, the affinity for hydrogenation increases with higher unsaturation in the fatty acid. This means that the linolenic acid would have the highest opportunity to react with hydrogen, while oleic acid would be reacted last.

Therefore, higher reaction temperature would be conducive for selective hydrogenation. This can be seen from the data presented in Table 7.8.

Review of the data in Table 7.8 and the graphs in Figs. 7.9 and 7.10 indicate that the lower-reaction temperature produced lower–*trans* fatty acid and higher-stearic acid.

7.5.9 Agitation

The effect of agitation has been shown by several researchers in their pilot plant studies. Theoretically, at higher agitation, especially higher high-shear mixing, one would increase contact between the catalyst, the triglyceride molecules, and hydrogen gas. This could facilitate hydrogenation of the linoleic, as well as the oleic acid. However, on an industrial scale, agitator speed demonstrates less dramatic effect on hydrogenation. For example, the AGR mixer, promoted by Praxair in the 1980s, showed improved dispersion but the oil processors were unable to obtain a noticeable improvement in the rate of hydrogenation.

7.6 CATALYST FILTRATION

The hydrogenated oil is filtered at the end of the reaction using pressure leaf filters or plate and frame filters. The oil is cooled down either in the reactor or drop tank. Alternatively, the oil from the drop tank is recirculated through a plate and frame heat exchanger, while the incoming oil is pumped through the same heat

TABLE 7.8 Impact of Reaction Temperature on Hydrogenated Oil Attributes Type of Oil Soybean; United Catalyst (Currently BASF Catalysts LLC); Type G135A

	Temp. 320°F (160°C); Pr. 310 kPa (45 psig); Ni 0.005%						Temp. 400°F (204°C); Pr. 310 kPa (45 psig); Ni 0.005%					
IV	C18:0	C18:1	C18:2	C18:3	*Trans*	SFC[a]	C18:0	C18:1	C18:2	C18:3	*Trans*	SFC[a]
130	3.3	23.0	54.4	8.2	**0.0**	0.0	3.2	22.9	54.4	8.2	**0.0**	0.0
125	3.7	29.6	49.8	6.4	**6.1**	0.0	4.1	27.5	51.0	7.1	**5.0**	0.0
120	3.9	36.0	45.1	4.9	**11.6**	0.4	4.7	32.4	47.2	6.0	**10.0**	0.1
115	4.0	42.2	40.5	3.6	**16.4**	0.6	5.1	37.5	43.0	5.0	**14.9**	1.4
110	4.0	48.0	35.8	2.6	**20.8**	1.0	5.5	42.6	38.7	3.9	**19.6**	2.5
105	4.2	53.3	31.1	1.8	**24.6**	2.0	5.8	47.6	34.2	3.2	**24.2**	3.5
100	4.4	58.1	26.6	1.1	**28.0**	3.5	6.3	52.4	29.7	2.4	**28.5**	4.7
95	4.9	62.2	22.3	0.7	**30.9**	5.7	7.0	56.7	25.2	1.8	**32.5**	6.3
90	5.8	65.5	18.2	0.4	**33.4**	8.7	7.9	60.6	20.9	1.2	**36.0**	8.5
85	7.1	68.1	14.4	0.2	**35.5**	12.6	9.3	63.7	16.7	0.7	**39.2**	11.4
80	8.9	69.7	11.0	0.1	**37.2**	17.5	11.0	66.1	12.9	0.3	**41.8**	15.2
75	11.4	70.2	8.0	0.0	**38.7**	23.5	13.3	67.5	9.5	0.1	**43.9**	20.2
70	14.6	69.7	5.5	0.0	**39.8**	30.8	16.3	67.9	6.6	0.0	**45.4**	26.5
65	18.6	68.0	3.5	0.0	**40.2**	39.4	20.0	67.1	4.2	0.0	**46.2**	34.4

The bold values indicate significant difference from the other conditions in comparison

All analyses are shown as percent.

[a]*At 70°F.*

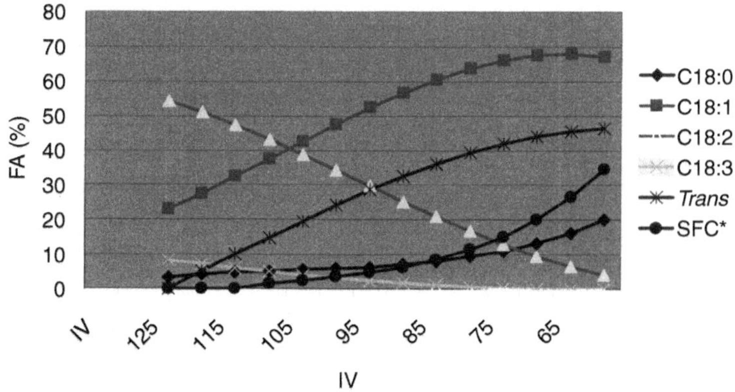

FIGURE 7.9 Hydrogenation curve for soybean oil. Temperature: 400°F; pressure: 45 psi; and catalyst: 0.005% (Ni). *SFC at 70°F.

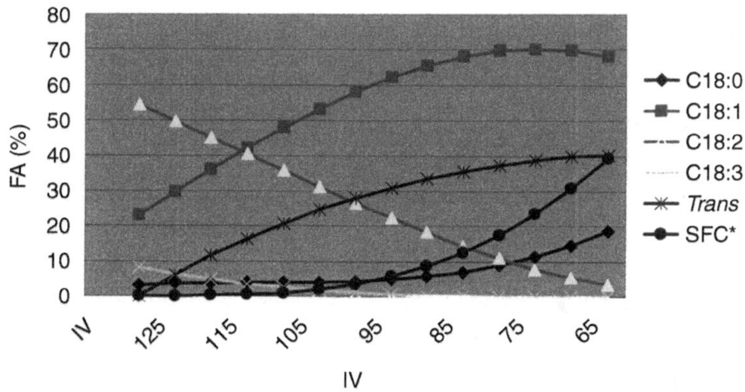

FIGURE 7.10 Hydrogenation curve for soybean oil. Temperature: 320°F; pressure: 45 psi; and catalyst: 0.005% (Ni). *SFC at 70°F.

exchanger. The reactor feed is further heated in a heater to raise the temperature to the initiation temperature for the reaction.

The pressure leaf filters for oil filtration can be vertical leaf and horizontal shell or vertical leaf and vertical shell. The filters need to be precoated as discussed in Chapter 6 on bleaching. The operating controls and procedures are very similar for bleaching, as well as for hydrogenation.

Plate and frame filters can oxidize the oil unless they have enclosed oil manifolds to protect the oil from excessive exposure to the air.

Like in bleaching, the oil is recirculated and checked for suspended solids. The oil is sent forward only after it meets the required clarity standard.

The hydrogenated oil are cooled and can be stored in tanks before being blended for product formulation or sent directly to the deodorizer as the case may be.

7.7 CRITICAL QUALITY PARAMETERS IN BATCH HYDROGENATION

Critical quality parameters for hydrogenated oils, their recommended limits, their significance, and consequences for noncompliance are listed in Table 7.9. The same quality standards apply for the oil from a continuous hydrogenation process.

7.8 *TRANS* FATTY ACIDS

Trans fatty acids are unsaturated fatty acids, which are formed during hydrogenation. In reality, these are the positional isomers of the *cis*-form, which is naturally present in vegetable oils (see Chapter 2).

Trans fatty acid has been one of the major subjects of discussion in the oil industry. Europe has been ahead of the United States in this respect because they recognized the ill effects of *trans* fatty acids in the diet and some countries have implemented regulations.

Trans fatty acids appear to behave the same way as certain saturated fatty acids (palmitic and lauric), by elevating the following constituents in human blood serum:

1. total cholesterol,
2. undesirable low-density lipoprotein, and
3. triglycerides.

Trans fatty acids also depress the level of desirable high-density lipoprotein in human blood serum.

There are several alternatives that are commercially applicable to reduce *trans* fatty acids in hydrogenated oil. These are:

1. Manipulation of the reaction conditions, such as reaction temperature, reaction pressure, and catalyst concentration, to achieve reduced *trans* fatty acids in the oil, using a suitable nickel catalyst.
2. Use of platinum or other precious metal catalysts.

7.8.1 Manipulation of the Reactor Conditions

Reaction conditions conducive to producing high- and low-*trans* fatty acid levels are shown in Table 7.10. These are:

- low-reaction temperature,
- high-reactor pressure,
- high agitation,
- low-catalyst concentration,
- high-catalyst activity, and
- low-selectivity catalyst.

TABLE 7.9 Critical Quality Parameters in Hydrogenation

Analysis	Recommended limits	Significance/consequence
Oil color	Varies with the type of oil • The hydrogenated oil color is lighter than the RB oil color (reactor feed), due to heat bleaching of the carotenes	• Darker color in the hydrogenated oil indicates presence of impurities in the RB oil, such as soap, high levels of trace metals, and/or phospholipids
FFA	• Hydrogenated oil generally has higher FFA than the RB oil (reactor feed) • The rise in the FFA must not be higher than 0.05% over that of the RB oil	• Higher FFA in the hydrogenated oil can be caused by longer than normal reaction time caused by any of the factors mentioned in the previous section, such as: • Poor-catalyst activity • Loss of catalyst activity due to poisoning • Poor-hydrogen gas supply • Poor-hydrogen gas distribution • Poor-hydrogen gas quality
IV	• Depends on the intended use of the hydrogenated oil	• The IV of most hydrogenated stocks are maintained at ±2 IV units from the target value
Solid content	• Depends on the intended use of the hydrogenated oil	• This is measured at various temperatures • The recommended ranges for the solids are: At 70°F (21.1°C) ±2 units At 80 °F (26.7°C) ±1.5 units At 92 °(33.3°C) ±1 unit At 104 °F (40°C) ±0.5 unit
Oil clarity after the catalyst removal	• The oil must be free of any suspended nickel catalyst and the filter aid	• Presence of nickel in the oil can be harmful because: • Nickel is a heavy metal and must be removed from the oil • Any nickel in the oil can reduce the stability of the shortening or margarine made with the oil • It can also reduce the shelf life of products made with the hydrogenated oil containing nickel • Presence of any filter aid can adversely affect the quality of the hydrogenated oil

(*Continued*)

TABLE 7.9 Critical Quality Parameters in Hydrogenation (*cont.*)

Analysis	Recommended limits	Significance/consequence
Melting point	• Depends on the intended use of the hydrogenated oil	• Controlling the melt point is important for: • Melting characteristics of margarine • Melting characteristics of shortening • Stability of cake batter and icing • Consistency of other products, such as coffee whitener and peanut butter stabilizer • Texture of cookies, crackers and other baked products

RB: Refined and bleached.

TABLE 7.10 Process Conditions in Producing High and Low Levels of *Trans* Fatty Acids for the Same End-Point IV

Process variable	Produces higher–*trans* fatty acids	Produces lower–*trans* fatty acids
Reaction temperature	High-reaction temperature	Low-reaction temperature
Reactor pressure (hydrogen gas pressure)	Low pressure	High pressure
Agitation	Low agitation	High agitation
Catalyst (nickel) concentration	High concentration	Low concentration
Catalyst activity	Low activity	High activity
Catalyst SR	High SR	Low SR

It must be understood that it is necessary to manipulate one or more of these process variables to obtain a reduced level of *trans* fatty acids in the hydrogenated oil. For example, a combination of high-reaction temperature (400°F), low-reactor pressure (45 psi), and low-nickel loading (0.005%) produced significantly higher–*trans* fatty acids for the same end point IV (65) in soybean oil as compared to when a lower temperature (220°F), higher pressure (90 psi), and higher-nickel loading (0.04%) were used. The results are shown in Table 7.11. The latter reaction conditions produced significantly lower–*trans* fatty acids when the reactor temperature was lower, while the operating pressure and nickel loading were much higher. The *trans* fatty acid level dropped further when a

TABLE 7.11 Manipulation of Reaction Conditions to Produce Lower–*Trans* Fatty Acids Type of Oil Soybean; United Catalyst (Currently BASF Catalysts LLC); Type G135A

	Temp. 400°F (204°C); Pr. 310 kPa (45 psig); Ni 0.005%						Temp. 220°F (104°C); Pr. 620 kPa (90 psig); Ni 0.04%					
IV	C18:0	C18:1	C18:2	C18:3	*Trans*	SFC[a]	C18:0	C18:1	C18:2	C18:3	*Trans*	SFC[a]
130	3.3	23.0	54.4	8.2	**0.0**	0.0	4.8	21.9	52.9	7.7	**3.1**	0.0
125	3.7	29.6	49.8	6.4	**6.1**	0.0	5.4	24.1	49.6	7.2	**5.1**	1.2
120	3.9	36.0	45.1	4.9	**11.6**	0.4	5.9	27.1	46.2	6.6	**7.1**	2.3
115	4.0	42.2	40.5	3.6	**16.4**	0.6	6.2	30.6	42.6	5.9	**9.2**	2.9
110	4.0	48.0	35.8	2.6	**20.8**	1.0	6.5	34.5	38.9	5.1	**11.2**	3.4
105	4.2	53.3	31.1	1.8	**24.6**	2.0	6.8	38.6	35.2	4.3	**13.2**	3.7
100	4.4	58.1	26.6	1.1	**28.0**	3.5	7.1	42.8	31.4	3.5	**15.2**	4.1
95	4.9	62.2	22.3	0.7	**30.9**	5.7	7.6	47.0	27.6	2.7	**17.2**	4.8
90	5.8	65.5	18.2	0.4	**33.4**	8.7	8.3	51.0	23.8	1.9	**19.0**	5.9
85	7.1	68.1	14.4	0.2	**35.5**	12.6	9.3	54.6	20.1	1.3	**20.8**	7.5
80	8.9	69.7	11.0	0.1	**37.2**	17.5	10.6	57.7	16.5	0.7	**22.5**	9.9
75	11.4	70.2	8.0	0.0	**38.7**	23.5	12.3	60.2	13.0	0.2	**24.1**	13.3
70	14.6	69.7	5.5	0.0	**39.8**	30.8	14.5	61.9	9.7	0.0	**25.6**	17.7
65	18.6	68.0	3.5	0.0	**40.8**	39.4	17.3	62.6	6.6	0.0	**26.9**	23.3

The bold values indicate significant difference from the other conditions in comparison
All analyses are shown as percent.
[a]*At 70°F.*

TABLE 7.12 Further Reduction of *Trans* Fatty Acids in Hydrogenated Soybean Oil Type of Oil Soybean; United Catalyst (Currently BASF Catalysts LLC); Type G135A

	Temp. 170°F (77°C); Pr. 1717 kPa (250 psig); Ni 0.11%					
IV	C18:0	C18:1	C18:2	C18:3	*Trans*	SFC[a]
135	4.0	20.7	55.8	8.0	**0.9**	—
130	5.2	22.3	52.8	7.6	**2.7**	0.0
125	6.1	24.3	49.7	7.0	**4.5**	0.0
120	6.9	26.8	46.5	6.4	**6.3**	1.3
115	7.6	29.7	43.2	5.7	**8.1**	1.9
110	8.3	32.7	39.9	4.9	**9.8**	2.5
105	8.9	35.9	36.6	4.2	**11.5**	3.0
100	9.7	39.2	33.2	3.4	**13.1**	3.7
95	10.5	42.4	29.8	2.6	**14.7**	4.5
90	11.5	45.4	26.4	1.9	**16.1**	5.6
85	12.7	48.2	23.1	1.3	**17.4**	7.2
80	14.3	50.6	19.8	0.8	**18.6**	9.2
75	16.1	52.6	16.6	0.3	**19.7**	11.9
70	18.3	54.0	13.4	0.1	**20.5**	15.3
65	21.0	54.8	10.3	0.0	**21.2**	19.5
60	24.2	54.9	7.4	0.0	**21.8**	—

The bold values indicate significant difference from the other conditions in comparison
All analyses are shown as percent.
[a]*At 70°F.*

reaction temperature of 170°F, a reaction pressure of 250 psi, and a nickel loading of 0.11% were applied.

The results are shown in Table 7.12. The *trans* fatty acid values from Tables 7.11 and 7.12 are plotted in Fig. 7.11.

7.8.1.1 Hydrogenation Under High Pressure

The impact of higher-reaction pressure has been illustrated in Table 7.7 earlier. The high-operating pressure illustrated in Table 7.6 is 25 bars. The batch hydrogenation reactors in the United States are designed for an operating pressure of 75–100 psi (approximately 7 bar max.) and the nozzles and flanges are designed for 150 psi (15 bar). Hydrogenation pressure of 200–300 psi (14–21 bar) would be required along with a lower-reaction temperature and a higher-nickel loading as discussed earlier to effectively reduce the *trans* fatty acid in the hydrogenated oil. This would require a reactor to be designed for an operating pressure up to 450–500 psi (67 bar).

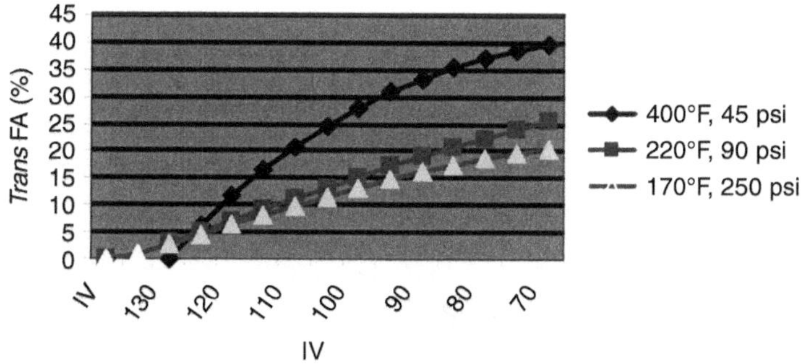

FIGURE 7.11 **Effect of hydrogenation temperature on the *trans* fatty acid level in hydrogenated oil.**

7.8.1.2 Economic Impact of High-Pressure Reactors

High-pressure hydrogenation can produce lower–*trans* fatty acid levels. This is a definite advantage in view of today's need for lower–*trans* fatty acids in hydrogenated oils. Looking at this potential solution for lower–*trans* fatty acid in products, one may become interested in changing the process to high pressure.

It must be remembered that this is not an easy solution that can be applied by the oil processors in the United States because the industry will have to make a number of changes in the process design for hydrogenation. Besides replacing the existing reactors with high-pressure ones at a tremendous cost, the various other required elements for this change are listed:

1. significantly higher cost for the reactor,
2. a heater before the filter to raise the temperature of the oil to 250–280°F (121–138°C) to decrease the viscosity of the oil and obtain a reasonable filtration rate,
3. larger filter or multiple filters because of the higher-nickel loading,
4. higher cost of depreciation for the new equipment, besides the high-pressure reactor,
5. higher cost of maintenance,
6. increased cost of catalyst,
7. higher oil loss in the spent catalyst,
8. higher labor cost for filter cleaning, and
9. higher cost of spent catalyst disposal.

7.8.2 Higher Cost of the Reactor

The typical cost of a standard reactor built for a capacity of 60,000 lb may be $250,000.00. A 500-psi reactor may cost as much as $850,000.00. The estimated installed cost for the standard reactor is $400,000.00, while that for the high-pressure reactor is $1,200,000.00.

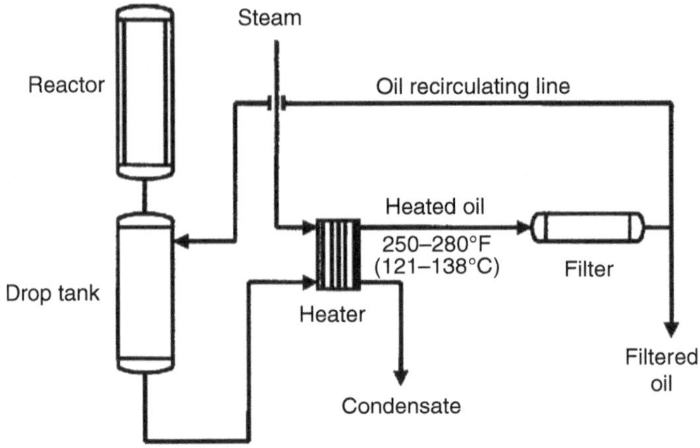

FIGURE 7.12 **Filter arrangement for high-pressure hydrogenation system.**

7.8.3 Heating Hydrogenated Oil before Filtration

The oil coming out of the reactor must be heated to 250–280°F (121–138°C) before it can be filtered. Fig. 7.12 shows the schematic diagram for the heating arrangement. This is essential to decrease the oil viscosity to obtain better oil flow through the filter.

7.8.4 Larger-Filter Area or Dirt Load Capacity

Nickel loading used at high pressure (250 psi, 1717 kPa) was nearly 3 times that used at moderate pressure (90 psi, 620 kPa) and 10 times that used in a typical hydrogenation process. With no changes in the filter design, the cycle time for the filter would be significantly reduced. Therefore, larger filters with possibly higher–dirt load capacity would be required. It might be necessary to install multiple filters to maintain the overall cycle time to maintain the productivity.

7.8.5 Higher Cost of Depreciation

The comparative depreciation cost for the high-pressure reactor is shown in Table 7.12. Using a 7-year straight-line method, the depreciation for the standard reactor is $57,000.00 compared to $171,000.00 for the high-pressure reactor.

7.8.6 Higher Cost of Maintenance

The typical cost of maintenance is 10% of the installed cost. Thus, the maintenance cost for the standard reactor is $40,000.00 per year and that for the high-pressure reactor is $120,000.00 per year. Owing to the high-pressure operation, the reactor requires greater attention to all seals and joints.

Besides the reactor, the high-pressure reaction system will need more filtering area, a heater, and certain other accessories. These will also add to the cost of maintenance.

7.8.7 Increased Cost of Catalyst

Normal nickel loading is 0.01%, while for the high-pressure reaction the nickel loading is 0.1%. This is an increase of 0.09% in nickel loading.

Typical catalyst contains 20% nickel. Thus, the catalyst loading for the two methods is:

Standard reactor	0.01/0.2 = 0.05 × 60,000/100 = 30 lb/batch
High-pressure reactor	0.1/0.2 = 0.5 × 60,000/100 = 300 lb/batch
Cost of catalyst	$8.50 per pound

The cost differential/batch = (300–30) × $8.50
= $2,295.00 per 60,000 lb
= $0.03825 per pound

7.8.8 Higher Oil Loss in the Spent Catalyst

The spent catalyst has approximately 33% oil absorbed in it. This oil is lost in the spent catalyst. Normally, 0.1% filter aid is used to improve filterability of the catalyst. Therefore, for a batch of 60,000 lb of oil, the amount of oil loss in the spent catalyst is:

Catalyst + filter aid = 300 (catalyst) + 60 (filter aid) = 360 lb
Oil absorbed = 360 × 0.33 = 118.8 lb
The estimated oil loss = 118.8 lb × $0.45 per pound = $53.5 per batch

7.8.9 Cost of Spent Catalyst Disposal

Spent nickel catalyst is banned from landfills in most countries. The spent catalyst must be sent to metal reclaimers, who recover the nickel. There is some cost involved in this. The cost can vary from region to region of the world. In the United States, it is necessary to pack the spent catalyst in Department of Energy (DOT)–approved drums for shipping to the metal reclaimers. The cost of a refurbished DOT approved drum is $30.00 and it can hold up to 400 lb of spent nickel catalyst. One could use the same drums in which they received the fresh catalyst for shipping the spent catalyst.

The reclaiming companies in the United States work with the following cost formula:

- Nickel credit — 40%–60% of recovereable nickel in the spent catalyst
- Processing fee — Variable, to be paid by the oil processor
- Shipping spent — Paid by the oil processor

7.8.9.1 Catalyst

The oil processor may recover some of the cost of catalyst, but in most cases, this simply offsets the cost of shipping. The oil processor may even incur some out of pocket expenses for shipping the spent catalyst to the metal reclaimer. It is possible to maximize the return by using the absolute minimum of the filter aid in the process to maintain high-nickel concentration in the spent catalyst.

7.8.9.2 Summary of All-Cost Elements for High-Pressure Hydrogenation Process

Table 7.13 lists the various cost factors related to the reactor that affect the price of oil hydrogenated under high pressure. This can amount to at least $0.04 per pound of oil.

TABLE 7.13 Cost Implications in High-Pressure Hydrogenation for Low–*Trans* Fatty Acids

	Low-pressure reactor	High-pressure reactor
Reactor capacity per batch	60,000	60,000
Cost of the reactor ($)	250,000.00	850,000.00
Cost of Installation ($)	150, 000.00	350,000.00
Installed cost ($)	225,000.00	625,000.00
No. of batches hydrogenated per day	4	4
Daily production (lb/day)	240,000	240,000
Annual capacity (300 days/year)	72,000,000	72,000,000
Actual production @ 85% production efficiency (lb/year)	61,200,000	61,200,000
Depreciation, using a 7-year straight-line depreciation ($)	57,000.00	171,300.00
Impact of depreciation ($/lb)	0.0009	0.0027
Maintenance per year (using 10% of the equipment)	40,000.00	120,000.00
Impact of maintenance cost ($/lb)	0.00063	0.0019
Catalyst (nickel loading) (%)	0.01	0.1
Actual amount of catalyst (20% nickel content) per 100 lb of oil	0.05 lb	0.50 lb
Cost of catalyst per pound of oil	$0.00425	$0.0425
Increased cost	—	$0.085 = $0.04 per pound of oil

TABLE 7.14 Hydrogenation Results with Platinum Catalyst

Basis of comparison	Nickel catalyst Nysosel 325	Platinum catalyst 5% Pt/C	Platinum catalyst 5% Pt/Al
Catalyst loading (lb of metal/100 lb of oil)	0.1 (1000 ppm)	0.01 (100 ppm)	0.01 (100 ppm)
Support	—	Carbon	Aluminum
Reaction temperature (°F)	284 (140°C)	140 (60°C)	140(60°C)
Reaction pressure (bar)	10	10	10
IV	100.5	101.2	101.8
Reaction time (min)	10	42	90
Stearic acid (%)	**6.8**	**12.8**	**17.6**
Linoleic acid (%)	1.7	3.0	4.2
***Trans* fatty acid (%)**	**18.8**	**8.5**	**5.4**

The bold values indicate significant difference from the other conditions in comparison
Source: Courtesy of BASF Catalysts LLC, GA, USA.

The additional cost of the larger filter, accessories, catalyst disposal, and loss of oil can be estimated and it would amount to some additional costs. Not all of those are shown in detail in Table 7.13.

7.8.9.3 Use of Platinum or Other Precious Metal Catalysts

Hydrogenated oil can be produced with low–*trans* fatty acid content using a platinum catalyst. These catalysts are highly reactive and the reaction can be carried out at very low temperatures compared to a nickel catalyst. Platinum catalyst can come with carriers (also called support), such as carbon or aluminum. Table 7.14 shows the reaction conditions and some specific analysis for soybean oil hydrogenated to an IV of 100 by using platinum catalyst and compares the data against those from the nickel catalyst (at higher-nickel loading).

It was observed that the platinum catalyst produced the reaction at significantly lower temperature (140°F, 60°C), compared to the nickel catalyst (284°F, 140°C), with lower–*trans* fatty acid but much higher-stearic acid in the oil. Reaction time for the platinum catalyst was much longer than with the nickel catalyst. However, the concentration of nickel was significantly higher than that of platinum.

7.8.9.4 Economics of Using Platinum Catalyst

Platinum catalyst offers the benefit of lower–*trans* fatty acids but it adds serious complexity to the operation, such as:

1. The oil processor needs to purchase a certain amount of platinum up front. This requires a large sum of capital for the vegetable oil refiner.

2. The spent catalyst must be recovered to nearly 100% of use to minimize cost. This necessitates the catalyst handling system that be designed with great care to prevent any losses in handling.
3. The spent catalyst is sent back to the catalyst company for the recovery of the platinum and converting it into a catalyst.
4. A certain amount of platinum in the spent catalyst is lost in the reclaiming process.
5. Number of uses of the catalyst.
6. Analytical cost.
7. Rerefining the metal.

On June 8, 2016

1. Spot market price of platinum was $1012 per troy ounce.
2. Spot market price of nickel was $3.91 per pound ($0.27 per troy ounce).

The price difference between platinum and nickel is obvious.

One can justify the use of platinum catalyst on paper by using the following assumptions:

1. cost of platinum,
2. cost of catalyst fabrication,
3. metal recovery from spent catalyst (30% oil content),
4. estimated metal loss in hydrogenation process = 15%,
5. number of uses = 10, and
6. metal loss in rerefining = 2%.

In actuality, the repeated use of the catalyst on a commercial scale has not been proven by any vegetable oil processing plant.

Loss of metal in the hydrogenation process has not been fully established because of the lack of data.

The loss of activity and change in selectivity of the platinum catalyst with multiple uses has not been documented in the oil industry. This adds another dimension to the lack of understanding of the behavior of this catalyst.

Additionally, it is very difficult to achieve perfect recovery of the spent catalyst in an existing vegetable oil plant for the following reasons:

- The oil filtration system in the existing plant is designed to handle the inexpensive nickel catalyst and is not designed for complete recovery of the expensive platinum catalyst.
- The number of reuses of the catalyst is unknown. No vegetable oil industry has used platinum catalyst for long enough to provide any meaningful information on the subject.
- The process requires high-pressure reactors.

Therefore, the use of platinum catalyst, although attractive for *trans* fatty acid reduction, has not been shown to be practical for a commercial vegetable oil refinery at this time.

The hydrogenation temperature with platinum catalyst is too low for filtration. The oil needs to be heated to 250–280°F (121–138°C) before it can be filtered, similar to what is shown in Fig. 7.12.

7.8.9.5 The Impact of Reheating of the Oil for Filtration

In both high-pressure nickel catalyst and platinum catalyst processes the oil from the reactor is required to be heated to a higher temperature as mentioned earlier. The oil in the reactor retains some dissolved hydrogen gas, and the catalyst surface also retains some amount of adsorbed hydrogen gas. The hydrogen gas in the oil and on the catalyst surface can react with the unsaturated fatty acids as the oil is reheated and there can be a small drop in the IV in the oil. The extent of such a reaction is not known.

7.9 SOURCES OF HYDROGENATION CATALYSTS

A couple of decades ago, there were several manufacturers of nickel catalyst for hydrogenation of vegetable oils. Many of these companies are no longer in existence. Forty years ago, names, such as Englehard, Synetix, and United Catalyst were household names among the oil processors. The Germany-based company BASF Catalysts LLC acquired Englehard Corporation and also acquired the rights to vegetable oil hydrogenation catalysts from the United Catalyst Co. and from some other catalyst companies. Johnson Mathey of the United Kingdom acquired the rights to the vegetable oil hydrogenation catalysts from Synetix. Thus, there are only two major companies in the world that supply nickel catalyst for hydrogenation of vegetable oils:

1. BASF Catalysts LLC
2. Johnson Mathey

There are two small companies in Japan (Nikki and Sakai) that supply nickel catalysts locally.

7.10 SELECTION OF HYDROGENATION CATALYST

Based on the many aspects of the hydrogenation process discussed earlier in this chapter, it should be clear that a catalyst for hydrogenation should be selected on the basis of the desired end product, as well as the process requirements. Generally, a catalyst is chosen based on the following criteria:

1. catalyst activity,
2. catalyst selectivity,
3. filterability of the hydrogenated oil,
4. physical integrity of the catalyst after use, and
5. cost.

7.10.1 Catalyst Activity

The hydrogenation catalyst must be of high activity to achieve the following:

- short initiation time,
- rapid turnover of a batch in the reactor,
- lowest amount of catalyst required for the production of a given batch of oil,
- minimum of side reactions to reduce the formation of some of the undesirable components in the hydrogenated oil, and
- reduced cost of hydrogenation.

7.10.2 Selectivity

Selectivity is important in the production of margarine and some other products, such as cocoa butter substitutes, that require sharp melting characteristics. Based on the previous discussions, it must be understood that these products tend to have high levels of *trans* fatty acids.

7.10.3 Filterability

The catalyst particles must be as small as possible to obtain maximum specific surface area (area in square inches per pound or square centimeter per gram).

A larger number for the specific surface area for the catalyst is desirable to maximize the reaction. However, a very small-particle size may slow down the filtration process, reducing the plant's productivity. The catalyst must not pass through the filter media causing quality issues.

Catalyst manufacturers design their products for maximum efficiency for both reactivity and filterability.

7.10.4 Physical Integrity

The catalyst for hydrogenation must be capable of withstanding the normal physical handling in the process without disintegrating because this would adversely affect the hydrogenation characteristics, filterability, and the overall productivity of the process.

7.10.5 Cost

The cost of the catalyst must be reasonable and still deliver the desired performance.

7.11 COMMERCIALLY AVAILABLE NICKEL CATALYSTS

Specific examples of commercial nickel catalysts and their applications in vegetable oil hydrogenation are listed in this section. The reader needs to understand that the catalyst manufacturers can mention the activity or selectivity of their catalysts; but the specific application of the catalysts depends on the user's knowledge and understanding regarding hydrogenation and the properties of the catalyst as well as those of the end product.

Table 7.15 lists the nickel catalysts from BASF Catalysts LLC for the hydrogenation of vegetable oils. The company emphasizes the following areas to characterize the catalysts:

1. filtration,
2. activity,
3. selectivity,
4. poison resistance, and
5. versatility.

Items 1–3 have been discussed earlier. Poison resistance is an important attribute that a catalyst must have to retain its activity.

Versatility indicates the catalyst's ability to hydrogenate various types of oils from vegetable, marine, and animal origin and also to produce different hydrogenated products, such as salad oil, margarine oil, shortening base, and fully hydrogenated oils.

Catalysts from Johnson Mathey for vegetable oil hydrogenation are listed in Table 7.16.

7.12 TROUBLESHOOTING THE HYDROGENATION PROCESS

Table 7.17 shows the typical process and oil quality difficulties or symptoms experienced in hydrogenation along with their probable cause or causes and the recommended solutions. It is beneficial for the process personnel to understand the cause and effect of these commonly occurring situations so that appropriate action can be taken in time to correct the problem.

7.13 HEAT RECOVERY IN HYDROGENATION

Hydrogenation is an exothermic reaction process. The reaction generates 1.65 Btu/lb of oil for every IV value drop. Therefore, the process has tremendous amount of heat energy. The oil processing industry has been recovering this energy for nearly three decades. Various engineering designs have been applied. Most of these used a heat-sink approach where hot-hydrogenated (filtered) oil was stored to preheat the incoming oil to the converter.

There were many drawbacks in this approach, such as:

- The hot oil stored for an extended period had higher-polymer content and reduced tocopherols content.
- The overall oil quality was compromised.

A new approach was published in 1994. Fig. 7.13 shows the schematic flow diagram for this heat-recovery system.

TABLE 7.15 Hydrogenation Catalysts From BASF Catalysts LLC

Catalyst	Application	Filterability	Activity	Selectivity	Poison resistant	Versatility	Active metal	Carrier
Nysosel 645	For partially hydrogenated oils with high spreads	Good	Superior	Superior	Acceptable	Acceptable	Ni	Coated in fully hydro fat
Nysosel 950	Similar to Nysosel 645	Superior	Good	Good	Good	Good	Ni	Coated in fully hydro fat
Nysosel 810	Suitable for partial and complete hydrogenation of marine and vegetable oils	Good	Good	Good	Good	Superior	Ni	Coated in fully hydro fat
Nysosel 325	Suitable for partial and complete hydrogenation of vegetable oils	Good	Superior	Superior	Good	Acceptable	Ni	Coated in fully hydro fat
Nysofact 120	Suitable for hydrogenating fatty acids of vegetable, marine, or animal origin	Superior	Good (for fatty acids <200°C) Superior (for fatty acids >200°C)	—	Superior	Acceptable	Ni	Coated in fully hydro fat
Nysosel 210	A sulfur-promoted catalyst suitable for producing steep melting curves in oils; suitable for making cocoa butter replacers	Good	Good	Superior for *trans* isomer formation	Acceptable	Acceptable	Ni	Coated in fully hydro fat

TABLE 7.16 Hydrogenation Catalysts From Johnson Mathey

Catalyst	Application	Activity	Selectivity	Poison resistant	Active Metal	Carrier
Pricat 9910	For partially and fully hydrogenated oils	Superior even in poor-quality oils	Medium	Good	Ni	Coated in fully hydro fat
Pricat 9920	Suitable for polyene-selective, partial hydrogenation for low-*trans* selective reactions	Good	Good	Good	Ni	Coated in fully hydro fat
Pricat 9908	A sulfur-promoted catalyst suitable for producing steep melting curves in nonlauric oils; suitable for making cocoa butter replacers	Good	Good	Good	Ni	Coated in fully hydro fat
Pricat 9936	For partial or complete hydrogenation; can be used in oils with poor quality, high FFA	Superior	Good	Good	Ni	Coated in fully hydro fat

TABLE 7.17 Troubleshooting Hydrogenation Process

Symptoms	Probable cause(s)	Recommended solution(s)
Reaction is slow	• Poor-catalyst activity • Old catalyst that lost its activity • Not enough catalyst added • Poor-quality RB oil	Add more catalyst or change the catalyst
		Use fresh catalyst
		Add more catalyst but determine the reason for the higher dosage
		Check the RB oil for quality and make sure it meets the following quality standards: Soap content 0 ppm Phosphorus Not >1 ppm Moisture Not >0.1% PV Not >2 FFA As low as possible Check refining and bleaching conditions and the crude oil quality if any of the aforementioned impurity is higher than as listed and correct the problem
	• Poor-hydrogen gas quality	Check hydrogen gas quality; the gas must meet the following quality standards: Sulfur compounds Trace Carbon monoxide Trace Carbon dioxide Trace Nitric oxide Trace Nitrous oxide Trace Moisture Trace Verify the operating conditions of the PSA system, gas dryer, and take corrective action if found necessary
	• Agitator impeller blade may have fallen off	Check the amperage reading of the agitator motor; shut down the reactor and repair the agitator
	• There is loss of hydrogen gas through the packing (or seal) of the agitator shaft at the top of the reactor	• Inspect the seal with an explosimeter to check for the gas leak and take appropriate action if the gas leak is found • Check if the agitator shaft is off-center and creating an escape for hydrogen gas • Estimate the theoretical amount of hydrogen required per batch using the estimate of 14.15 $ft.^3$/1000 lb of oil/IV drop (or 883.3 L of hydrogen gas/1000 kg of oil/IV drop)

Melting point is too low at the desired end-point IV	Poor-activity catalyst	Use catalyst with good activity
	• Reaction temperature is higher than the specified target	Check and correct reaction temperature
	• Reactor pressure is too low	Increase reactor pressure
	• Poor agitation	Check and fix agitator
	• Poor-gas dispersion	Inspect hydrogen gas sparger, clean it or replace it if it is plugged
	• Not enough hydrogen gas	Check the gas flow and correct it if necessary
	• Possible cross-contamination in the: • Converter • Drop tank • Cooler	Eliminate the source of cross-contamination by: • Identifying the contaminant via chemical analysis • Make a thorough inspection of the system and clean it of contaminants • Apply measures to prevent such cross-contamination in the future
Filtered oil does not meet the clarity standard	There may be a tear on the screen	Shut down, clean the filter, and repair any tear or replace the torn screen
	Precoating may not have been done properly	Verify the integrity of the precoat as it has been discussed in Chapter 5 on bleaching
	The "O" ring may be damaged or the nozzle at the bottom of the screen or the socket on the manifold is worn and/or damaged	Shut down, clean the filter, check the "O" ring, the nozzle, and the socket and take the necessary action
Slow filtration of the hydrogenated oil	All of the factors causing slow reaction can produce colloidal nickel causing blinding of the screen	Check and correct any situation that may have caused the slow reaction
	The precoating is not done properly causing premature blinding of the screens	Check precoating
	The overall pressure control on the filter is not set correctly or not working correctly	Correct the problem
	Check the filterability characteristics of the catalyst	Change catalyst if necessary to improve filterability
	Excessive reuse of the catalyst that has potentially affected the particle size distribution	Change to fresh or less used catalyst

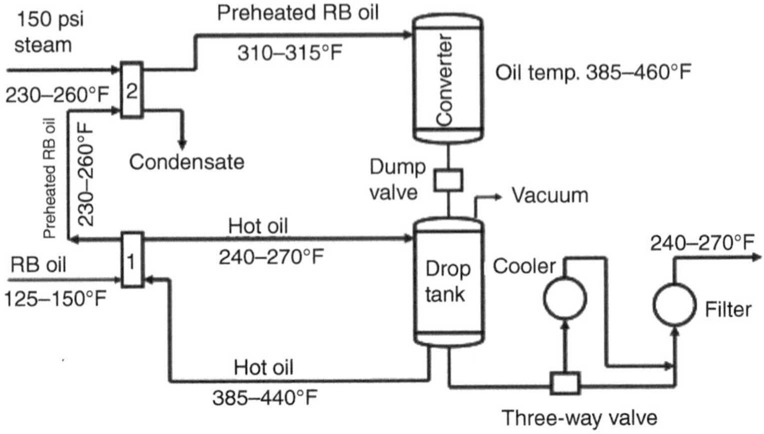

FIGURE 7.13 Schematic diagram for hydroheat recovery system.

1. In this process there are two plate and frame heat exchangers that are used for the heat recovery.
2. These are APV R-57 heat exchangers with Viton Gasket in grooved construction, which can stand up to 440°F. The gasket material is not exposed to the fluid media.
3. The reacted oil is cooled down to 440°F in the converter and then dropped into the drop tank.
4. The hot oil is circulated through the plate and frame heat exchanger.
5. Simultaneously, the RB oil is passed through the same heat exchanger, which gets heated to 230–260°F.
6. The preheated RB oil then passes through the second plate and frame heat exchanger where it is heated to 310–315°F by steam before it enters the converter.
7. The oil loses some heat as it enters the converter because it loses the dissolved moisture due to the vacuum, but it gets heated quickly by the steam coils inside the converter.
8. The reaction begins by the time the converter gets filled.
9. At the end of the reaction the oil is cooled down to 440°F using the internal cooling system in the converter.
10. The oil is dropped into the drop tank as soon as the oil temperature reaches 440°F.
11. The hot oil in the drop tank reaches 240–270°F at the end.
12. The oil from the drop tank is filtered, cooled, and stored or sent to the deodorizer as needed.

Since this work was done, Alfa Laval introduced gasket-less plate and frame heaters. With this heat exchanger no precooling of the hydrogenated oil is needed.

This process achieved a significant amount of heat recovery from the hydrogenated oil and improved the productivity of the plant by 25% without the use of a hot-oil sink.

READING REFERENCES

Allen, R.R., 1967. J. Am. Chem. Soc. 44, 466–467.

Allen, R.R., 1982. Bailey's Industrial Oil and Fat Products, fourth ed., vol. 2 . John Wiley and Sons (Chapter 1).

Allen, R.R., Kiess, A.A., 1956. J. Am. Oil Chem. Soc. 33, 335–359.

Bodman, J.W., James, E.M., Rini, S.J., 1950. In: Markley, K.S. (Ed.), Soybean and Soybean Products. Interscience, NY (Chapter 17).

Buchenal, J.J., 1915. US Patent 1,135,351.

Conen, J.W.E., 1976. J. Am. Oil Chem. Soc. 53, 382–389.

Geuking, W.A., 1999. Latin American Oil Chemists' Society Congress, 24–27 October, 1999.

Gupta, M.K., Farr, W.E., 1994. Inform. 5, 1238–1244.

Hasman, J.M., McLaughlin, P. D., 1994. Presented at the International Canola Short Course at the POS Pilot Plant. 3–8 October, 1994, Saskatoon.

Hoffman, H.P., Green, C.E., 1938. Oil and Soap, pp. 16236–16238.

Kaufmann, H.P., 1935. Studienauf dem Fettegebiet. Verlog Chemie G.M.B.H, Berlin, (pp. 234–251).

Normann, W., 1903. British Patent #1,515.

Richardson, A.S., Knuth, C.A., Milligan, C.H., 1924. Ind. Eng. Chem. 16, 519–522.

Sabatier, P., Catalysis in Organic Chemistry, translated by E. E. Reid. Van Nostrand, New York, 1922.

Tumer, H.V., Fegue, R.O., Cousins, E.R., 1964. J. Am. Oil Chem. Soc. 41, 212–214.

Chapter 8

Deodorization

8.1 INTRODUCTION

Deodorization is the final step in vegetable oil processing in an oil refinery. The next step is product storage, packing, and shipping. Therefore, the product leaving the deodorizer, at the final stage, should be ready for shipment. All finished product standards must be met: otherwise the product will be either reprocessed or might be returned by the industrial users for unsatisfactory quality, or there would be complaints on the consumer brands. In any case, it is a costly affair for the company to respond to these customer complaints. Additionally, persistent customer complaints can lead to long-term loss of customers.

The finished product standards are established by the company based on the following criteria:

1. Typical industry standards for the commodity products.
2. Specific quality standards set by the individual companies to meet the following:
   a. customer needs,
   b. product shelf life (taking product warehousing and distribution into account).
3. To offer the clients superior-quality products compared to the competition.
4. Finished products specified by certain industrial users.

Making good-quality deodorized oil requires not only proper operation of the deodorizer, but goes all the way back to crude oil. The deodorizer cannot correct the quality issues inherent to the crude oil or bad oil quality due to poor oil handling from refining to hydrogenation steps at the plant. Therefore, the plant supervision must pay attention to the entire process if good-quality product is expected out of the plant.

8.2 PURPOSE OF DEODORIZATION

Refined and bleached (RB) oil is steam distilled under high vacuum to make the oil palatable by removing the odoriferous compounds from the deodorizer feed oil. In reality, the process of deodorization performs numerous other functions as listed below:

Practical Guide to Vegetable Oil Processing. http://dx.doi.org/10.1016/B978-1-63067-050-4.00008-8
Copyright © 2017 AOCS Press. Published by Elsevier Inc. All rights reserved.

1. Reduces free fatty acid (FFA) to <0.05%, preferably <0.03%.
2. Reduces the red and yellow color in the refined and bleached (RB) oil and makes it lighter by heat bleaching in the deodorization process which decolorizes the carotenoids at high temperature under vacuum.
3. Removes the odoriferous compounds, such as aldehydes, ketones, hydrocarbons, lactones, alcohol, etc. produced from decomposition of the oils.
4. Reduces peroxide value (PV) to zero. At the same time, the anisidine value (pAV) increases.
5. The oil loses a significant portion of the natural antioxidants (mostly tocopherols and some sterols).
6. Any residual trace metals, picked up by the oil after bleaching, are reduced via citric acid treatment (chelation or scavenging process). This is an essential step and is not a substitute for the bleaching step.
7. There is some increase in the amount of polymers, conjugated dienes, or other oil decomposition products.
8. There can be a very small but detectable increase in the *trans* fatty acid content in the oil, depending on the deodorizer temperature.

8.3 DESCRIPTION OF THE DEODORIZATION PROCESS

Deodorization of refined and bleached oil is carried out under vacuum and at an absolute pressure of 1–6 mm of mercury in the United States, depending on the type of vacuum system on the deodorizer. Modern deodorizers in Europe and the United States are operated at 2–3 mm of mercury or even lower vacuum (1 mm) via a special ejector design to be discussed in Chapter 14. The various steps involved in the deodorization process are outlined as follows:

1. The refined, bleached (RB), and possibly hydrogenated (RBH) oil, meeting all oil quality standards as discussed in Chapter 6 on bleaching and Chapter 7 on hydrogenation, is deaerated at a temperature of 185–195°F (85–90°C) under the same vacuum as the deodorizer.
2. The oil is then heated to a temperature of 480–490°F (249–254°C), under the same vacuum as the deodorizer for heat bleaching, which reduces the red and the yellow color from the carotenes in the oil.

   For palm oil deodorization (not red palm) the temperature can be higher, such as:

   - Heat bleaching 446–509°F (230–265°C).
   - Deodorizing at a temperature as high as 518°F (270°C).
3. The oil is then steam distilled under very low pressure using dry saturated sanitary stripping steam injected at the bottom of the oil bed in the deodorizer. The temperature is normally maintained below 500°F (260°C). At temperatures above this, oils like soybean, sunflower, cottonseed, corn oil, and low linolenic soybean oil polymerize. The oil also exhibits higher levels of *trans* fatty acids after deodorization. However, the higher temperature is needed to deodorize physically refined oils because of the higher FFA content.

4. The combined effect of the low vacuum and the stripping steam produces the bland-tasting light-colored oil, which meets consumer acceptance.
5. The deodorized oil is cooled to less than 290°F (143°C). Fifty percent citric acid solution is added under vacuum before the oil is pumped through an external cooler. Citric acid decomposes at temperature higher than 290°F (143°C) producing a number of compounds that are not effective metal chelators like citric acid. Therefore, addition of citric acid at higher temperatures is not recommended.
6. The oil is cooled down to about 260°F (127°C) inside the deodorizer under vacuum before it is pumped through an external cooler. Below 250°F (121°C), there is some condensation of the steam into the oil. Therefore, the oil temperature is maintained slightly higher 260°F (127°C) as mentioned earlier.
7. The temperature of the oil after the final cooler depends on the type of oil. For example:
   a. Liquid soybean, liquid cottonseed, sunflower, safflower, corn, and palm olein oil must be cooled down to <100°F (38°C) and not exceeding 110°F (43°C).
   b. Palm oil and palm stearine must be cooled down to <120°F (49°C).
   c. Hydrogenated products must be cooled down to a temperature not exceeding 10°F (5°C) above the complete melt point.
8. The cooled oil is stored under nitrogen protection.

8.4 OPERATING PRINCIPLES OF DEODORIZATION

Operating principles for deodorization are complex but can be expressed in a simple equation as shown below:

$$\mathrm{Ln}\left(\frac{X_F}{X_O}\right)=\left(\frac{P\times L}{P_O\times g}\right) \tag{8.1}$$

where X_O = initial mole fraction of the component that needs to be removed, such as FFA, aldehydes, ketones, etc.; X_F = target mole fraction of the same component in the deodorized oil; P_O = vapor pressure of the pure component to be removed (FFA, aldehydes, ketones, etc.); g = moles of stripping steam; P = operating pressure in the deodorizer (mm of mercury); L = moles of oil being deodorized.

8.4.1 Interpretation of the Previous Formula

The object of deodorization is to maximize the removal of the undesirable component. Therefore, the value of X_F must be as low as possible.

In other words, *minimize* the ratio, X_F/X_O.

This also means that the value of the ratio $\frac{PXL}{P_O X_g}$ must be at its minimum.

This can occur only when:

1. The operating pressure, P, is low in the deodorizer (Conversely, the vacuum in the deodorizer is high).
2. The amount of oil, L, is at its minimum.
3. The vapor pressure of the pure component P_O is high (high volatility). This can be attained at higher deodorizer temperature.
4. Moles of stripping steam, g, is at its maximum without affecting the value of P (operating pressure). Higher amounts of g can exceed the vacuum ejector's capacity and increase the value of P. This is because the ejector is designed for the maximum amount of stripping steam in use.

This brings us to the basic premise that the deodorizer must be operated at:

1. The lowest possible operating pressure (or highest vacuum).
2. Maximum operating temperature that can be maintained without damaging the quality.
3. Maximum amount of stripping steam achievable without affecting the vacuum on the system.
4. Minimum amount of oil in the deodorizer without sacrificing the productivity.

8.5 CRITICAL CONTROL POINTS FOR THE DEODORIZING PROCESS

The critical control points for deodorization are:

1. incoming oil quality,
2. deaeration,
3. heating the oil for deodorization,
4. operating pressure (vacuum),
5. operating temperature,
6. amount of stripping steam,
7. batch size or oil flow rate,
8. citric acid addition, and
9. cooling the deodorized oil before storage.

8.5.1 Incoming Oil Quality

This is critical for the deodorizer to process the oil under standard operating conditions and deliver high-quality products. Feed oil quality plays a significant role in determining the performance of the deodorizer. Feed oil (RB) quality to the deodorizer varies in terms of FFA and Lovibond color, depending on the type of oil, but they all must be low in certain trace impurities without exception, such as:

• Phosphorus (in seed oils)	<1 ppm
• Chlorophyll	<30 ppb
• Iron	<0.2 ppm

• Calcium	<0.2 ppm
• Magnesium	<0.2 ppm
• Nickel (from hydrogenation)	Trace (<0.5 PPM)

Note: The feed oil to deodorizer must also have an upper limit as specified by the deodorizer manufacturer.

Palm oil or palm olein will not be able to meet the above standards on phosphorus on a regular basis because the oil is normally refined by physical refining process and the typical phosphorus in the refined oil is 3 ppm and can be higher.

It is important to note that the deodorizer cannot reduce chlorophyll. Therefore, if the feed to the deodorizer is high in chlorophyll, the deodorized oil will remain green after deodorization.

Phosphorus cannot be reduced in the deodorizer either. However, there is another issue with the deodorizer feed that relates to high phosphorus. This involves the following phenomena:

Some amount of hydrolysis of the oil takes place during deodorization. The FFA produced in the oil by the stripping steam is removed immediately in addition to the FFA in the incoming oil. This is why the deodorized oil always has a low FFA. However, if the phosphorus, which is natural emulsifier present in the oil at high level, some additional amount of hydrolysis takes place and the FFA in the deodorized oil can reach a higher than normal equilibrium value seen at lower phosphorus levels in the RB oil. This phenomenon has been observed in deodorizing corn oil or cottonseed oil with high phosphorus content where the deodorizer could not reduce the FFA in the oil below 0.07% (standard is 0.03%–0.05%). Some similar experience has been reported by oil processors handling palm oil, sunflower oil, and soybean oil. High phosphorus content of the oil is most likely the driver in those cases.

The deodorizer vacuum system is designed for a certain amount of condensable (FFA and other volatile oil decomposition products plus the stripping steam) and noncondensable gas (dissolved air) per hour. If the FFA in the deodorizer feed oil is high, the following actions would be necessary in order to reduce the FFA in the oil:

- increase the residence time in the deodorizer,
- increase the deodorizer temperature without damaging the oil, and
- increase both residence time and the deodorizer temperature.

This reduces the deodorizer throughput and increases oil loss in the distillate. The net result is higher cost for the products.

8.5.2 Deaeration of the Oil Before Heating It for Deodorization

The oil must be deaerated before it is heated to heat bleaching or deodorization temperature. The purpose is to remove the dissolved oxygen from the oil to prevent oxidation and formation of oxidative polymers and also to remove the dissolved moisture from the oil. This is accomplished through the following steps:

- The oil is heated to 185–195°F (85–90°C) and then distributed through a distribution ring and nozzles into a deaerator which is maintained at the same vacuum as the deodorizer.
- In a batch deodorizer, the vacuum is pulled before the oil is heated to any higher temperature.
- In a semicontinuous deodorizer, this is done in the deaerator tray maintained at the same vacuum as the deodorizer.
- A continuous deodorizer uses a separate deaerator vessel under full vacuum as the deodorizer.

Deaeration temperature and vacuum are both very critical for the final oil quality. Dissolved air under poor vacuum and at a temperature of 285°F (140°C) or higher produces oxidative polymers in the oil. This process is aggravated if the deodorizer feed oil contains high phosphorus due to poor refining and bleaching. High oxidative polymers have the following negative impacts:

- Produce poor flavor stability in the deodorized oil.
- Cause rapid fouling of the heat bleacher and deodorizer system.

Without the deaeration step, the oil must not be heated further for deodorization because this will seriously oxidize or even scorch the oil and make it unusable.

8.5.3 Heating the Oil for Deodorization

Heating of the oil before deodorization is a very critical step. The oil must be deaerated as mentioned earlier. Seed oils high in polyunsaturated fatty acids can get seriously polymerize at 300°F (149°C) in the presence of air. Most of these are oxidative polymers and are very harmful to oil flavor stability. Even oils like palm oil or palm olein must be deaerated before heat bleaching in order to protect the oil against oxidation. This is why deaeration is critical.

Deaeration of the oil has another important consequence for palm oil or palm olein. Beta carotene and other carotenes in these oils are heat bleached (except for red palm oil) in the heat bleacher and the deodorizer. The presence of oxygen oxidizes these compounds, causing color fixation that cannot be removed in deodorization. The oil remains darker after deodorization. This also increases the yellow color of shortening when measured on the Hunter Color machine.

8.5.4 Operating Pressure (Vacuum)

The process of deodorization is carried out at a low absolute pressure and at high temperature. It must be clear from the discussion in the previous section that the temperature of the oil must be raised in order to increase the vapor pressure of the components, such as FFA and other volatile compounds. The absolute pressure in the deodorizer must be lower than the combined vapor pressure of these components so they will distill from the oil.

Low operating pressure (vacuum) in the deodorizer must be maintained at all times. A higher than normal operating pressure (lower vacuum) reduces the ability of the deodorizer to remove the odoriferous compounds from the oil. Poor vacuum in the deodorizer can arise from:

1. poor performance of the vacuum ejector,
2. a possible air leak in the system, and
3. higher than designed amount of stripping steam.

In either case, there is more air in the system that can oxidize and polymerize the oil. Using excess amounts of stripping steam may exceed the vacuum system's capacity for the condensable vapor. This can result in poor vacuum and, hence, produce poor-quality oil because of the overloaded vacuum system.

Therefore, it is recommended that the vacuum be maintained at or below the maximum operating pressure recommended by the manufacturer of the deodorizer.

8.5.5 Operating Temperature

As stated earlier, the temperature of the oil must be sufficiently high to volatilize compounds like FFA, aldehydes, ketones, and other volatile oil decomposition products in order to obtain good flavor in the deodorized oil.

On the other hand, very high deodorizer temperature can oxidize and polymerize the oil. Therefore, the oil must always be deodorized at the minimum possible temperature that produces the best oil quality, including flavor. Sometimes, it is necessary to increase the deodorizer temperature in order to remove the odoriferous compounds from the oil. One must always remember that this temperature must never be excessively high. In such cases it is advised that the plant supervisor check the quality and the history of the deodorizer feed stock and take corrective action to fix it because the deodorizer cannot and should not be used to correct any fundamental issue with the oil quality prior to deodorization.

The typical operating pressure and temperature conditions for deodorization are shown in Table 8.1. One can see in Table 8.1 that the oil temperature used for deodorization in Europe is lower than that in the United States and, at the same time, deodorization in Europe takes nearly twice as long compared to that in the United States. Deodorization time for physical refining in Europe is nearly 3 times longer than that used for chemically refined oils in the United States and twice as long for chemically refined oils than that used in Europe for the same.

8.5.6 Amount of Stripping Steam

The stripping steam is injected into the oil in the deodorizer. This serves several purposes:

TABLE 8.1 Typical Deodorizer Conditions

Deodorizer condition	Chemically refined oil (USA)	Chemically refined oil (Europe)	Physically refined oil (Europe)
Temperature	460–500°F 238–260°C	440–460°F 230–240°C	440–480°F 230–250°C
Pressure mm of mercury	3–6	2–3	2
Stripping steam, (% of oil) (2 Torr pressure)	0.5–2	0.5–1	1–2
Deodorization time (min)	20–30	40–60	60–90
FFA (%)	0.03–0.05	0.03–0.05	0.03–0.05

Source: From the Presentation of Wim De Greyt, De Smet, Tunis, 2004.

- It creates agitation in the oil, which helps remove the volatile matters from the oil.
- The steam expands under the reduced pressure, which increases the specific volume of the steam many times, increasing the specific surface area. This enhances the contact between steam, oil, and the volatile components in oil.
- Owing to the expanded volume, the steam can remove the volatile matter more effectively.

In deodorizing, the FFA is removed quite readily. It takes additional time to reduce the red color and distill some of the odoriferous compounds.

Vaporization of the volatiles is highest at the surface of the oil. Therefore it is very important that the entire mass of the oil reach the surface as many times as possible during deodorization to maximize the vaporization process. This is why it is essential to inject stripping steam into the oil through a sparge ring at the bottom of the oil with a mammoth pump (an eductor) to circulate the oil to bring all of the oil to the surface. Eductor hats are also used to enhance the effect of splashing of the oil with the stripping steam.

The amount of stripping steam varies with the type of deodorizer. For example:

- A batch deodorizer requires 3%–4% steam in older designs; newer designs uses <2% steam.
- A semicontinuous deodorizer uses 1%–2% steam.
- A continuous deodorizer uses 0.25%–1.0% steam.
- A thin film deodorizer requires 0.3%–0.6% steam.

8.5.7 Batch Size or Flow Rate

Batch size in a batch system is determined by the production requirements. However, building a batch deodorizer larger than 45,000 pounds (20,450 kg) is not practical.

Eq. 8.1 and the subsequent discussions in Section 8.5 indicate that the amount of oil in the deodorizer must be at its minimum in order to achieve maximum reduction of the volatile matter in the oil. Given that all conditions remain within the operating limits, a batch deodorizer must not be overfilled, or the semicontinuous or a continuous deodorizer must not be operated at higher than the designed flow rate.

8.5.8 Citric Acid Addition

Citric acid is added to the deodorized oil at the cooling stage of the deodorization process. Citric acid acts as a chelating agent to complex with trace metals like iron, calcium, and magnesium. The following are the required conditions for citric acid addition:

- Citric acid dosage 50 ppm (of the oil).
- Citric acid addition temperature <290°F (143.3°C).
- Deodorizer condition under normal vacuum with stripping steam.

At higher temperature, citric acid decomposes, leaving very little or no beneficial effect on the oil.

Some commercial antioxidants containing citric acid are used by many manufacturers. This makes it easier for the oil processors to add these additives into the oil. However, the majority of the processors do not use any high-shear mixing. Some even add the additives in the trucks and argue that the agitation in the truck during transit disperses the additives into the oil. The author finds this to be "wishful thinking," especially when it comes to mixing two immiscible liquids in such disproportionate ratios to be uniformly distributed in the oil from the movement of the truck. A properly designed citric acid addition system is highly recommended for all deodorization operations.

8.5.9 Cooling Deodorized Oil

The deodorized oil must be cooled before storage. This should also be done with care and taking into consideration of the type of oil being processed. Oils high in polyunsaturated fatty acids must be cooled down to a temperature of 300–375°F (149–190°C) under vacuum and with stripping steam, in order to prevent the formation of undesirable flavor in the oil.

Cooling the oil directly through an external cooler may produce unacceptable flavor in the oil. On the other hand, using stripping steam at a temperature below 250°F (121.1°C) in the deodorizer may increase moisture content in the deodorized oil.

Citric acid addition to the oil in the deodorizer makes it necessary to cool the oil as discussed earlier.

The recommended temperature for deodorized oil before storage is discussed in Section 8.3.

This step is carried out in external coolers under the positive pressure of the deodorizer discharge pump. A sealless pump is better suited for this step in order to prevent any air leakage into the oil during cooling before it is stored.

8.6 DEODORIZED OIL QUALITY

Deodorized oil is evaluated for certain physical and chemical attributes as well as for organoleptic acceptability.

8.6.1 Physical Attributes

The deodorized oil must have the following characteristics:

- odorless,
- have a clean taste with no unpleasant aftertaste,
- must be light in color, and
- Lovibond color must meet the company standards.

8.6.2 Chemical Attributes

The deodorized oil must meet the chemical standards as shown in Table 8.2.

8.6.3 Organoleptic Attribute—AOCS Method Cg-2-83 (09)

Chemical attributes, such as PV, pAV, conjugated dienes, AOM, and OSI are good tests for the oil but they do not always correlate to the flavor of the freshly deodorized oil or oil in storage. This is because the deodorized oil may contain some volatile compounds at less than a 1 ppb level that still can affect the flavor but cannot be detected in the standard analysis of the deodorized oil. The oil may exhibit a slight off flavor with bitter aftertaste in the fresh oil when the polymer level is high. Unacceptable flavor in the oil becomes noticeable after storage. Product made with the questionable-flavor oil may have acceptable flavor initially, but it deteriorates rapidly during storage.

Therefore, testing flavor of the oil is an important part of determining oil quality, and it is essential that the plant personnel are trained to do so and they also conduct some storage stability tests.

8.6.4 Significance of the Deodorized Oil Quality Standards

It would be appropriate to go over the deodorized oil quality and discuss the significance of each of the quality standards in terms of deodorizer performance or other factors that could be driving these specific quality parameters. This is discussed in Table 8.3.

8.7 TYPES OF DEODORIZERS

There are several deodorizer manufacturers, but all deodorizers fall under the following three categories:

TABLE 8.2 Desired Quality Standards for the Fresh Refined Bleached and Deodorized (RBD) Oil

Analysis	Desired standard	Maximum standard	Am Oil Chemists' Society Method (AOCS)
FFA (%)[a]	0.03	0.05	Ca 5a-40 (12)
PV (mEq/kg)	0	<0.5	Cd-8b-90 (11)
para Anisidine value, pAV (AVU)	<4	6	Cd-18-90 (11)
Conjugated dienes (%)	Trace	<0.5	Ti-1a-64 (09)
AOM (h)	Depends on oil type	—	Cd-12-57 (93)
OSI (h)	Depends on oil type	—	Cd-12b-92 (13)
Polymers	<0.5	<1	Cd-22-91 (09)
Phosphorus, P (ppm)[b]	<0.5	<1.0	Ca-20-99 (09)-ICP
Iron, Fe (ppm)	<0.2	<0.5	Ca-17-01 (09)-ICP Ca-18-79 (09)-AA
Calcium, Ca (ppm)	<0.2	<0.5	Ca-17-01 (09)-ICP Ca-18-79 (09)-AA
Magnesium, Mg (ppm)	Trace	<0.5	Ca-17-01 (09)-ICP Ca-18-79 (09)-AA
Monoglyceride (%)	Nondetectable	Trace	Cd-11b-91(09)-capillary GLC Cd-11b-96(09)–HPLC-ELSD
Diglyceride (%)[c]	<0.5	<1.0	Cd-11b-91(09)-capillary GLC Cd-11b-96 (09)–HPLC-ELSD

These numbers can be significantly lower in palm oil and palm olein processed by some companies in Malaysia.
[a]*Can be up to 0.5% in many commercial palm oil or palm olein.*
[b]*Can be 3–5 ppm in most commercial palm oil or palm olein.*
[c]*Can be 3%–10% in most commercial palm oil or palm olein.*

1. batch
2. semicontinuous
3. continuous

8.7.1 Batch Deodorizers

Batch deodorizers were originally introduced in the oil industry during the middle of the 19th century and the actual units were used around the 1890s. In these units, the oil (especially coconut oil) was heated and direct steam

TABLE 8.3 Significance of the Deodorized Oil Quality Standards

Quality attribute	Desired standard	Maximum standard	Discussion
FFA (%)	0.03	—	• It is possible to obtain <0.02% FFA in the deodorized oil routinely. • It is not possible to obtain <0.005% FFA because this is the equilibrium FFA in deodorization with steam stripping.
FFA (%)	—	0.05%	A higher FFA content may imply any or all of the following: • Very high FFA in the deodorizer feed. • Vacuum, temperature, stripping steam, or other issues with the deodorizer. • High level of phosphorus in the deodorizer feed. • Chances are that the high FFA oil will have shorter shelf life because of the presence of other possible impurities left behind in the oil.
PV	—	0.5	• A higher PV value indicates poor deodorizer vacuum.
pAV	—	6	• A higher pAV value indicates high PV in the deodorizer feed or poor deodorizer vacuum.
Conjugated dienes (%)	—	0.5	• A higher conjugated diene value implies oxidation of the oil, before, during, or after deodorization. • Generally it indicates poor handling of the oil in process.
Polymers (%)	—	1	A higher polymer value implies: • Heating of the oil in presence of air forming oxidative polymers. • Poor deodorizer vacuum forming oxidative polymers. • Very high deodorizer temperature even under good vacuum forming thermal polymers.
Phosphorus (ppm)	—	1	High phosphorus in the deodorized oil implies: • Crude oil was obtained from heat and moisture damaged oilseeds. • Inadequate caustic refining. • Incomplete bleaching of the oil.

Iron (ppm)	—	0.5	A higher iron content in the deodorized oil implies: • Poor or incomplete bleaching. • Citric acid addition system in deodorizing is malfunctioning.
Calcium (ppm)	—	0.5	A higher calcium level in the deodorized oil implies: • Crude oil obtained from poor quality oilseeds. • Poor or incomplete refining and bleaching.
Magnesium (ppm)	—	0.5	Same as for high calcium.
Monoglyceride (%)	—	Trace	High level of monoglyceride in seed oil implies: • Lipase activity during seed storage at high temperature and humidity. • Caustic overdose in refining. • Use of stronger than the recommended strength of caustic in refining.
Diglyceride (%)	—	1.0	Same as for high monoglyceride.

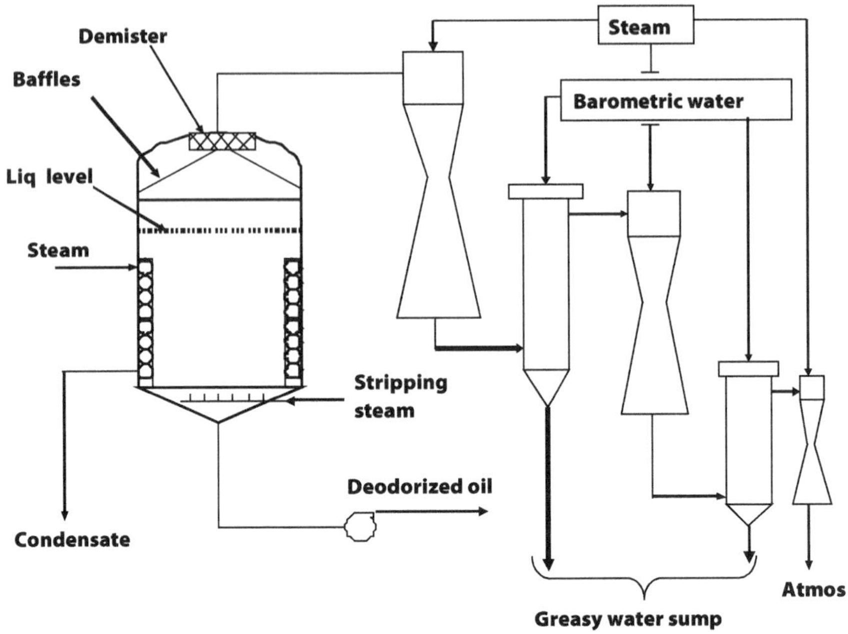

FIGURE 8.1 **Schematic diagram for Batch Deodorizer.**

sparge was used. The batch deodorizer with a vacuum system and superheated steam stripping was introduced in the United States in the late 1800s. The first high-temperature batch deodorizer with a vacuum system and made from nonoxidizing material was developed in France in the early 1900s. Several design modifications have been made to batch deodorizers since that time and are still used in many oil-processing plants around the world. Some modern plants are using batch deodorizers made from stainless steel for vegetable oils and specialty products. A schematic diagram for a batch deodorizer is shown in Fig. 8.1.

8.7.2 Typical Operating Steps in a Batch Deodorizer

In this process, the oil is deodorized in several stages as outlined:

1. The vacuum is turned on.
2. A batch of oil is charged into the deodorizer through a distributor ring with nozzles.
3. The oil is heated by using either steam or thermal fluid.
4. The steam is at a pressure of 450 psi (32.7 kg/cm^2).
5. The deodorizer oil temperature should be set at 250°F (121.1°C) for deaeration.
6. The deaeration step can take 30 min or a little longer.
7. A small amount of stripping steam is injected to agitate the oil for deaeration.

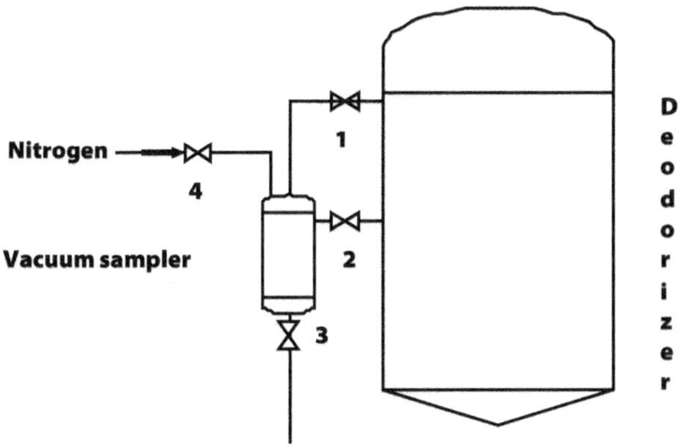

FIGURE 8.2 **Vacuum oil sampler.**

8. The volume of the stripping steam is increased after deaeration and as the oil temperature rises. The amount of stripping steam at this time is generally 3% (2%–4% in the older designs and <2% in the newer designs).
9. Deodorizer temperature is maintained at 460–480°F (238–249°C).
10. The oil is deodorized for a predetermined amount of time, which is established from prior experience. The end point is established from the satisfactory quality of the deodorized oil.
11. The oil is cooled down to <290°F (143°C) and 50 ppm of citric acid is added into the oil to chelate (scavenge) trace metals, such as iron, copper, etc.
12. A deodorizer sample is collected by using a specially designed sampler as shown in Fig. 8.2.
13. Stripping steam is shut off when the oil temperature reaches 250°F (121.1°C) otherwise there will be condensation of steam in the deodorized oil if the stripping steam is left on while the oil temperature drops below 250°F (121.1°C).
14. The oil is cooled down in an external cooler and handled as described in Section 8.3.
15. Antioxidant is added, sometimes before the oil is discharged from the deodorizer.
16. Alternatively the antioxidant can be added as the oil is transferred to storage. An inline mixer, such as a static mixer can be used. However, a high shear mixer is recommended.

8.7.3 Vacuum Sampler

The schematic diagram for the vacuum sampler is shown in Fig. 8.2. It is a very small pressure vessel connected to the side of the deodorizer as shown. There are four connections to the sampler. The step-by-step operation is described as follows:

1. Initially check that all four valves are closed.
2. Using thermally protected gloves and face shield, open valve #3 with a bucket underneath to collect any oil residue from the vacuum sampler.
3. Leave valve #3 open.
4. Slowly open valve #4 and let some nitrogen flow through the sampler.
5. Close both valves #3 and #4.
6. Open valve #1 to bring the sampler to the same vacuum as the deodorizer. It will take only a second or two to equalize the vacuum between the two vessels.
7. Close valve #1.
8. Open valve #2 for a few seconds to allow the oil from the deodorizer to flow into the sampler.
9. Close valve #2.
10. Crack open valve #4 very slowly to let nitrogen flow into the sampler to break the vacuum, and then close it.
11. Hold a thermally insulated safe sample collector at the bottom of the drain line.
12. Slowly open valve #3 and very slowly crack open valve #4 again.
13. Collect the oil in a safe sample collector.
14. Close valves #4 and #2.
15. Chill the oil immediately in the laboratory.
16. Bubble nitrogen through it for 5–10 min while it is being chilled.
17. Analyze the sample.
18. Make sure all four valves on the vacuum samplers are fully closed after collecting the sample.

8.7.4 Semicontinuous Deodorizer

The semicontinuous deodorizer was introduced in 1948 in the United States by Girdler (subsequently Votator and now Desmet Ballestra). The original deodorizer was designed by A. E. Bailey. This provided the vegetable oil industry a large boost in productivity. The design of a semicontinuous deodorizer includes some intriguing features as discussed here.

There are five trays (in the original design) or six trays (in the newer design) that are vertically stacked. These trays perform several functions, such as:

- deaeration
- heating
- deodorizing
- cooling

A schematic drawing of the modern Desmet Ballestra semicontinuous deodorizer is shown in Fig. 8.3.

1. Oil is measured and deaerated in an external measuring tank using the same vacuum system as that of the deodorizer.

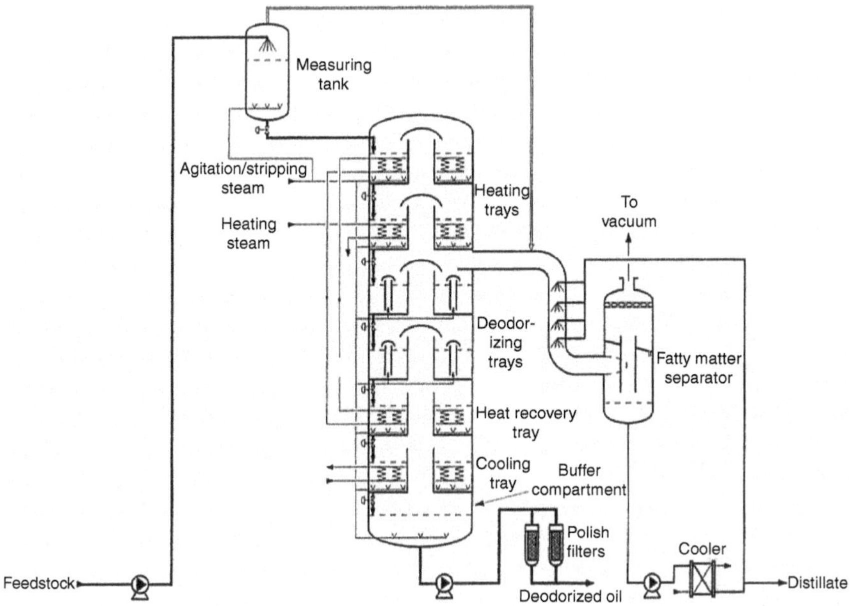

FIGURE 8.3 Schematic diagram for semicontinuous deodorizer.

2. The oil is automatically discharged into the first tray when it goes empty. Here the oil is heated further with steam stripping. The typical heating coils used in a semicontinuous deodorizer are shown in Fig. 8.4.
3. The oil drains into the second tray, which is also used for heating and heat bleaching.
4. The oil drains into the third tray. The oil drains into the next tray after the predetermined residence time.
5. The third and the fourth trays are for deodorizing. One can see the mammoth pumps for steam stripping and oil recirculation.
6. The fifth tray is used for heat recovery from the hot oil by the oil in the first tray.
7. The final tray is used for precooling the oil, and for the addition of citric acid, and possibly other additives.
8. The oil drops to the buffer tank at the bottom of the vessel before it is pumped out of the deodorizer.
9. The oil is cooled to a temperature as discussed in Section 8.3.
10. The oil is saturated with nitrogen and stored under nitrogen protection, to be discussed later.
11. The oil passes through a polish filter before it goes to through the external cooler.
12. The vapors leaving the deodorizer enter the indirect-contact type condensing ejector, where the volatiles, such as fatty acids and the entrained triglycerides are captured in liquid form.

FIGURE 8.4 **Internal coils for heating and cooling in trays.**

13. The fatty acids are collected at the bottom of the fatty matter separator, cooled through an external cooler, and then returned to the vapor line coming from the deodorizer to make the initial contact with the vapor before the separator.
14. The uncondensed vapor leaves the separator at the top and goes to the vacuum system.
15. Most modern ejectors have four stages. Some combination of liquid ring vacuum pumps (or dry vacuum pumps) and steam ejectors is used. Liquid ring vacuum pumps cannot deliver the very low pressure required by modern ejectors on a reliable basis.
16. The fatty acids are collected in stainless steel tanks and sold as a by-product.
17. Dowtherm was used as the heating medium. This required a Dowtherm boiler for each deodorizer. The Dowtherm was heated by natural gas either in a liquid boiler or in a Dowtherm vaporizer. Dowtherm vaporizers are still in use, except the new thermal medium is different.
18. Steam spargers are located at the bottom of the coils.

8.7.5 Advantages of Semicontinuous Deodorizers

There are several advantages of this type of deodorizer over either batch or continuous deodorizers. The primary advantages are:

1. Higher productivity than the batch deodorizers.
2. The stock changes do not pose any serious threats of cross-contamination of the products, unless there is a breakdown of the control system or there is a human error.
3. Uses much less energy than the batch deodorizer.
4. Deodorized oil quality is good.
5. Offers good opportunity for energy recovery. Can recover 40%–65% of the total energy.

The older Votator design had a double shell. The trays were placed in the inner shell. The idea was to protect the oil from any air leak from the bottom of the deodorizer. In case of a bottom air leak the air would not come in contact with the oil inside the trays. The disadvantage was that the deodorizer had a tendency to accumulate oil at the bottom of the outer shell, which was referred to as shell-drain. The outer shell in the older design deodorizers was made of carbon steel. This made the shell-drain very dark from the reaction with the fatty acids in the oil and the carbon steel of the shell. Plants experienced leaks in the outer-shells quite often, requiring expensive shutdowns and repairs. Later, the design was replaced by Votator with a single-shell, stainless steel construction, which was further improved by Desmet Ballestra as shown in Fig. 8.3.

A deodorizer design was introduced by Desmet Ballestra in early 1970s called a multiple stock deodorizer (MTD). This is a unique concept in the sense that the oil is deodorized in six semicontinuous individual deodorizers built inside one shell looking almost like a giant batch deodorizer. The oil actually moved continuously through the cells. This deodorizer did not receive widespread acceptance in the United States, primarily because of the advancements that were made to semicontinuous deodorizers during the same time frame and later. In addition, the MTD design was complex and certain areas were hard to reach for repairs.

8.7.6 Continuous Deodorizers

Continuous deodorizers were introduced to the oil industry much later. Continuous deodorizers from several manufacturers are available. They differ in their physical features, mode of internal oil flow, type of packing, or trays, but the basic approach is the same for all of these designs. The oil is continuously fed to the deodorizer with deaeration, and internally they are provided with the capabilities for heat bleaching and deodorization to produce good-quality oils. There are minor as well as some significant differences in physical designs between the manufacturers, but the end results are always good when proper deodorization procedures are followed and the deodorizer feed oil is of good quality. The latest innovation in the design for continuous deodorizers utilizes the principle of maximum interface between the oil and the stripping steam. Two varieties have been well recognized in vegetable oil processing, namely:

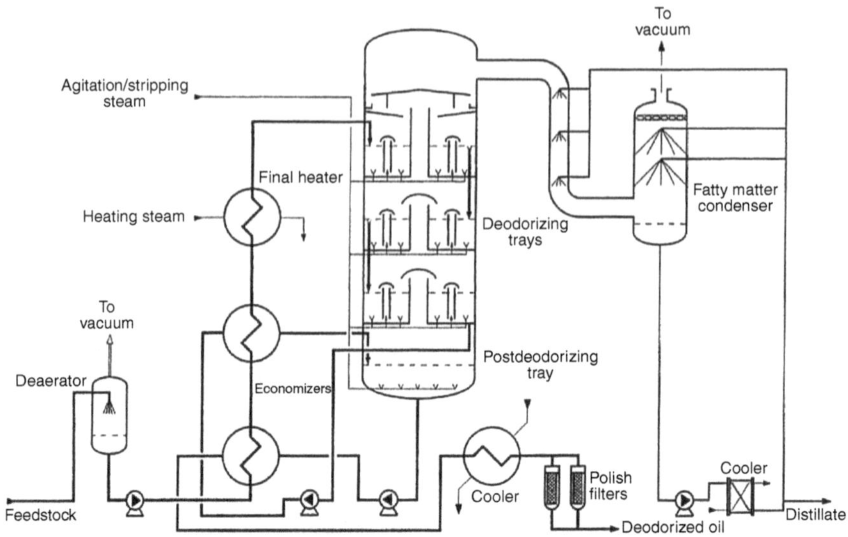

FIGURE 8.5 Schematic diagram for continuous deodorizer.

1. short path distillation by Alfa Laval and
2. packed column deodorizers made by several manufacturers.

Fig. 8.5 shows the schematic diagram for one of the continuous deodorizers from Crown Iron Works. This is a single-shell construction with the following features:

1. The oil is deaerated in an external deaerator.
2. Deaerated oil is then pumped through economizers using the hot oil from the deodorizer.
3. A steam heater then brings the oil to the deodorization temperature.
4. There are three trays in this deodorizer stacked vertically.
5. The hot deaerated oil enters the top tray and goes down the next two while getting heat bleached and deodorized with the help of stripping steam.
6. The hot deodorized oil leaving the last tray is passed through one of the two economizers to preheat the deaerated feed.
7. The hot oil returns to the bottom of the deodorizer shell, which is also referred to as the postdeodorizing tray. Here the oil undergoes further deodorization under vacuum with stripping steam.
8. The oil from the postdeodorizing tray is pumped through the economizer which first heats the deaerated oil. Then it is further cooled, filtered through polish filters, and stored under nitrogen.
9. It is highly recommended that an inline gas diffuser be used to inject nitrogen gas into the deodorized oil as it leaves the deodorizer.
10. The combination of nitrogen diffusion and nitrogen blanket in the storage tank protects the oil from oxidation.

8.7.7 Advantages of Continuous Deodorizers

The continuous deodorizer offers several advantages over batch or semicontinuous deodorizers. Some of the advantages are listed as follows:

1. The production rate is the highest among all three types.
2. Heat recovery is very high through the use of external heat economizers. Over 80% of the total heat in the oil can be recovered.
3. The unit cost of production is lowest among all three types of deodorizers.

8.7.8 Disadvantages

The continuous deodorizer is not for frequent stock changes. This is because the system consists of a number of heat economizers that retain a substantial volume of oil. This residual oil gets mixed in with the oil that follows in the process due to a high degree of back-mixing in the system. It takes a long time to get pure stock to come out of the deodorizer after any product changeover. Fig. 8.6 shows the extent of back-mixing that can occur during product changeover in a continuous deodorizer.

In this example the deodorizer was processing Product A. Product B was pumped behind Product A and samples were collected every 5 min at the exit of the deodorizer. The samples were analyzed and the amounts of product A and B were determined in every sample and were plotted in Fig. 8.6.

One can see that the oil leaving the deodorizer was 100% Product A for up to 20 min. During the same period, the concentration of Product B in the sample was 0%.

Product leaving the deodorizer at 50 min was made of 50% Product A and 50% Product B. It took 90 min for the oil at the end of the final oil cooler to reach 100% Product B.

Thus, one could continue to send Product A into its original storage tank for 20 min and then divert the deodorizer to a divert tank for another 70 min.

If the two products are not compatible, then the entire 70 min of production would have to be used in formulating some other product. In addition, the other

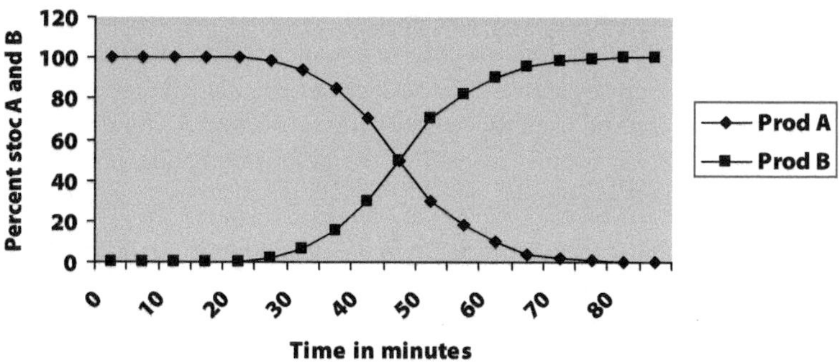

FIGURE 8.6 Cross blending during product changeover in continuous deodorizer.

product will have to be formulated and deodorized. This would add cost to the new formulated product.

One could follow Product A with a compatible product and allow a certain amount of cross-mixing between the two products and not have to formulate a third or a fourth product.

Sometimes, the first product (Product A) may be required to meet a certain cold test. In that case the second product (Product B) cannot have any solid component. In this case the deodorizer system may be required to be flushed with liquid oil before Product B can be introduced. This means that there could be potentially two intermixed stocks on hand that will have to be disposed of in the proper manner. In addition, the liquid oil must be compatible with both Product A and B in order to meet the regulatory requirements for labeling.

The time for stock change can be reduced somewhat by the following means:

1. Select heat exchangers with minimum oil holding capacity.
2. Use an internal packing design that can minimize the back-mixing of stocks.
3. Optimize the plant layout, locations of the heat exchangers, and economizers relative to the deodorizer.

It is clear from the previous discussions that the deodorizer system must be evacuated well at shut down for sanitation; otherwise there can be high oil loss. The heat economizers must be installed in such a manner that the oil can be drained well before introducing the caustic solution.

Therefore, the disadvantages can be summed as follows:

1. Less flexible for stock change.
2. High level of cross-contamination at stock changeover.
3. Potential for high oil loss during sanitation.

8.7.9 Residence Time Distribution in a Continuous Deodorizer

This is a way to establish the product changeover time for a deodorizer. This must be done as soon as a new continuous deodorizer is installed. The deodorizer manufacturer cannot provide the needed information without knowing the residence time in the entire system and the estimated oil held in the system. The amount of the interface is a function of the actual installation of the system. It is essential for the plant to establish the residence time distribution in a continuous deodorizer if it is to be used for deodorizing more than a single stock or two incompatible stocks. A simple procedure for determining the residence time distribution is outlined as follows:

1. A stable oil-soluble tracer is injected into the oil stream at the deodorizer feed.
2. A sample collected at the regular sample port is and analyzed for the tracer.
3. The tracer will not appear in the sample initially, but then it will appear.
4. The concentration of the tracer will go through a peak and then eventually it will reach zero.

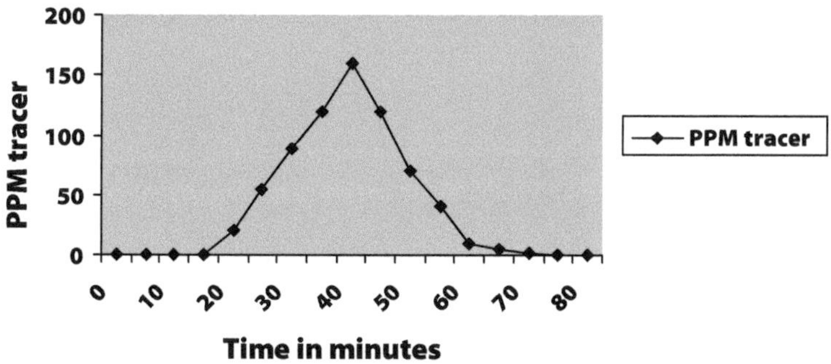

FIGURE 8.7 **Residence time distribution.**

5. Fig. 8.7 shows a plot developed by using chlorophyll as a tracer.
6. Using the flow rate through the deodorizer, one can calculate the amount of the intermixed product.
7. In this case, the amount of the intermixed product was:
   a. The rate of deodorizer flow in pounds per minute × (80 − 17).
   b. At 45,000 lb/h = (45,000/60) × 63 = 47,250 lb.
   c. At least this much intermixed product will be generated in this particular deodorizer at every product changeover.

Alfa Laval's soft column design has overcome this issue. Unlike all other random-packing packed column deodorizers, the soft column deodorizer can be drained quickly and completely to have a clean changeover. However, one must drain all the heat economizers and heat exchangers at the same time to minimize cross-contamination. This, of course, does not fully address the issue with the cold test for the oil if it has to follow a previous stock containing solids.

In this deodorizer, the deaerated and heated oil trickles down the packed column, which offers very high surface contact between the oil and steam a facilitating removal of the volatiles. The oil accumulates in the retention vessel under vacuum to complete heat bleaching of the oil.

This deodorizer received the attention of the palm oil industry. The seed oil processors were somewhat apprehensive of the short distillation time and long retention thereafter. It took somewhat longer for the seed oil processors to overcome their apprehension over the design.

It has been stated earlier that most of the back-mixing occurs from the heat exchangers and the random packing. Since random packing does not occur in this deodorizer, a fair amount of back-mixing may still exist in the heat exchangers. The residence time distribution in this unit should be studied because back-mixing may not be permissible between all products.

A comparison between the three types of deodorizers—batch, semicontinuous, and continuous—is shown in Table 8.4. The discussion in Table 8.4 highlights the distinctiveness of each type of deodorizer.

TABLE 8.4 Comparison Between the Three Groups of Deodorizers

Item	Batch	Semicontinuous	Continuous
Suitability	Suitable for: • Discrete batches of product • Specialty products • Can be used for making emulsifiers and interesterified products	• Suitable where frequent product changeover is needed • Not suitable for making emulsifier or interesterified products	Suitable for: • Continuous production of large volume of product with minimum number of product changeovers • Not suitable for making emulsifier or interesterified products
Production size	Can be any size but it becomes uneconomical over batch size of 6000 US gallons	Can be any size but not less than the minimum volume of oil needed to cover the heating or the cooling coils in the trays	Usually large volume of production, 20,000 US gallons or larger batch size for minimum cross-contamination
Stripping steam Usage [Pressure (Pr)—2 Torr]	• High, 3%–4% of the batch size by weight (old design) • 1%–2% (new design)	1.25%–1.5% of the oil flow by weight	• 0.8%–1.2% of the oil flow • 0.0.4%–0.7% (for packed column)
Production rate	Low, normally 7–8 h per batch: regardless of the batch size	Normally 4–6 times of that of the batch deodorizer	Normally 5–8 times of that of the batch deodorizer
Energy recovery	• Normally none • Can be 25%–35% through heat economizer	• Relatively high • Can be up to 65% of the total heat input by using internal and external heat recovery	• High • Can be up to 85%–90% of the total heat input by using external heat economizers
Cost of deodorizing per number or kilogram	• Very high	• Normally 30%–40% of the batch process	Normally 10%–20% of the batch process
Risk of product cross-contamination	• Low risk	• Relatively low but it is higher than the batch process	• High risk
Versatility for use	Versatile. Can be used for: • Deodorizing • Making emulsifiers • Making interesterified products	Not versatile. Can be used solely for deodorization	Not versatile. Can be used solely for deodorization

Table 8.5 is devoted to quality issues and troubleshooting of deodorizer operation. These are some of the issues that the plant may experience from time to time during operation. The author believes that this can help the readers identify the cause of oil quality issues and implement solutions in their operation.

In this chapter, the author did not include the pictures of many deodorizers. The names of most of the deodorizer manufacturers are listed as follows for reference.

Company	Location (Country)
Crown Iron Works Co.	USA
S. A. Extraction Desmet Ballestra N.V.	Belgium
Lurgi GmbH	Germany
Masiero Industrial S.A.	Brazil
Andreotti Impianti SpA	Italy
Harburg Freudenberger GNBH	Germany
Oiltek Sdn. Berhad	Malaysia
Alfa Laval	Sweden
LurgiGmbH	Germany
Fratelli Gianzza SpA	Italy
Lipico	Singapore

8.8 VACUUM SYSTEM FOR DEODORIZER

The importance of high vacuum (low operating pressure) for deodorization has been discussed in various sections in this chapter and in detail in Chapter 15.

The latest trend in vacuum ejector design is the freeze condensing unit which significantly reduces the size of the steam ejector and steam consumption. This system is discussed in Chapter 15.

The surface condensing vacuum ejector system was common in Europe before it expanded to the United States. This system also complements the physical refining (see Chapter 5) because palm oil, palm olein, coconut oil, and palm kernel oil contain high FFA, and physical refining is most suited for these oils. A vacuum system with a surface condenser or freeze-condenser is very appropriate for the process. Although both surface condensing and freeze-condensing are discussed in detail in Chapter 15, a surface condensing design using a cleaning technique, offered by EquirepSA of Spain, has some unique capability. Fig. 8.8 shows the schematic diagram for the EquirepSA condensing system with plate heaters.

The system was developed to reduce the odor produced by the greasy water. Cold process water from the cooling tower cools the vapor, which travels down the tubes inside the surface condensers. The fatty material can accumulate inside the tubes and obstruct the flow of vapor, causing poor vacuum. With a specially designed spray device a small amount of dilute caustic (12.5% NaOH) solution is introduced at the top of the tubes near the tube plate, which goes down the tube and forms a thin film of soap. This keeps the tube surface clean.

The normal pH of the greasy water is around 6. The soap maintains the pH in the greasy water at 11. The condensed water from the vapor dilutes the soap. More caustic solution is automatically sprayed into the tubes at the top.

TABLE 8.5 Trouble Shooting Deodorizer Process

Symptom	Probable cause	Recommended solution
FFA > 0.05%	• High absolute pressure in the deodorizer (poor vacuum). • Low oil temperature. • Oil flow rate is high. • Not enough stripping steam. • Poor distribution of stripping steam. • Too much sparge steam affecting the vacuum. • Incoming oil has high FFA.	• Correct ejector problem as outlined in Table 14.4. • Increase the oil temperature by 5°F (2.5°C) and check FFA. • Reduce oil flow rate. • Increase stripping steam flow. Make sure it does not raise the operating pressure in the deodorizer (reducing the vacuum). • Clean steam sparger if the steam distribution appears uneven. • Reduce sparge steam flow. • Investigate the cause for high FFA in the deodorizer feed and correct it. Increase deodorization time by reducing the size of the batch in a batch deodorizer or increase residence time in a semicontinuous deodorizer or reduce flow rate in a continuous deodorizer.
Freshly deodorized oil has PV > 0.	Poor vacuum due to: • Ejector issue. • Air leak into the deodorizer.	Identify the cause for poor vacuum. • Correct vacuum ejector issue. • Identify the source of air leak and correct it.
Deodorized oil is too dark.	• Deodorizer feed oil has poor quality and contains one or more of the following impurities: 1 High soap. 2 High phosphorus. 3 High iron or nickel. • Old or abused incoming oil with dark color. • No or low citric acid in the deodorized oil. • Reflux of the distillate into the batch deodorizer due to high level in the catchall. • Deodorizer coil is not fully covered by the oil in a batch or a semicontinuous deodorizer or in continuous deodorizers where heating coils are used. • Deodorizer temperature is too high. • Deodorized oil is not cooled properly in the deodorizer and also through the external cooler to the correct temperature.	• Correct the problem in the RB oil, whether it is in the refining, or in the bleaching step. • Rectify the issue. • Check and make sure that the correct amount of citric acid is being added. The citric acid addition temperature is too high. • Empty the catchall and set up a regular emptying schedule for the catchall accumulation. • Check and verify the oil level in the deodorizer. Also check the level control to make sure that it did not drift. • Reduce the deodorizer temperature and make sure that the FFA and oil flavor are satisfactory after reducing the temperature. • Make sure the oil cooling is carried out properly in the deodorizer under vacuum with steam sparge and also check if cooler is working properly. If not rectify the issue.

Deodorized oil has poor flavor.	• Deodorizer feed oil is of poor quality. • Poor vacuum. • Deodorizer temperature is low. • Insufficient stripping steam. • Stripping steam distribution is poor. • Deodorizer needs a caustic wash.	• Correct the deodorizer feed oil quality issue and reduce the oil flow in a semicontinuous or a continuous deodorizer. Increase deodorization time in a batch deodorizer. • Correct the vacuum issue. • Raise the deodorizer temperature by a few degrees, recheck the oil flavor. • Increase the amount of stripping steam without affecting the vacuum. • Check and clean the steam sparger. • Shut down and caustic wash the deodorizer.
Deodorized oil shows high level of polymers.	• Deodorizer feed oil is oxidized. • The deodorizer vacuum is poor. • The deodorizer temperature is too high.	• The deodorizer feed oil is high in PV, possibly high pAV and polymer. Correct the deodorizer feed oil oxidation issue. • Fix the vacuum issue. • Reduce deodorizer temperature. Recheck the deodorized oil quality.

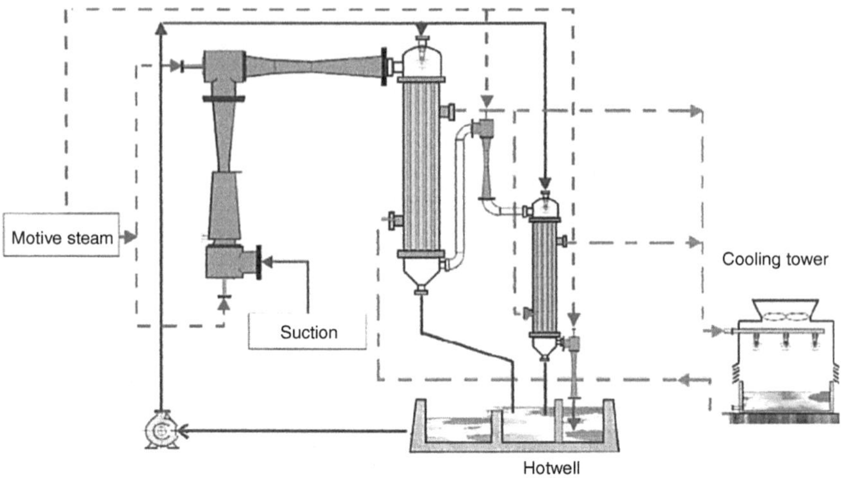

FIGURE 8.8 Surface condenser with plate heaters by EquirepSA, Spain.

Maintaining the pH of the greasy water is critical. Care must be taken to make sure the caustic solution addition system is working properly. The tubes can get plugged up and the deodorizer can have a very high operating pressure if the fatty acid is allowed to build up.

An automatic pH monitor with an alarm could provide the safeguard. The flow of caustic solution can stop for many reasons, the most likely of which could be:

- Failure of the caustic pump.
- Caustic solution supply tank is empty.
- Failure of the control system.
- The spray nozzles are plugged up.

8.9 PERIODIC CLEANING OF THE DEODORIZER

A deodorizer must produce clean-flavored oil. One of the reasons for poor flavor in the freshly deodorized oil is that the deodorizer is not clean. The unsaturated fatty acids and even some triglyceride molecules can oxidize and polymerize inside the deodorizer. Polymer deposits can be found on the heating coils, steam bubble caps, mammoth pumps, packed columns, deodorizer trays, walls, and demisters. Polymers at low concentration can produce unacceptable fresh oil flavor. In addition, the flavor can deteriorate very rapidly if the dimer or polymer content is over 1% in deodorized oils.

This is why it is recommended that all deodorizers be caustic washed once a year or more frequently. Packed column deodorizers require more frequent caustic cleaning. Cleaning the deodorizer is a long and tedious process, but it is absolutely necessary. The lye wash setup would be similar for a batch deodorizer and a semicontinuous unit. A continuous deodorizer requires a somewhat different setup for lye washing. A specific caustic wash (sometimes called lye wash) procedure should be established by the plant personnel.

A generalized step-by-step lye washing procedure is outlined as follows.

1. Shut down the deodorizer.
2. Shut off the heating and stripping steam.
3. Shut off the cooling water.
4. Blow the entire system with nitrogen to push the residual oil and collect it in the appropriate tanks.
5. Open all drain valves to let any residual oil out of the system. Sometimes a scavenger pump can be used.
6. Close the drain valves.

8.9.1 Batch Deodorizer

1. Fill the deodorizer with cold water covering the coils.
2. Add a sufficient amount of caustic solution totaling approximately 5% strength.
3. A stronger caustic solution does not clean the vessel any better.
4. Heat the water to 180–190°F (82–87°C).
5. Use stripping steam to agitate the water.
6. Leave the small stage ejector on if the pressure in the vessel begins to rise. Most of the time it is not necessary to turn on the ejector.
7. Leave the caustic solution with the steam on overnight.
8. Shut off the steam the next day and drain the water with the proper amount of phosphoric acid added to the water to meet the guidelines of the local municipality.
9. Refill the deodorizer with cold water. Agitate with stripping steam.
10. Add a certain amount of phosphoric acid to neutralize the residual caustic in the vessel.
11. The pH of the water must match that of the city water, and not necessarily 7.0. The pH should preferably be measured by a pH meter and not by litmus paper.
12. Let the vessel cool down.
13. Take down the demisters and clean them in a vat of light caustic solution in hot water.
14. Rinse and treat the demisters with phosphoric acid.
15. Sometimes it is better to replace the demister pad when the deodorizer is cleaned once a year. It is not very easy to clean the demister pads well.
16. The deodorizer must be sprayed with deodorized oil as soon as the vessel is dry. This step is not so critical if the unit is built with 304 or 316 stainless steel.

The deodorizer is ready for reuse.

8.9.2 Semicontinuous Deodorizer

The semicontinuous deodorizer is lye washed in the same manner as a batch unit. The deaerator vessel and each tray are handled as with a batch deodorizer. It is a good idea to replace the demister pads if they look black.

8.9.3 Continuous Deodorizer

A continuous deodorizer requires recirculation of hot caustic solution through the system. This requires a separate tank to supply the deodorizer with hot caustic solution. Fig. 8.9 shows a schematic diagram for a caustic wash arrangement for a continuous deodorizer.

The step-by-step procedure is outlined as follows:

1. Fill the tank with water.
2. The capacity of the tank should be 30% more than the total capacity of the deodorizer system.
3. A centrifugal pump with a capacity of 100–200 gallons per minute (gpm) with proper NPSH calculated on the basis of water temperature of 200°F (93.3°C) is used for pumping water through the system.
4. Caustic is added to make a 5% solution in the tank.
5. The water in the tank is heated with low-pressure steam. A self-actuated temperature control valve could be used for temperature control with a thermodynamic steam trap.

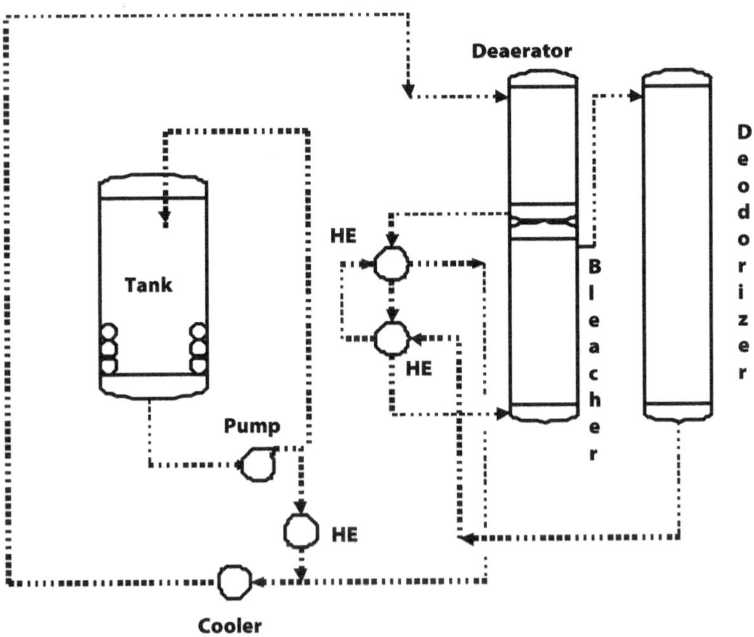

FIGURE 8.9 Schematic diagram for caustic wash system.

6. The caustic solution from the tank is recirculated for an hour while it is being heated.
7. After the water reaches a temperature of 180–190°F (82–87°C), the water is pumped through the deodorizer system.
8. All control valves are kept open with manual override.
9. The isolation valve to the ejector system is kept closed.
10. The caustic solution is circulated for 12–24 h. The temperature of the caustic solution is maintained.
11. Neutralize the caustic with phosphoric acid and circulate it for 2–4 h.
12. Drain the liquid from the system.
13. Open all the drain lines at the low spots as before and drain the residual water.
14. Turn on the vacuum system.
15. Fill the deodorizer with startup oil.
16. Heat the oil slowly to 250°F (121.1°C) and recirculate it until full vacuum is established.
17. Heat the oil to full temperature and start deodorizing.

Chapter 9

Finished Product Storage and Handling

9.1 INTRODUCTION

Liquid oils and the bases for solid shortening, margarine, and pourable liquid shortening from the deodorizer are collected and stored in tanks. The products are then shipped as:

1. Bulk products in tank trucks, drums, rail cars, ISO tanks, or totes for the food manufacturing industry.
2. Consumer products in bottles or cans.
3. Food service or restaurant products in totes, drums, or cans.

Finished products, after deodorization, are stored in tanks under nitrogen protection. The tanks are maintained at certain temperatures that are specific to the products. Packaged products are stored in the warehouse before shipping. In addition, some special conditions apply for storing pourable liquid shortening. All of these will be discussed in detail in this chapter.

9.2 TRANSFER AND STORAGE OF DEODORIZED PRODUCTS IN TANKS

Freshly deodorized products are analyzed to ensure the finished product standards have been met. These standards vary between products but the common themes between the products are good flavor, low FFA, low PV, and light color. Table 9.1 lists the typical finished product attributes for various types of products.

As discussed in Chapter 8 on deodorizing, the product is reprocessed if one of the finished product standards is not met.

In addition to the eight attributes listed in Table 9.1, the product may have added antioxidants. These antioxidants can be synthetic or natural. The product in the storage tank must be analyzed for the antioxidant content to make sure that the required amount of the antioxidant has been added. Based on the analysis, the plant may have to add some additional amount of antioxidant to meet the

Practical Guide to Vegetable Oil Processing. http://dx.doi.org/10.1016/B978-1-63067-050-4.00009-X
Copyright © 2017 AOCS Press. Published by Elsevier Inc. All rights reserved.

TABLE 9.1 Typical Finished Product Standards Checked After Deodorization and Storage

Type of product	FFA (%)	PV (mEq/kg)	Lovibond color red/yellow	Cold test (h)	Solid content index	Pourability (g/30 s)	Flavor grade	Citric acid (ppm)
Margarine base	<0.05	<0.5	X	—	X	—	8	20–50[a]
Solid shortening	<0.05	<0.5	X	—	X	—	8	20–50
Pourable shortening	<0.05	<0.5	X	—	X	X	8	20–50
Salad oil	<0.05	<0.5	X	X	—	—	8	20–50

[a]*Added amount is 25–50 ppm. There is some possible loss in the deodorizer.*

minimum level required in the product as per company's standard or as required by the customer.

9.3 DEODORIZED OIL STORAGE TANK

In the United States, deodorized seed oils are generally stored in carbon steel tanks. The transfer lines are also made of carbon steel. However, palm oil, palm oil fractions, coconut oil, palm kernel oil, and palm kernel oil fractions must be stored in 304 or 316 stainless steel tanks and all transfer lines and the accessories must also be made of stainless steel. Some plants use stainless steel tanks for storing all oils.

The oil in the deodorized oil storage tank should be properly protected against oxidation by nitrogen blanket.

9.3.1 Components of the Deodorized Oil Storage Tank

The schematic diagram for deodorized oil storage tank is shown in Fig. 9.1. The features of a deodorized oil storage tank are listed as follows:

1. The line loading the tank with deodorized oil is extended to the floor of the tank.
2. The tank has a side entering mechanical agitator with a low-level cutoff switch (LLS). The purpose of the low level cut off will be discussed in Chapter 15.
3. The tank has a vacuum vent valve at the top. The purpose of this valve is discussed later in this section.
4. There is a rupture disc at the top of the tank. This feature is also discussed later in this section.

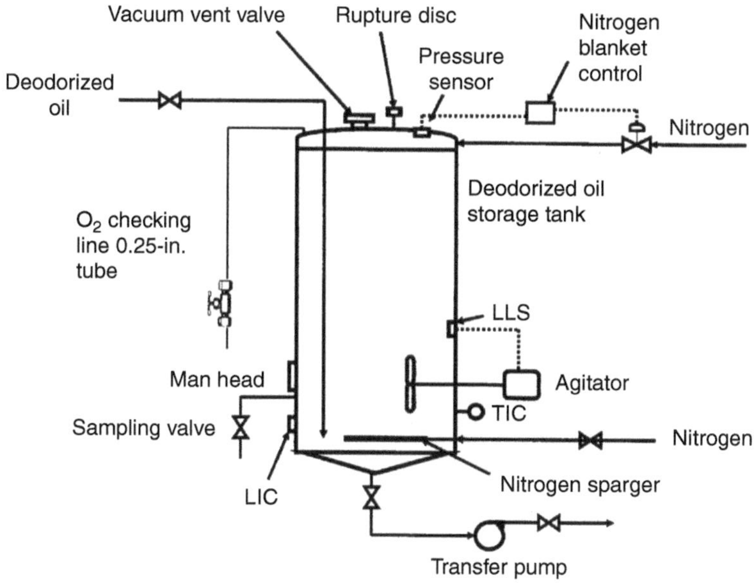

FIGURE 9.1 Schematic diagram for Deodorized Oil Storage©.

5. A nitrogen control device regulates nitrogen flow into the tank to maintain a maximum oxygen content of 0.5%.
6. The tank has a temperature indicator controller (TIC).
7. A level indicator controller (LIC) with high level cut off, which shuts the oil inlet valve to the tank and diverts the oil from the deodorizer to a similar storage tank.
8. A ¼-in. tube coming down from the top of the tank where the headspace oxygen content can be measured with a hand-held oxygen meter.
9. A sampling line and a valve are used for collecting samples for analysis.

9.3.2 Nitrogen Blanketing

The term *nitrogen blanketing* is misunderstood by many. This is discussed as follows:

1. Some believe that nitrogen gas is heavier than air. Therefore, the nitrogen-blanketed tank has a layer of nitrogen gas lying on top of the oil protecting it from coming in contact with the air.
2. The blanket moves up or down as the oil level in the tank changes.

In reality, there is no such thing as the oil being protected under a layer of nitrogen gas. This is because:

1. Nitrogen gas is lighter than air.

2. The air we breathe is a mixture of 79.1% nitrogen and 20.9% oxygen.
3. When nitrogen gas is released into the atmosphere or inside a tank, it takes a fraction of a second for it to completely dissipate into the environment in which it is released.

In addition, there is a common fear among people that a nitrogen-blanketed tank can be a serious health hazard when it is located inside a building (even if the building is well ventilated) because nitrogen can fill the whole building and kill the workers.

This belief is incorrect, and this is why:

1. The air we breathe contains 20.9% oxygen.
2. Human life can be sustained at 19.5% oxygen content in the air.
3. It will take 6.7% of nitrogen by volume to reach 19.5% oxygen level in a confined space without ventilation.
4. No building should be built without ventilation and it takes a lot of nitrogen to reduce the oxygen content in a well-ventilated building. In reality, it is virtually impossible to reach this condition.
5. It is a different story when it comes to *nitrogen-blanketed tanks*. One must never enter a nitrogen-blanketed tank without following the precautionary measures discussed in Chapter 16.

Note

Gases heavier than air, such as carbon dioxide, methane, hexane vapor, etc., can form a layer near the floor of the building and can be very hazardous.

Nitrogen-blanketed tanks are maintained under a very low positive pressure of nitrogen. This is normally ½–2 in. of water. Blanketing is accomplished through a control system that is capable of putting nitrogen gas into the tank as the oil is being pumped out and a partial vacuum is created in the tank. The system consists of a few critical components.

9.3.2.1 *Pressure Sensor*

As the oil is pumped out of the tank, a partial vacuum is created inside the tank by the suction of the oil discharge pump. A pressure sensor senses the pressure at the tank headspace and sends a signal to the control device to open a control valve to let fresh nitrogen enter the tank.

9.3.2.2 *Vacuum Vent Valve*

This is mounted at the top of the tank. It is a breather valve and performs two functions to protect the tank:

1. When the tank is being filled with oil, the nitrogen in the headspace is pushed upward. This creates a positive pressure inside the tank (much higher than

the nitrogen blanket pressure). The vacuum vent valve opens to allow the nitrogen gas from within the tank to escape into the environment.

2. On the other hand, when the oil from the tank is pumped out, the partial vacuum created by the oil unloading pump opens up the nitrogen control valve to allow nitrogen gas to fill the tank head space. If the nitrogen control valve fails to open or if there is no nitrogen gas supply, the vacuum vent valve opens and fills the tank head space with air to prevent the tank from collapsing due to the vacuum created by the oil discharge pump.

> **Note**
>
> Every tank has a loading and unloading pump. One of these pumps may have higher pumping capacity. Therefore, the vacuum vent valve must be sized to allow the flow of nitrogen gas, matching the volumetric flow of oil through the pump with larger capacity.

9.3.2.3 *Rupture Disc*

This is also mounted on the top of the tank. It protects the tank from overpressure during oil loading. As mentioned earlier, the vacuum vent valve releases the nitrogen gas from the tank as the oil level rises. If the vacuum vent valve fails to open, high pressure is created inside the tank and it can match the discharge head of the oil pump at the deodorizer discharge. These tanks are not designed to withstand high pressure and are built with weak tops. Nitrogen-blanketed tanks do not have an atmospheric vent. Therefore, if the vacuum vent valve fails to open while the tank is being filled with oil, the high pressure built inside the tank can blow up the tank top.

The rupture disc is set to break at a certain pressure and open up the passage to release excess pressure from within the tank. This protects the tank from rupturing.

9.3.2.4 *Checking Head Space Oxygen*

A hand-held oxygen meter is used to check the oxygen content in the tank headspace. There are many portable instruments available on the market, but the author has used one brand of instrument that has worked very well. The particulars are given as follows:

Oxygen Meter, Model 8108-O
North American Manufacturing Co.
4455 East 71st. Street
Cleveland, Ohio 44205

One must check the percentage of oxygen in the tank headspace once per day. The following steps are recommended if the oxygen content in the tank head space is >0.5%:

1. Open the manual valve to the nitrogen sparger at the bottom of the tank and allow the nitrogen gas to bubble up the bed of oil in the tank.

2. The exact duration for sparging will depend on the value of the oxygen content in the tank headspace.
3. Sparging is stopped after some time (typically 15 min) and the oxygen in the headspace is checked.
4. Sparge the tank with nitrogen for some additional time if the oxygen content in the tank headspace is still >0.5%.
5. Sparging is stopped if the oxygen content has dropped <0.5%.

In many modern plants, oxygen in the tank headspace is monitored automatically. An alarm sounds if the oxygen content is >0.5%. Nitrogen is sparged manually. The alarm stops when the oxygen in the headspace is 0.5%.

9.3.2.5 Nitrogen Sparger

These are made of sintered metal with micro pores. Normally, there is more than one sparger and they are installed inside the tank at the bottom such that they cover the entire cross section of the tank.

9.3.3 Temperature Indicator Controller

This can be local or with remote control from the control panel. The recommended temperatures for oil storage are:

Oil type	Maximum temperature
Seed oils, palm oil, palm olein, coconut oil	<115°F (46.1°C)
Hydrogenated seed oils	No higher than 10°F (5°C) above the complete melt point of the stock
Hydrogenated palm, palm kernel oil	
High melting fractions of palm, palm kernel oil	

9.3.4 Agitator

A side-mounted agitator is needed for hydrogenated or higher melting stocks to keep the product well mixed. A heating coil using low-pressure steam is used in the tank (not shown in Fig. 9.1). The agitator also prevents any scorching of the stock on the heating coil surface. A low level switch disables the agitator motor at a predetermined oil level (above the agitator blades) in order to protect the agitator shaft and the seal.

Also not shown in Fig. 9.1 are the three or four baffles in the tank. The baffles are placed at every 120 degree or at every quadrant and run vertically along the wall of the tank. The width of each baffle is one twelfth of the diameter of the tank. The baffles prevent churning of the oil inside the tank and break any vortex.

9.4 LOADING FINISHED OILS IN TRUCKS

Most bulk shipments of liquid oils are made by tank trucks; others are made by rail cars. The capacity of a tank truck can vary worldwide. The author has encountered tank trucks with 6 MT up to 30 MT capacities. It is necessary to follow certain protocol before the tank truck is loaded with the oil. These are:

1. The tank truck, upon arrival must be checked inside and outside to confirm that it is clean.
2. A tank truck must have a cleaning certificate that clearly describes the following:
   a. the description of the previous load and it must be an oil that is compatible with the current oil to be loaded
   b. method of cleaning and the date and location of cleaning
3. The tank truck must not have any foreign or objectionable odor.

Note

A tank truck must always be filled to its capacity with a little space at the top. The oil inside the truck suffers a great deal of movement as the truck starts or stops. The oxygen from the air in the tank truck headspace gets absorbed in the oil. This may not be detectable upon its arrival at the destination but may show as higher PV in the oil after storage. This is why a partial load in a truck should be avoided.

The tank trucks have different features, such as:

- Some have *two or three* compartments so more than one type of oil can be transported at a time or less than a full truckload of oil can be transported. However, they are hard to find.
- Some are insulated to protect the higher melting stocks from solidifying in the transit.
- Some have heated jackets to maintain higher temperature during extreme cold weather. This is not always safe because the temperature control is not always good.
- Some have two heating coils inside that can be used to melt the solidified product upon arrival by using low-pressure steam and thermodynamic traps.

The following steps are recommended for physically loading the truck with deodorized oil after the loading truck has been found satisfactory.

1. Get a fresh (representative) sample from the deodorized oil storage tank and analyze it for finished product standards.
2. Depending on the product, this may include FFA, PV, Lovibond color, flavor, and the melt point.
3. The truck is ready to be loaded if the analysis of the oil is satisfactory.
4. The truck must be bottom loaded which is the preferred way.
5. It can be loaded from the top with the loading boom extended to the floor of the truck to minimize entrainment of air into the oil.
6. The oil from the deodorized oil storage tank passes through an in-line nitrogen diffuser, which has a sintered metal diffuser along the axis. This saturates the oil with micro size nitrogen bubbles as the oil passes through the diffuser.

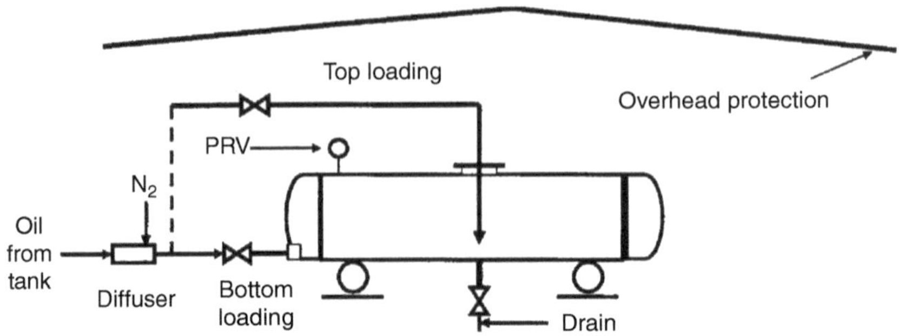

FIGURE 9.2 Oil loading station at Oil Processor©.
Note:
- Do not purge the empty truck with nitrogen. This consumes too much nitrogen.
- Bottom-fill the truck with oil saturated with nitrogen. This displaces most of the air from the truck.
- Some of the nitrogen, dissolved in the oil, is released and also displaces the air from the truck.
- Purge the headspace with nitrogen after the truck is filled.
- Close the hatch. Apply seals.

7. The oil rises from the bottom and some of the residual dissolved air is expelled which collects at the top and gets purged out of the tank truck.
8. The tank truck is filled to the maximum liquid level as indicated on the truck.
9. A representative sample is taken from the tank truck as an official loading sample and analyzed. A duplicate sample is sent with the truck driver to be delivered to the buyer and a third sample is saved at the plant in a cool place for 3 months for any future reference.
10. The hatch is closed and locked.
11. Nitrogen gas is purged for an additional time (5–10 min), which purges more of the residual air collected at the top of the oil through the pressure relief valve (PRV).
12. The truck is sealed and is released for delivery if the truck sample meets the finished oil quality standards.

Fig. 9.2 shows the typical tank truck loading station and the proper procedure for the same.

9.5 UNLOADING FINISHED OIL FROM TANK TRUCKS

Sometimes an oil processing plant may have to bring in fully processed oil from another source. The truck inspection procedure is similar to that described in the previous section but there are some differences, such as:

1. The exterior of the truck must be clean. The truck can be rejected (in USA or many other countries) if it is not clean on the outside.

2. There are several seals located at strategic locations on the truck. These seals must be in place and not tampered with. If a seal is missing, the plant must not unload the oil until the issue on the missing seal is resolved between the company's purchasing personnel and the oil supplier.
3. Once the decision is made to unload the oil, the seal at the top hatch is broken.
4. The inside of the truck is inspected for:
   a. appearance of the oil
   b. any unusual or foreign smell
   c. presence of any foreign object in the truck
5. Oil samples are taken with the help of a zone sampler at three heights, namely top, middle, and bottom of the truck, and mixed in a clean container for analysis only if the previous inspection is satisfactory.
6. The oil is analyzed for:
   a. FFA
   b. PV
   c. Flavor
   d. Lovibond color (if required)
   e. Presence of any antioxidant if the end user requires has to declare no antioxidants in their finished products
7. The oil is unloaded if the analysis is satisfactory.
8. The oil passes through a basket strainer. An in-line nitrogen diffuser, as it is unloaded.
9. The strainer basket is checked before and after unloading the truck to inspect any foreign object that might have escaped the visual inspection. Sometimes glass bottles in the truck are hard to see from the top.
10. If any foreign object is discovered in the strainer after unloading the oil, it must not be released for use until the foreign object is identified and it is deemed harmless for human health.
11. The unloaded oil must be isolated if the foreign object is glass or something injurious to human health.
12. The storage tank and the associated piping and accessories must be cleaned to ensure that the foreign material has been completely eradicated from the system.

Fig. 9.3 shows the schematic diagram for unloading a tank truck and oil receiving facility.

9.6 PACKAGED PRODUCTS STORED IN THE WAREHOUSE

Various products are stored in the warehouse and shipped to the customers.

The packaged products can be:

1. Salad oil in bottles packaged in cardboard cases.
2. Salad oil in gallon jugs packaged in cardboard cases.
3. Salad oil in 5 gallon plastic containers with cardboard outer package.

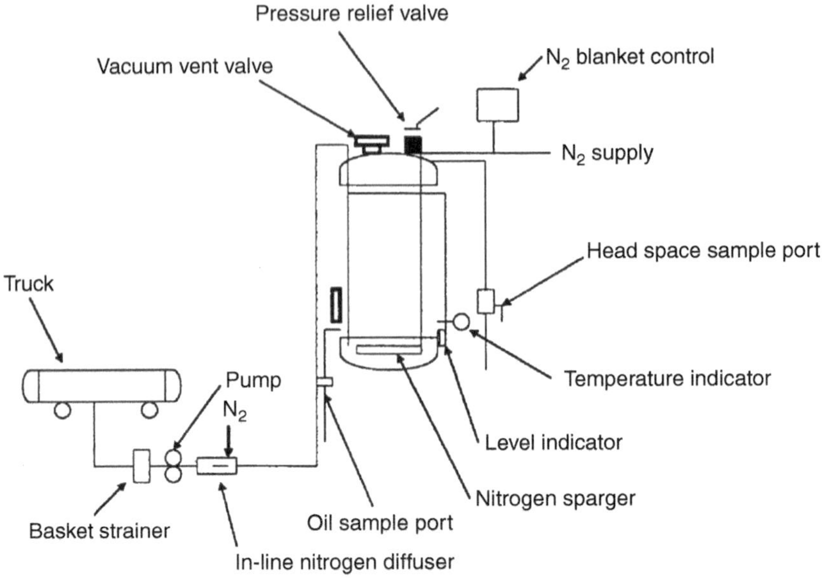

FIGURE 9.3 Schematic diagram for Oil Unloading©.

4. Salad oil in 55 gallon drums.
5. Plastic shortening in cubes (plastic bags inside cardboard cases).
6. Pourable shortening in bottles, 1 and 3 gallon containers packaged in cardboard cases.
7. Pourable shortening in 55 gallon drums.
8. Stick or tub margarines.

Products 1–7 must be stored in a cool warehouse at 70°F (21.1°C) with the temperature not to exceed 85°F (29.4°C).

All margarine products must be stored in a refrigerated storage at <45°F (7.2°C).

The crystal structure of the plastic shortening is damaged if the product is exposed to a high temperature where it gets partly melted and then allowed to resolidify. The resultant crystal structure in the product is very different from that in the original product. This can be quite harmful for the shortening if it is used in making dough, cake batter or icing at a bakery. The bread roll dough can be sticky and the cake batter can be more dense than normal. It will also perform poorly in icing, producing lower icing volume and poor structural stability.

If the pourable shortening is partly melted, there is a separation of the liquid, which floats to the top and the solids settle at the bottom.

Therefore, it is critical to maintain the products at the lowest possible temperature. Unfortunately, in many warehouses in the United States, Latin America, Asia, and Africa the warehouse temperature can be high. This is why many companies have to make summer and the winter formulations for solid shortenings. The summer formula uses higher amounts of high melting fractions in the product so it can withstand the higher ambient temperature during storage and shipment. However, summer and winter formulas for pourable shortening are not feasible.

9.7 MAINTAINING PRODUCT QUALITY IN THE WAREHOUSE

The product quality in the warehouse should be monitored until it is shipped. It is advisable to implement a warehouse quality standard similar to that for the product as packed. This is important because the industrial users of packaged shortening have a shelf life requirement. On the other hand, the quality of the packaged shortening will decline in the warehouse. Similarly, bottled liquid oil also loses flavor during storage. This can pose some additional challenges to the plant quality assurance management. Following guidelines can be followed to develop a program that would allow the plant to make a decision whether or not the product in the warehouse could be shipped if it has been stored there for several weeks or months.

9.7.1 Consumer Products

- Through storage studies, one can establish the stability of the product under normal storage conditions.
- The code date is established in this manner and is used on the label (especially for consumer products).
- At the same time, the company needs to determine how long a consumer product stays on the store shelf.
- Thus, with the information of the market data and the storage stability data, the company can determine whether or not the consumer product would exceed its code date and if it would become unacceptable to consumers even when it is on the store shelf.
- The company can establish the quality criteria on consumer products based on the age of the product in the warehouse and the time remaining on the code date.
- If the quality of the product is unacceptable or marginal and it is anticipated that the product would become unacceptable by the time the consumer buys it, it would be appropriated to reprocess the product. This will add cost to the product. This will be discussed in Chapter 13.

Quality criteria for the consumer product should be:

- Flavor, PV, FFA, and Lovibond color for liquid oil.
- Flavor, PV, Lovibond color, Hunter color, or other tests to check the appearance or color of packaged shortening for consumers.

9.7.2 Industrial Products

Similar to the previous approach, one can establish the quality standard and code date for industrial shortening and industrial margarine.

There is another factor that should be considered: industrial users make shelf stable products that must meet the shelf life standard set by their own quality standards. These users are critical about the production date on the packaging because of the projected shelf life requirements for their own products. They prefer to use the freshest product from the supplier.

Therefore, it is advantageous for the plant to have proper stock rotation. However, it always helps when the product is shipped out of the company warehouse as quickly as possible after production, reaching the end-user as soon as possible.

The warehouse quality parameters for the industrial shortening and margarine would be the same as shown for the shortening under consumer products earlier.

9.8 SHIPPING OF PACKAGED PRODUCTS

Packaged products in cases, 55 gallon drums, and totes are shipped in trucks. The inside of the truck can reach a temperature of 145°F (62.8°C) or even higher in the summer. All plasticized shortening and pourable shortening can suffer serious damage in the truck. Insulated or refrigerated trucks can alleviate this issue. With a refrigerated truck, one can maintain a temperature of 75°F (23.9°C) and protect the product during transit.

Delivering the products in proper conditions is the oil processor's responsibility, and one must always use the best means to protect the product during storage, loading, and delivery.

Chapter 10

Fundamentals of Fat Crystallization Related to Making Plastic and Pourable Shortenings

10.1 INTRODUCTION

Polymorphism or crystal type is a unique property that defines the physical consistency of solid fats or pourable liquid shortenings. Oil technologists, food technologists, and chemical engineers have studied the crystal behavior of various fats and fat blends to learn about their intrinsic properties. Food scientists have put this knowledge into making various food products. Margarine and shortening made from hydrogenated vegetable oils have been in use for nearly a century. Margarine made from animal fats was invented in France during the 19th century. Shortening made from animal fats was instituted in Europe and the United States at the same time. The first shortening made from hydrogenated cottonseed oil was introduced in the United States by Procter & Gamble Co. in 1912 with the brand name Crisco.

Although these products were made in the 19th and early 20th centuries, the understanding of the crystal behavior in fats started in the early 20th century. This field of study is still continuing, and advanced knowledge in this area is enabling us to understand shortening and margarine formulation and processing techniques. Fat crystallography of palm oil, palm kernel oil, and their fractions have also been studied extensively in the past decades.

A better understanding of the polymorphic behavior of fats was beneficial to the oil technologists in formulating shortenings applicable to products, such as bread, dinner rolls, puffed pastries, cakes, pastries, icing, frozen dough, frozen cakes, ice cream, and many others that the consumers are enjoying today. These products require fat crystals of specific polymorphic behavior, which is achieved by proper selection of the hard stock and the appropriate crystallization process in each case.

Practical Guide to Vegetable Oil Processing. http://dx.doi.org/10.1016/B978-1-63067-050-4.00010-6
Copyright © 2017 AOCS Press. Published by Elsevier Inc. All rights reserved.

10.2 FAT POLYMORPHISM

The most common types of fat crystals in edible shortening and margarine products belong to the following three categories:

1. alpha
2. beta prime
3. beta

10.2.1 Alpha Crystals

Alpha crystals are first formed when the blended fat containing hard stock (saturated triglycerides) or hard stock (by itself) is chilled rapidly. Alpha crystals can be characterized as:

1. Randomly oriented crystals with an approximate diameter of 5 μm.
2. Transparent or translucent in appearance.
3. Loosely packed with an ill-defined hexagonal subcell structure.
4. Having the lowest melting point among all the three forms of crystals.
5. Extremely unstable and readily transformed to beta prime or beta form.

Transformation from alpha to beta prime or beta can occur under ambient conditions. This is carried out via controlled temperature conditions in the crystallization process for manufacturing shortening or margarine.

10.2.2 Beta Prime Crystals

The beta prime crystals can be described as:

1. Tiny, needle-shaped, cross-linked crystals that form three-dimensional structures.
2. Large amounts of liquid oils can be held in the interstitial spaces between the crystals.
3. The crystals are typically 1-μm long. The alternating fatty acid chain axes are oppositely oriented.
4. Crystal diameters range from 3.8 to 4.15 μm, which is smaller than the alpha crystals. The crystal size and shape is strongly influenced by processing and packing procedure applied. These numbers are for comparison purposes.

Fig. 10.1 shows a simplistic pictorial concept of a beta prime crystal.

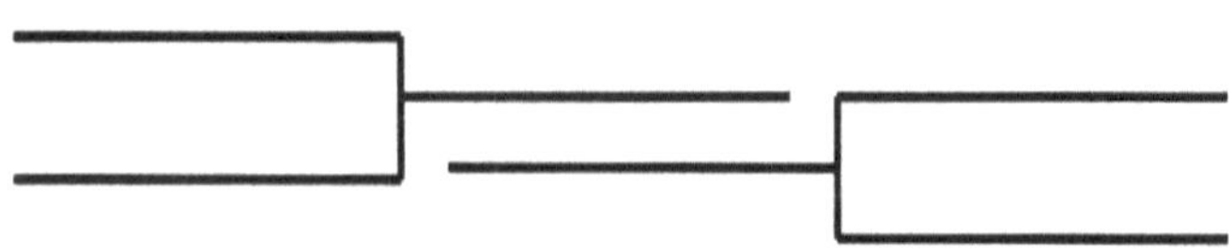

FIGURE 10.1 **Schematic pictorial concept of beta prime crystal.**

10.2.3 Beta Crystals

Beta crystals are formed from beta prime crystals. This transformation can take place naturally with storage of the product or through thermal manipulation of the beta prime crystals. Transformation from alpha to beta with a very short duration in the beta prime state can occur when:

1. The saturated triglyceride (hard stock) is strongly beta stable.
2. The crystallization of product containing beta stable hard stock is cooled at a slower rate.
3. When the triacyl glycerol (also referred to as TAG) or the oil molecule population is very uniform and homogeneous in terms of carbon chain length and saturation.

Beta crystals exhibit the following characteristics:

1. The crystals do not form three-dimensional structures like beta prime crystals.
2. Unlike beta prime crystals, these fat crystals cannot hold liquid oil in the interstices.
3. The crystals are 25–50 μm long and can grow up to 100 μm in length.
4. The crystals have a diameter of 4.2 μm.
5. The melting point is the highest among the three types of crystals.
6. The fatty acid chain axes are oriented in the same direction.

Fig. 10.2 shows a simplistic pictorial concept of a beta crystal.

10.2.4 Melting Points of the Three Polymorphic Phases

It was briefly mentioned earlier that the three polymorphic phases have different melting points and that the beta crystals have the highest melt point of the three types of fat crystals. The following example will illustrate the point.

Crystals obtained from tristearine, where all three glyceride links are occupied by stearic acid, demonstrated the following melting points.

Polymorphic state	Melting point, °F (°C)
Alpha	129.2 (54)
Beta prime	147.2 (64)
Beta	163.4 (73)

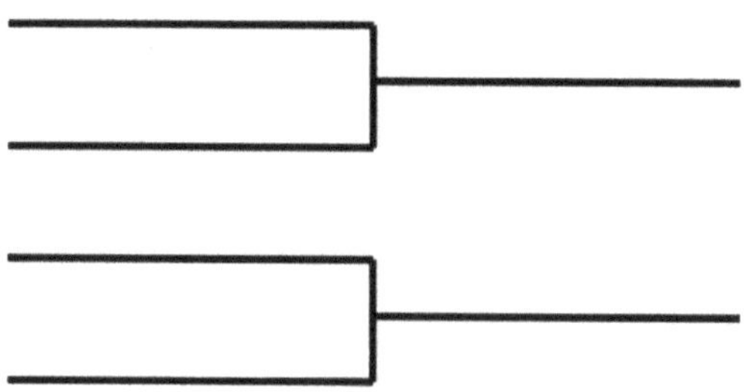

FIGURE 10.2 Schematic pictorial concept of beta crystal.

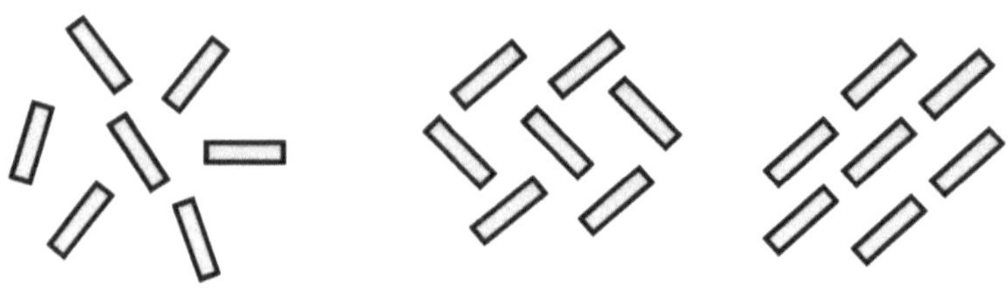

FIGURE 10.3 **Conceptual crystal orientation patterns of three polymorphic forms (alpha, beta prime, and beta).**

10.2.5 Crystal Packing Pattern of Alpha, Beta Prime, and Beta Crystals

Alpha crystals are loosely packed and appear to be packed in a random fashion. Beta prime crystals are cross-linked, while the beta crystals are unidirectional. Fig. 10.3 shows the conceptual patterns of these crystals.

10.3 TRIGLYCERIDE STRUCTURE

A shortening base is a mixture of different triglycerides. These are usually made by blending the following components:

1. A liquid or lightly hydrogenated oil with a low-melt point.
2. An intermediate melt point fraction that is somewhat more hydrogenated.
3. A high-melting fraction that is fully hydrogenated (hard stock).

From the standpoint of polymorphic behavior, there are four fundamental types of triglycerides found in the aforementioned blends, namely:

1. Triunsaturated triglycerides, where all three positions on the glycerol molecule are esterified with unsaturated fatty acids.
2. Diunsaturated fatty acids, where one of the unsaturated fatty acids has been replaced by a saturated fatty acid.
3. Disaturated triglycerides, where there is only one unsaturated fatty acid on the triglyceride molecule.
4. Trisaturated triglycerides, where all three fatty acids of the molecule are saturated. These are commonly referred to as hard stocks.

Fig. 10.4 shows a pictorial view of these triglyceride molecules.

10.3.1 Fatty Acid Distribution in Trisaturated Triglycerides and Their Polymorphic Properties

Fig. 10.5 shows the polymorphic behavior of the trisaturated triglyceride molecules. Molecules, such as SSS and PPP are beta stable, whereas PSP and PSS are beta prime stable. As a rule, the fully hydrogenated fats exhibit the following

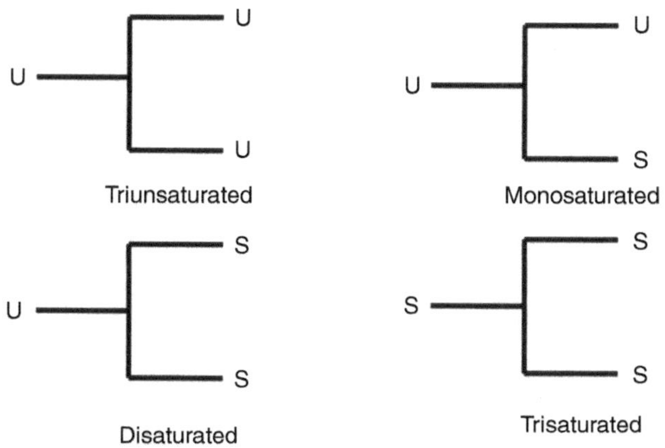

FIGURE 10.4 Four basic triglycerides. Triunsaturated, diunsaturated, disaturated, and trisaturated. *S*, Saturated fatty acid; *U*, unsaturated fatty acid.

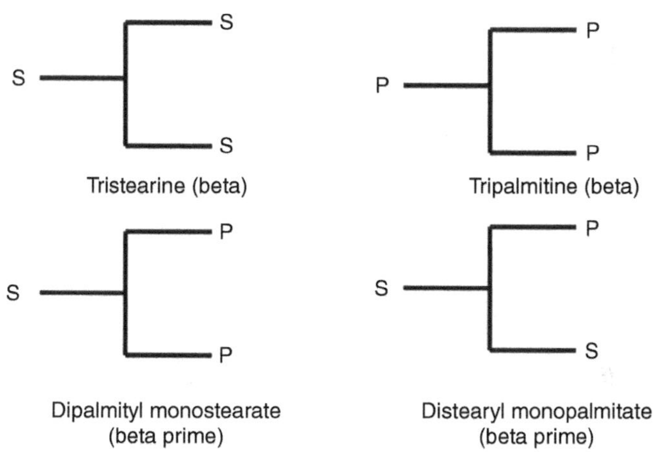

FIGURE 10.5 Types of trisaturated triglycerides. *P*, Palmitic acid; *S*, stearic acid.

polymorphic behavior. A hard stock made from TAG with wide differences in fatty acid chain length is beta prime stable. The following examples show that rapeseed hard stock is beta prime stable, while canola hard stock is beta stable. This is because the rapeseed hard stock contains triglycerides with a much higher content of behenic acid compared to the canola hard stock.

Fig. 10.6 shows the polymorphic behavior of the disaturated triglyceride molecules. Here PUS is the only molecule that is beta prime stable; the other three molecules—SSU, SUS, and PUP—are beta stable. If there is a high population of PUS, such as in cocoa butter, the system will tend to be beta.

The beta tendency of tristearine is extremely strong. Therefore, even a small amount of this hard stock added to a beta prime shortening can very rapidly

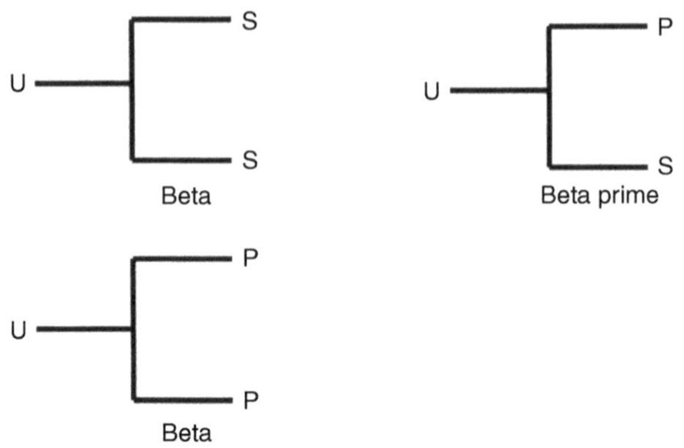

FIGURE 10.6 **Types of trisaturated diglycerides.** *P*, Palmitic acid; *S*, stearic acid; *U*, unsaturated fatty acid.

transform the polymorphic state of the shortening from beta prime to beta under suitable conditions. The beta prime to beta transition in a shortening is maximum when it is held at a temperature 2°F (1°C) below the complete melting point of the shortening for some time.

Composition of the hard stock can be established from the carbon number on the hard stock. This method is well-known in the oil industry. Table 10.1 shows the carbon numbers, their percentages, and the polymorphic behavior for hard stocks obtained from various commonly used oils in the industry.

Although not shown in Table 10.1, fully hydrogenated corn oil, sunflower oil, and lard are beta tending, while partially hydrogenated soybean corn or sunflower oil with high–*trans* fatty acid content is beta prime tending. Table 10.2 shows the polymorphic behavior of trisaturated triglyceride molecules in the oils. The more heterogeneous the TAG population and/or fat population, the greater the tendency toward beta prime stability.

TABLE 10.1 Carbon Numbers in Various Hard Stocks and Their Polymorphic Behavior

Oil type	Carbon #48 PPP (%) (Beta)	Carbon #50 PSP (%) (Beta prime)	Carbon #52 PSS (%) (Beta prime)	Carbon #54 SSS (%) (Beta)	Dominant polymorphic form
Soybean	0.2	3.3	27.6	66.7	Beta
Canola	—	1.6	11.6	28.3	Beta
Cottonseed	0.9	13.6	43.5	40.5	Beta prime
Palm	8.4	40.0	41.9	10.7	Beta prime
Tallow	7.5	21.0	44.9	24.5	Beta prime

TABLE 10.2 Properties of C16 and C18 Triglyceride Molecules

	Melting point (°C)			Stable polymorphic forms
Triglyceride	**Alpha**	**Beta prime**	**Beta**	**Comments**
SSS	54.9	64	73.1	Exhibit all three forms Beta stable
SPS	51.8	—	68.5	Only beta or fleeting beta prime form
PSS	50.6	61.1–65	65.2	Beta prime from melt Beta from solvent Stable in both beta prime and beta form
PSP	46.5	68.6	—	Beta prime stable
SPP	47.4	57.7–61.7	62.7	Beta prime and beta stable
PPP	44.7	56.6	66.4	Exhibit all three forms Beta stable

10.3.2 Summary of the Rule of Thumb on the Polymorphic Behavior of Triglyceride Molecules

- Saturated triglycerides have the strongest driving force to form crystals.
- Symmetrical trisaturated fat molecules are strongly beta tending.
- Nonsymmetrical trisaturated fat molecules are strongly beta prime tending.
- High-*trans* fatty acids make the fat beta prime tending.
- Heterogeneous distribution of fatty acids in the hard stock makes it more beta prime stable.

10.4 FAT CRYSTALLIZATION

Trisaturated triglycerides crystallize in a fat blend (shortening or margarine base) when it is cooled down. The cooling process removes the heat of crystallization and produces fat nuclei, which then grow into fat crystals under proper conditions. This process can take place when:

1. the cooling occurs naturally under ambient condition, or
2. the cooling is conducted under controlled chilling process.

Addition of certain emulsifiers, such as monoglycerides and diglycerides, can modify the crystallization process.

For making shortening and margarine products, the fat blend is always chilled under controlled conditions. The trisaturated triglycerides form nuclei. The nuclei are then allowed to grow under controlled conditions.

The amount of nuclei formed depends on the rate of cooling. A rapid chilling process produces nuclei in larger numbers. However, an extremely fast chilling process can make the fat blend supercooled, and the nuclei do not appear until the temperature rises and the system comes to a thermal equilibrium.

In the crystallization process, the trisaturated triglyceride molecules dictate the polymorphic behavior of the product, and its polymorphic stability depends on the type of saturated fatty acids and their distribution on the triglyceride molecule.

Polymorphic Behavior of Fully Hydrogenated Oils (Hard Stocks)

Type of hard stock	Beta prime stable	Beta stable
Cottonseed	Yes	—
Palm	Yes	—
Tallow	Yes	—
Rapeseed	Yes	—
Soybean	—	Yes
Sunflower	—	Yes
Canola	—	Yes
Corn	—	Yes
Lard	—	Yes

10.4.1 Sequence of Events in Controlled Crystallization Process

The following sequence of events occurs in the controlled fat crystallization process:

1. Through the rapid chilling process, the nuclei are formed from the trisaturated triglycerides and some from the disaturated triglycerides.
2. These nuclei form alpha crystals.
3. The alpha crystals are then converted to beta prime crystals fairly rapidly even at the low temperature. This conversion depends on the following factors:
    a. The fatty acid distribution in the trisaturated triglyceride molecules.
    b. The rate of cooling.
4. Under suitable conditions the beta prime crystals are converted to beta.
5. The transformation of beta prime to beta crystals depends on the following factors:
    a. The composition and fatty acid distribution in the trisaturated triglyceride molecule.
    b. The rate of cooling.
    c. Storage temperature and storage time of the finished product.
    d. Some beta prime products slowly get converted to beta with time. Therefore, if a shortening is intended to remain in the beta prime phase, it would be necessary to conduct a storage study on the product over its code date to make certain that the product is still predominantly beta prime. It must be noted that all beta prime products will eventually have beta crystals at higher and higher levels under normal storage.

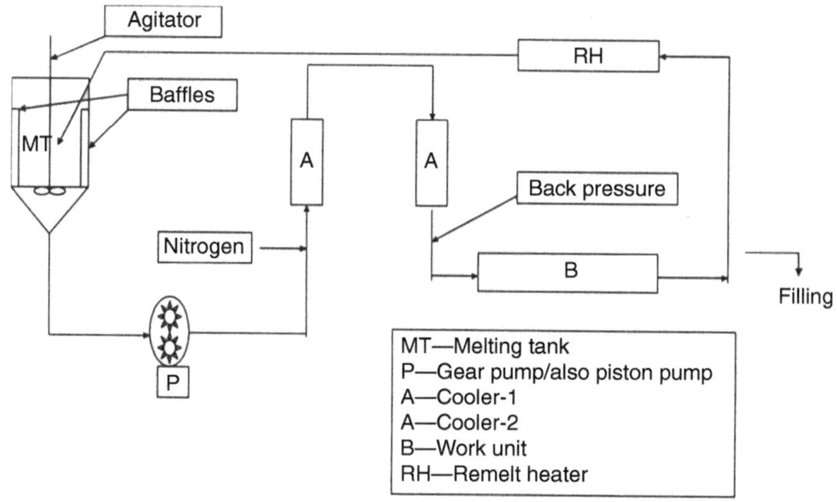

FIGURE 10.7 **Schematic diagram for a typical fat crystallization process.**

10.4.2 Typical Crystallization Process for Making Shortening

Fig. 10.7 shows the schematic diagram for a typical fat crystallization process for making shortening. It consists of the following components:

1. A melting tank with a top-entering agitator, three/or four baffles, and a heating coil or a heated jacket.
2. A transfer pump with variable frequency drive (VFD) motor control for controlling the flow into the crystallizer system.
3. Crystallizer (unit A), either one or two.
4. A work unit (unit B).
5. A back pressure valve after unit A.
6. A remelt heater.
7. All piping is made of 304 stainless steel and is heated by hot-water jackets.

10.4.3 Process Description

1. The shortening base in liquid state is held in the heated and agitated tank (MT). The temperature of the melted shortening base is maintained at 15–20°F (8.3–11.1°C) above its complete melt point.
2. The melted product (shortening or margarine) is pumped through one (or two or more) scrape-wall coolers (A), where the product is chilled down rapidly to a very low temperature, using brine ammonia or a Freon chilling system.
3. The temperature of the product at the exit of unit A is controlled within ±2°F/1°C.
4. The scrape-wall coolers have rotating shafts that have fixed blades along the horizontal axis of the shaft. These blades continuously scrape the inner wall of the chiller to facilitate the formation of nuclei.

5. The shaft rotates approximately at 300–330 rpm. The speed varies with the type of product, as well as the diameter of the freezer unit.
6. A back pressure of 300–350 psi (21.9–25.5 kg/cm^2) is applied at the outlet of unit A. This pressure can vary with the type of product being made.
7. Atomized nitrogen is injected into the oil as it enters unit A when aerated and a white-looking product is desired.
8. The product then passes through a work unit (B), which contains a set of fixed and a set of rotating fingers across the shaft.
9. The shaft typically runs at 600–700 rpm. Again, shaft speed can vary with the size of the cooler (A) and the type of product being made.
10. In some manufacturing processes, a second back-pressure valve is placed after unit B.
11. There is a rise in the product temperature in the B unit, indicating crystallization of the fat.

10.4.4 What Happens to the Product?

1. The saturated triglycerides form fat nuclei.
2. The amount of nuclei formed depends on the following factors:
   a. The solid content of the product.
   b. The degree of cooling of the product.
   c. The rate of cooling of the product.
3. The nuclei then pass through the short passage between the cooler (A) and enter the work unit (B). This distance must be minimum to prevent any crystal formation prior to unit (B).
4. This is where several events occur, such as:
   a. The nuclei formed in the cooler (unit A) begin to form fat crystals in unit B, as can be confirmed by the temperature rise in the product in unit B.
   b. The product attains the smooth texture and consistency as it leaves unit B.
   c. There is an optimal distance or residence time between the freezer and worker unit that provides sufficient time for crystal development, aggregation, and networking to happen prior to mixing in the worker unit; the finished product texture is affected by the level of crystal development prior to the working unit.

10.4.5 Primary and Secondary Crystal Bonds

During the passage of the product through unit A and unit B, a number of important events take place as discussed:

1. Fat crystals form three-dimensional structures during crystallization.
2. The crystals exhibit two types of bonds, known as:
   a. the *primary* bond and
   b. the *secondary* bond

3. The crystals vary in their size and shape.
4. The crystals grow together due to the *primary bond* and not as single individual crystals.
5. The crystals can agglomerate with a few points of contact, creating a branched intertwining mass and interstices due to the *secondary bond*. This produces the three-dimensional network in the crystallized fat system.

10.4.6 Primary Bonds

There are some significant contributions made to the shortening structure by the primary bonds. These are listed:

1. These bonds are very important for obtaining the smooth consistency in shortening and margarine.
2. These bonds are strong and exhibit the following properties:
   a. The bonds can be destroyed by mechanical force.
   b. These bonds cannot be restored once they are destroyed by mechanical force. Therefore, it is recommended not to put in excessive mechanical work in the unit (B).

10.4.7 Secondary Bonds

These are somewhat weaker than the primary bonds but provide very important physical properties to the shortening or margarine structure. The characteristics of the secondary bonds are listed:

1. These are weaker than the primary bonds.
2. Mechanical force can break these bonds, but the bonds are restored once the source of mechanical force is removed.
3. Extreme and prolonged mechanical force can permanently destroy these bonds.

10.4.8 Utilizing the Properties of the Primary and the Secondary Bonds

The object of making shortening in the crystallization process is to obtain the product with the desired consistency and plasticity. This is accomplished by directing the process to optimize the utilization of the properties of both primary and secondary bonds in the product.

1. The amount of nucleation must be maximized in the scrape-wall heat exchanger (unit A) by rapid chilling of the melted shortening feed.
2. This is achieved by applying a very high–temperature differential to the feed as soon as it enters the unit A by using brine or Freon.
3. Crystal growth (due to the primary bond) is minimized in unit A by applying the mechanical force (scrapers).

4. The residence time in unit A is controlled by the product flow rate to prevent crystal growth in the unit.
5. The agglomerates and some of the branched crystals (formed due to the secondary bonds) are broken in the work unit (unit B) to provide the smooth texture to the product.

10.4.9 Factors Determining the Physical Properties of Crystallized Fats

The main factors controlling the physical properties, such as consistency, hardness, etc., in a shortening are listed as follows:

- temperature
- concentration of solids (solid content)
- processing condition
- crystal size
- crystal distribution
- crystal shape
- interparticular force (van der Waal's force)
- mechanical work on the crystal
- polymorphic state of the crystals
- crystal dispersion
- secondary crystallization can occur after packaging and storage unless the product is properly tempered

10.4.10 General Rules of Fat Crystallization

The fundamental guidelines for fat crystallization are listed in Table 10.3. It is important for the plant personnel to have a good understanding of these principles.

10.4.11 Critical Process Variables for Fat Crystallization

Table 10.4 shows the critical process variables for the fat crystallization process, the recommended conditions, and the consequence of noncompliance.

10.4.12 Discussions on the Crystallization Process

The following comments are very important for the fat crystallization process (Fig. 10.7).

1. The feed must be completely melted and fed to the freezer at a temperature 15–20°F (8.3–11.1°C) higher than the melting point of the shortening mix. This is to ensure complete melting of the hard stocks. Incomplete melting of the hard stock will produce softer than normal consistency in the product.

TABLE 10.3 General Rules of Fat Crystallization

Higher amount of solid in the blend	Produces harder shortening
Slow cooling rate	Produces larger crystals and harder shortening, gets harder in storage
Rapid cooling	Produces smaller crystals, smoother, and softer product
High throughput	The product has softer initial consistency but becomes hard in storage
Slow throughput	Larger amount of nuclei, may start crystal growth in the unit A and the piping leading to the unit B, and may even cause plugging of the system
Low agitation in the unit A	Produces shortening that becomes hard in storage
High agitation in the unit A	The product is softer
Low agitation in the unit B	The product may appear lumpy
High agitation in the unit B or in the tempering tank	The product may lose its matrix, it may not regain the matrix
Use of back pressure	• Back pressure (throttling) at the end of the unit A produces smoother consistency in the product, and dissolves nitrogen in the product • Sometimes an additional throttle valve is used after the unit B

2. A higher temperature differential between the feed and the refrigerant produces more nuclei, but a very high temperature of the feed beyond the limit discussed earlier will significantly increase the energy cost for the refrigeration.
3. The flow rate must be in accordance with the design of the process (units A and B).
4. Exceeding the designed flow produces soft initial product, which becomes hard later in storage.
5. A rate lower than the recommended flow causes crystallization before unit B and can freeze up the system.
6. Incomplete melting of the product returned from unit B produces prenucleation and softer consistency product than the normal product.

10.4.13 Establishment of Crystal Matrix

It is important to know if the shortening leaving unit B will develop the proper crystal matrix within a reasonable time. This is critical for the following reasons:

1. The shortening might be packaged in cans. These cans go through a number of handling steps before they are placed in the cardboard cases. The

TABLE 10.4 Critical Process Variables in Fat Crystallization

Process condition	Operating condition	Consequence of noncompliance
The incoming feed temperature	Typically >150°F (65.6°C)	To ensure that all saturated triglycerides are completely melted, otherwise the desired consistency for the shortening will not be achieved
Cooler (unit A, also called the freezer) outlet temperature	• Commonly referred to as FOT • The outlet temperature must be controlled at the target ±2°F (1°C) • This varies, depending on the amount of solids and the type of product	• Higher FOT will produce larger crystals; softer initial product, after tempering will be firmer than the target • Lower FOT will produce smaller crystals; firmer initial product will be softer than the target after tempering
Temperature at the exit of unit B	• The outlet temperature of the product from the unit B varies with the type of product • This can range from 6–20°F (3.3–11.1°C) higher than the FOT	• The higher outlet temperature at the exit of unit B indicates the release of heat of crystallization • A lower than the typical temperature indicates lack of crystallization; this means fewer nuclei produced in the unit A for some reason
Remelt	All returned product must be completely melted	Incompletely melted returned shortening will make the product softer due to: • prenucleation which makes the product softer • crystal memory (due to the secondary bond) can alter the consistency of the product
Agitator shaft speed (unit A)		
Shortening and margarine	1200–1300 scrapes/min	• A higher rpm may permanently destroy the primary bonds
Industrial margarine	1000–1100 scrapes/min	
Puff pastry	800–1000 scrapes/min	
Total residence time in the system	1.5–2 min	• A longer residence time in the unit A may start the crystallization process in it or in the piping between units A and B; this will clog up the system • Crystallization process must not start before the unit B

FOT, Freezer outlet temperature.

shortening must develop the proper crystal matrix so it can withstand the physical handling of the packaging equipment without the can lids getting smeared by the shortening.

2. In a bakery, the shortening from unit B goes to a tempering tank before it is pumped into the mixers. Here the shortening stays for 1–2 h for the development of the crystal matrix before it is added into the mixers for making dough. An inadequately developed crystal matrix or the shortening that has been overworked in the process will produce soft and sticky dough.
3. In the sandwich cookies, inadequately developed crystal matrix will cause oozing of the filling material and smear the wrapper.
4. In shortening packaging, the product from the unit B goes to the filler, where it is packaged in cans, cubes, pails, etc., and then sent to tempering (to be discussed later). The crystal matrix must be fairly developed before the product is packaged to avoid smearing of the product on the lids of the cans or the pails.
5. In the case of stick or industrial margarine, the crystal matrix is allowed to be developed in a former tunnel (tube) before it goes to the packaging machine.

Therefore, it is important, at least during the development stage of a shortening formula or a newly installed crystallizer system, that the crystal matrix development in the shortening be properly studied. This is very important for a bakery where a fat crystallizer is used. A new shortening formula should be checked for its crystal matrix development property. The following method can be used to conduct this test:

1. A penetrometer with a light-weight, wide-angle cone is used (Fig. 10.8). The picture shows an automatic penetrometer, but a manual penetrometer could be used with no loss of accuracy.
2. The fresh product from unit B (before tempering) is tested with this light-weight, wide-angle cone.
3. Several samples are collected and the time of collection is marked on the samples.
4. The tip of the cone is lowered to touch the surface of the product, and the cone is suddenly released.
5. The cone sinks into the product and eventually comes to a stop.
6. The distance traveled by the cone is a measure of the point at which the crystal matrix is being formed.
7. The penetration value is initially high, but it becomes lower and constant after a certain time.
8. The time to reach the steady-state value of penetration is a measure of the rate of setting of the matrix. A longer time means a slow-setting rate for the product. Fig. 10.9 shows the concept behind this test and it also indicates that in the test case the steady state was reached 20 s after the product left unit B.

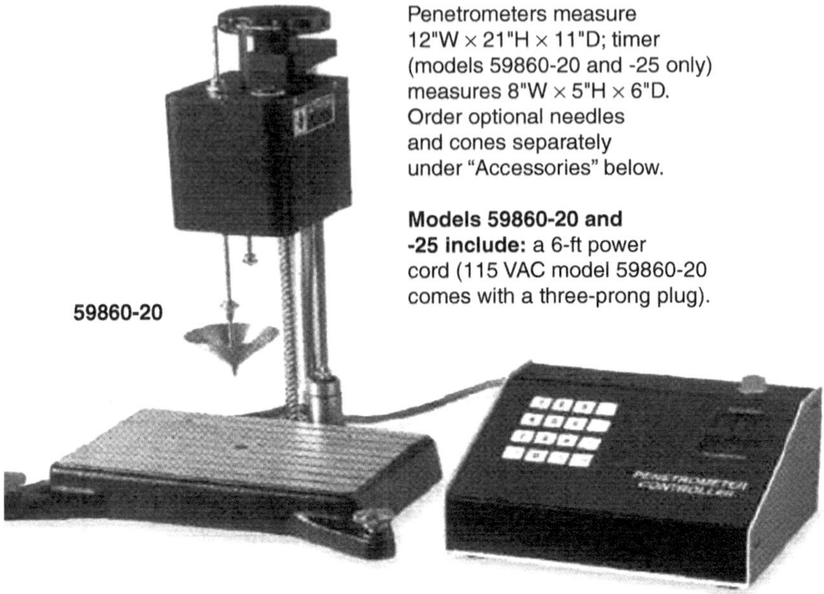

FIGURE 10.8 **Cone penetrometer to measure crystal matrix setting.**

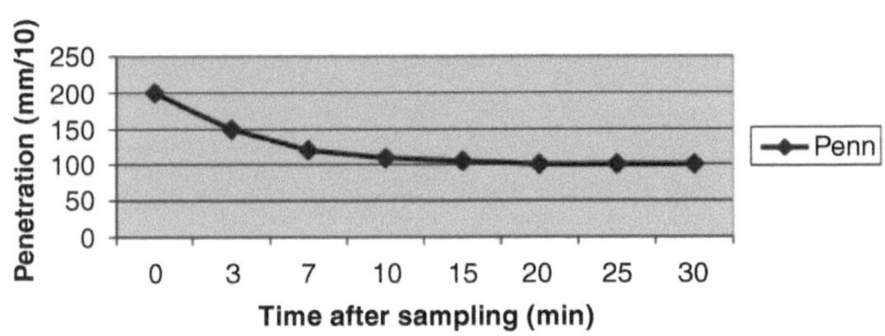

FIGURE 10.9 **Measurement of setting of a crystal matrix with a cone penetrometer.**

10.4.14 Purpose of Tempering

This is an important step in fat crystallization for making shortening. In this process, the product coming out of unit B and packaged in cans or cubes, undergoes an equilibration process. Some of the trisaturated triglycerides may have been supercooled in the crystallization process. On the other hand, some of the disaturated triglycerides might have completely crystallized at the same time.

The tempering process allows the supercooled trisaturated triglycerides to recrystallize from the supercooled state and some of the crystallized disaturated triglycerides may return to the liquid state. In reality, this is a very complex process and requires further discussion.

1. In the manufactured packaged shortening (cans or cubes), the product is stored in a constant temperature room for 24–48 h for tempering of the product after packaging.

2. The product is allowed to come to equilibrium, where any amount of supercooled trisaturated triglyceride is allowed to recrystallize. At the same time, some of the disaturated triglycerides may go into the liquid state from the pseudocrystal form.
3. Lower-tempering temperature produces a finished product with softer consistency.
4. Higher-tempering temperature produces a product with harder consistency.
5. Untempered product may be soft initially but continues to get harder in storage. In the untempered product hardening starts soon, within days of production.

10.4.15 Comments on Tempering of Shortening Made and Used at a Large Bakery

At the bakery, a cake shortening is tempered in the process in a tempering tank with a slow-moving agitator scraping the crystals from jacketed walls of the tempering tank. This allows the crystal matrix to be formed before the shortening is put into the dough mixer. The typical tempering time for the shortening is:

1. 1–2 h for shortenings made from seed oils.
2. 1.7–2 times longer for shortening made from 100% palm oil fractions.
3. This time can be reduced by using a certain amount of fully hydrogenated cottonseed oil.

10.4.16 Tempering Procedure

This is done to all packaged shortening used for making pastries and icing to obtain the desired crystal matrix.

1. Tempering of packaged solid shortening is done in an enclosed room maintained at 85°F (29.4°C) in the United States. The room has one sliding door with good seals to prevent air leakage. This temperature is chosen based upon the average ambient temperature of the country.
2. The product is stacked in cases on wooden or plastic pallets and stored in the tempering room.
3. The products are stacked two-pallets high and there should be plenty of space between the product stacks and the ceiling.
4. There are heating and temperature control devices along the walls that allow the room temperature to be maintained on target.
5. Air circulation fans should be placed at various strategic locations in the room for good air circulation around the pallets.
6. To prevent moisture condensation on the packages, the humidity must be controlled where the ambient humidity is very high. This is because there are many geographic areas where the ambient temperature is 100°F (37.8°C) with a relative humidity of 90%. This air will deposit some moisture when cooled to 85°F (29.4°C) without the humidity control. Therefore, the

humidity should be reduced to <60% at 85°F (29.4°C) to prevent any mold growth in the tempering room.

10.4.17 Benefits of Tempering Shortening

There are several benefits derived from tempering of the solid shortening:

1. Tempering temperature is carefully chosen to obtain the desired crystal matrix in the shortening.
2. In the absence of this step, the crystal structure in the product continues to change with time and temperature during storage and distribution.
3. The consistency of the shortening changes even in properly tempered shortening when it is exposed to high or low temperature. However, there is always some recovery of the consistency as the product is returned to the original storage temperature.
4. Properly tempered shortening exhibits plasticity over a wider range of temperature, as well as fluctuations in the storage temperature.

Figs. 10.10–10.12 illustrate the impact of tempering temperatures on solid shortening. These figures also demonstrate the impact of the storage temperature on the fresh consistency, as well as the plasticity of the shortening.

Three temperatures were chosen for tempering the product. These were:

1. 85°F (29.4°C), which is the average warehouse temperature in the United States.
2. 90°F (32.2°C) and 70°F (21.1°C) were chosen to demonstrate the impact of tempering temperature on the consistency of the same product.

In this test:

1. The product was tempered at these temperatures for 48 h.
2. The products were then placed at 50, 60, 70, 80, 90, and 100°F overnight to equilibrate at these storage temperatures.
3. Penetrations were measured at the various storage temperatures.

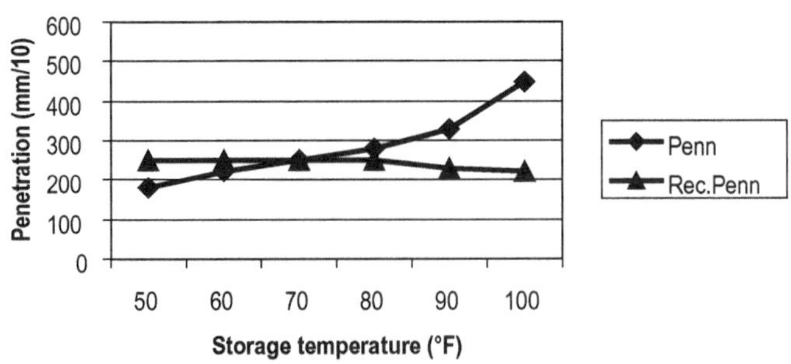

FIGURE 10.10 Tempered at 85°F (29.4°C).

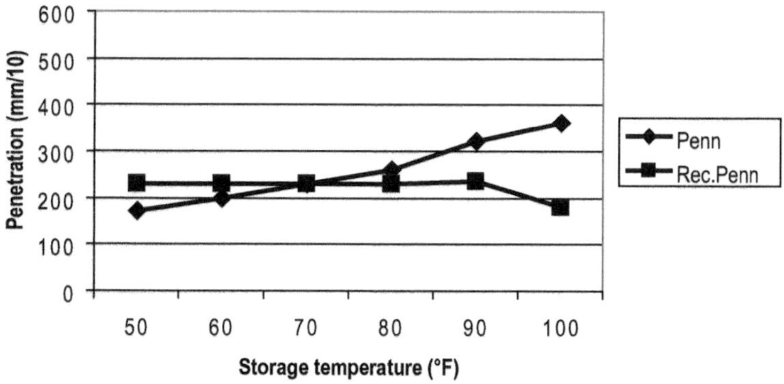

FIGURE 10.11 Tempered at 90°F (32.2°C).

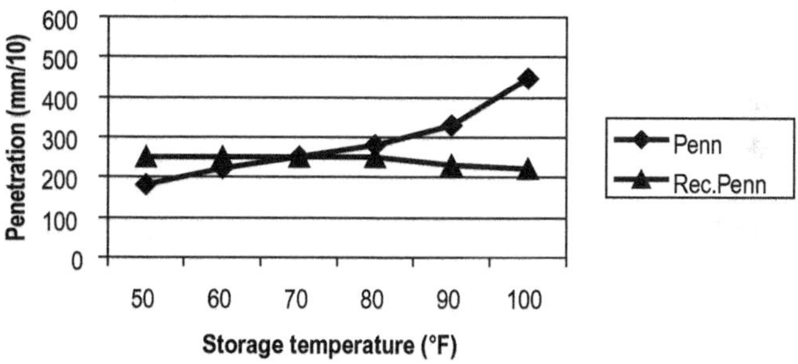

FIGURE 10.12 Tempered at 70°F (21.1°C).

4. All of the product samples were then removed from the respective storage rooms and placed at 70°F overnight to equilibrate.
5. Penetrations were measured on all samples. This is designated "recovery penetration."

The fresh and recovery penetration data are shown in Table 10.5 and the plots are shown in Fig. 10.10 (85°F), Fig. 10.11 (90°F), and Fig. 10.12 (70°F).

Compared to the sample tempered at 85°F, the other samples produced very different results, described further.

10.4.17.1 Sample Tempered at 70°F (21.1°C)

The following observations were made on the sample tempered at 70°F (21.1°C)

1. The sample had softer consistency (higher-penetration values) at all storage temperatures.
2. The recovery penetration indicated that the same sample had a shorter range of plasticity (the recovery penetration remained flat over a narrower temperature range).

TABLE 10.5 Penetration and Recovery Penetration Data on Shortening Tempered at Different Temperatures

Temperature °F	Tempered at 70°F		Tempered at 85°F		Tempered at 90°F	
	Penetration (mm/10)	Recovery penetration (mm/10)	Penetration (mm/10)	Recovery penetration (mm/10)	Penetration (mm/10)	Recovery penetration (mm/10)
50	230	300	180	250	170	230
60	270	300	220	250	200	230
70	**300**	**300**	**250**	**250**	**230**	**230**
80	350	270	280	250	260	230
90	420	250	330	230	320	235
100	540	220	450	220	360	180

The bold numbers signify penetration and recovery penetration at 70°F. This is because 70°F is a significant temperature for shortening storage and use.

10.4.17.2 Sample Tempered at 90°F (32.2°C)

1. The sample had harder consistency (lower-penetration values) at all storage temperatures.
2. The recovery penetration indicated that the same sample had a wider range of plasticity (the recovery penetration remained flat over a wider-temperature range).

10.5 CHARACTERIZATION OF FAT CRYSTALS

Crystalline fat products are characterized in many ways, such as:

1. hardness
2. consistency
3. plasticity
4. structure
5. spreadability
6. pourability
7. polymorphic phase

10.5.1 Hardness

Hardness in solid shortening is measured by a penetrometer, shown in Fig. 10.13.

The penetrometer is generally manually operated. Automatic units are also available (Fig. 10.8). A solid cone of specific weight is dropped on the clean

FIGURE 10.13 Penetrometer.

surface of the shortening from a fixed height above the surface. The distance traveled by the cone into the product is a measure of its hardness.

The texture meter does the same thing with an instrument that measures the force required for the needle to penetrate through a given distance into the shortening.

10.5.2 Consistency (Smoothness/Graininess)

The texture-meter can perform the test, where a probe is inserted into the product and the product is moved laterally. A smooth product produces a steady straight line on the recorder chart. A grainy product produces peaked or jagged lines, indicating the resistance experienced by the probe due to grainy or lumpy product.

10.5.3 Plasticity/Spreadability

A penetrometer is used to determine plasticity in a given shortening. This can be measured as was discussed on the data presented in Table 10.5 and Figs. 10.10–10.12.

10.5.4 Structure

The structure of a shortening can be measured by a texture meter called TA.XT21, made by Texture Technologies. There are other instruments that can also measure the texture in a shortening.

10.5.5 Pourability

This relates only to pourable shortening and not solid shortening. The method will be described under the section : Pourable Shortening (Section 9.1).

10.5.6 Polymorphic Phase

- This is normally done with the X-ray diffraction method (see results of X-ray diffraction results on fat crystals in Fig. 10.14).
- The polymorphic appearance can also be determined by the electron microscope (Fig. 10.15)

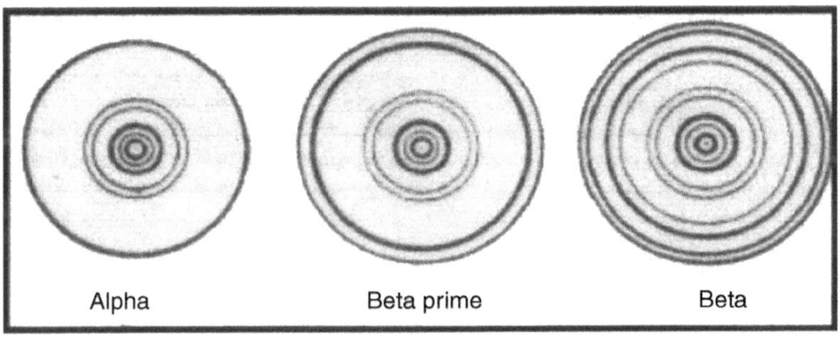

FIGURE 10.14 **X-Ray diffraction pattern for tristearine crystal.**

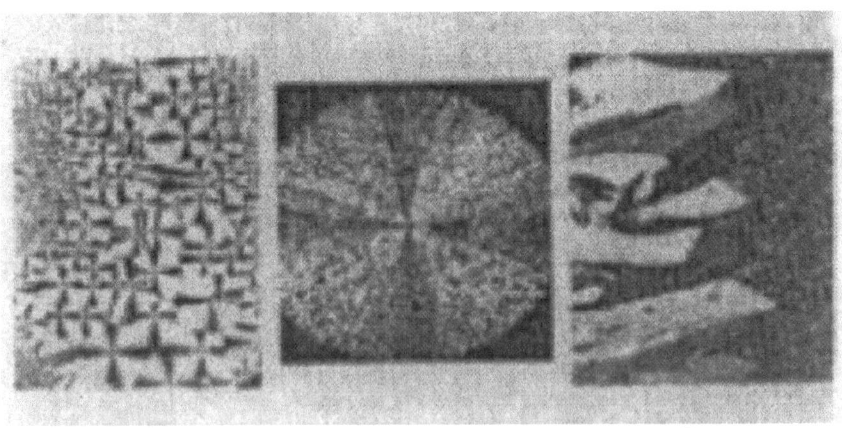

Alpha Beta prime Beta

FIGURE 10.15 **Electron microscopic view of fat crystals under polarized light.**

- Differential thermal calorimeter (DSC) can also be used to determine the predominant polymorphic form of crystals present in a shortening. It has been described in Table 10.2 that each polymorphic phase of fat crystal has a specific melting point. In addition, each requires a distinct amount of latent heat of fusion. The DSC method can determine the presence of any polymorphic phase of a fat through melting and cooling curves controlled at a certain rate for cooling or heating.

10.6 PALM OIL IN SOLID SHORTENING

Palm oil shortening takes longer to get the crystal matrix established compared to shortening made from hydrogenated seed oils. It has been mentioned earlier that the tempering time requirement is 1.7–2 times longer for palm oil shortening for bakery plants compared to that for shortening made from seed oils. It is understood that the following factors are responsible for this difference.

1. The crystal growth process is slow in palm oil shortening.
2. The transition time from alpha to beta prime phase is longer in palm oil than in the hydrogenated seed oils.
3. Commercial palm oil can contain 3%–10% diglycerides. This is one of the major factors for slow crystal growth in the palm oil shortening.
4. A longer residence time is required for palm oil shortening in the crystallizer system. However, at a very low–flow rate, crystal growth from the nuclei begins after unit A, and in the early part of the work unit (unit B). This can be detrimental to proper crystallization of the shortening made with palm oil:
   a. Slow throughput develops fine crystals by destroying the primary bonds.
   b. High throughput delays the crystallization process and continues into the tempering stage and beyond, resulting in larger crystals that produce harder consistency in the shortening.

Palm oil shortening requires extra A and B units to compensate for the slower rate of the crystal matrix formation. The end product works perfectly in all applications. Certain icing formulae can show lower volume when made from 100% palm oil.

Fig. 10.16 shows the schematic flow diagram for the process for palm oil shortening.

10.6.1 Improving Crystallization Rate in Palm Oil Shortening

The crystallization rate in the palm oil shortening can be improved significantly by using one of the following options:

1. Using partially hydrogenated palm oil in the formula.
2. Using a small amount of fully hydrogenated palm oil in the formula.
3. Using a small amount of fully hydrogenated cottonseed oil in the formula.

The rate of transformation from the alpha to beta prime in the palm oil products is found to be as follows:

Hydrogenated palm oil → Palm oil → Palm stearine

10.7 ISSUES WITH THE INTERESTERIFIED PRODUCTS

Interesterification of fully hydrogenated cottonseed oil, palm oil, or palm stearine with liquid oil has become the subject of great interest in reducing *trans* fat in shortening. These products have a very small amount of trisaturated triglycerides left after the interesterification process. As a result, the product takes longer to form the crystal matrix and behaves more like an-all palm oil shortening. The process unit looks very much like the one shown in Fig. 10.16.

10.8 VERY HIGH–HARD STOCK CONTENT

In an attempt to reduce the *trans* fat, some oil manufacturers have formulated products with very high (15%–16%)—hard stock with the rest made of liquid oil. This product does not behave like the regular seed oil shortening in the freezer. A standard shortening becomes softer and yields when stress is applied to it. This is known as the thixotropic property of the shortening.

The very high–hard stock blend shortening does not exhibit thixotropic behavior in the crystallization process. Contrary to the standard shortening, these blends exhibit dilatent fluid behavior. This means the shortening becomes

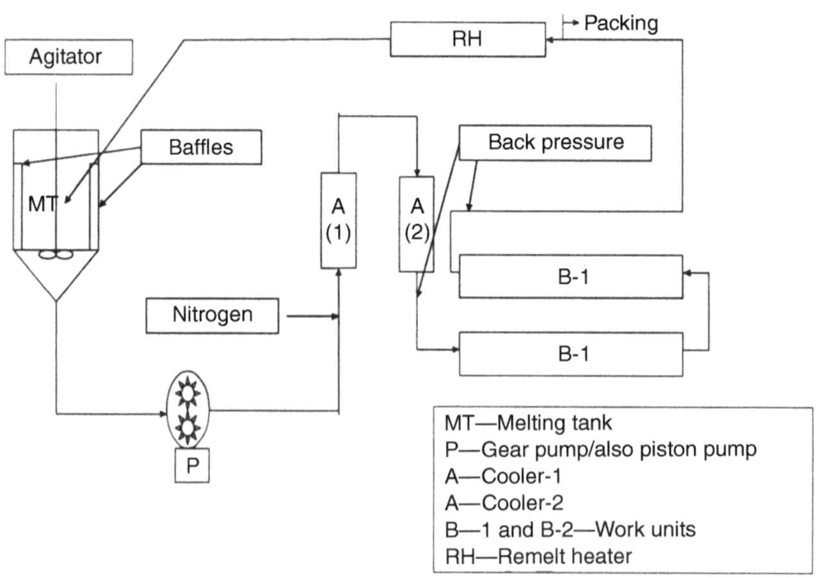

FIGURE 10.16 Schematic flow diagram for processing palm oil shortening.

harder as more work is applied on it. Therefore, the shortening requires much larger A and B units to accommodate the significantly lower-flow rate and also a lower rpm of the work unit to prevent work hardening of the shortening during processing.

10.9 POURABLE LIQUID SHORTENING

10.9.1 Product Description

This is an opaque fluid product made from hydrogenated vegetable oils. The product can remain fluid at room temperature for a long time.

10.9.2 Special Properties

The liquid shortening offers the following distinct advantages:

1. It can be poured out of a can or a bottle.
2. It can be squeezed out of a soft bottle.
3. It can be easily poured in a cup for recipe preparation.
4. It can be easily mixed with dry ingredients.

Comparison between the solid and the liquid shortenings are shown in Table 10.6, showing the suitability of their applications.

10.9.3 Formulation

Pourable shortening is made from a mixture of fully hydrogenated fat (hard stock) and liquid oil using a crystallization process with a somewhat similar crystallizer as for the solid shortening, but in many ways it is quite different in

TABLE 10.6 Comparison of Solid and Liquid Shortenings

Areas of comparison	Solid shortening	Liquid shortening
Consistency	Solid	Fluid (pourable)
Applicability	Baking	Baking
	Cakes	Bread
	Pastries	Soft cookies
	Cookies	Nutri bars
	Crackers	
	Bread	
	Pie Shells	
	Frying	Frying
	Heavy-duty frying	Heavy-duty frying

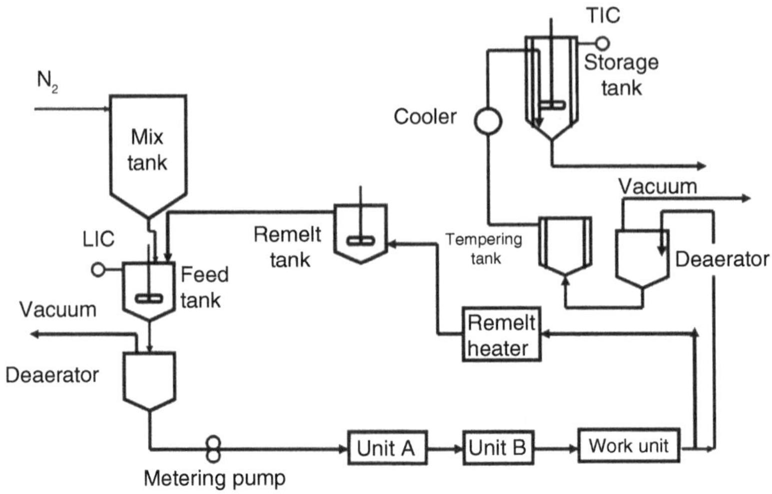

FIGURE 10.17 **Schematic flow diagram for making pourable shortening.** LIC, Level Indicator Controller; TIC, Temperature Indicator Controller.

principle from the process for making solid shortening. The schematic process flow diagram is shown in Fig. 10.17.

Table 10.7 shows the various types of oils that can be used to formulate the liquid shortening.

10.9.4 Polymorphic Phase

1. This is a beta stable product.
2. Small crystals of the high-melting fraction in beta phase are suspended and uniformly distributed in the liquid fraction.

TABLE 10.7 Formulation of Pourable Liquid Shortening

Solid fraction	Liquid fraction
Fully hydrogenated oil (IV < 8) Soybean oil Canola oil Sunflower oil Corn oil Mid-oleic sunflower oil High-oleic sunflower oil Low-linolenic soybean oil Low-linolenic canola oil Mid-oleic, high oleic Canola oil Amount of solid fraction 2%–8%	Partially hydrogenated oil Soybean oil (IV 104–110) Canola oil (IV 86–94) Liquid oil (unhydrogenated) Sunflower oil Mid-oleic sunflower oil High-oleic sunflower oil Corn oil Palm olein Low-linolenic canola oil High-oleic canola oil Amount of liquid fraction 92%–98%

IV, Iodine value.

Beta prime phase crystals, such as cottonseed or palm oil hard stock, will make the product nonpourable.

10.9.5 Processing Steps for Making Pourable Liquid Shortening

(Fig. 10.17)

1. The blend of the solid and the liquid fractions is transferred from the mix tank to the feed tank.
2. The mix temperature must be sufficiently high to melt the hard stock and produce a high-temperature differential (DT) between the oil and the refrigerant temperature to produce small crystals.
3. The mix is deaerated under vacuum in the deaerator vessel, before the oil mix enters unit A.
4. The nuclei formed in unit A then enter unit B to produce crystals.
5. A work unit is added after unit B to provide some additional time for crystallization and breaking any agglomerates.
6. During start-up, the product leaving the work unit is recycled back to a remelt tank where the product is heated to at least 10°F (5°C) above the complete melt point of the mix.
7. The residence time in the remelt tank is at least 45 min.
8. The mix is returned to the mix tank for recycling.
9. Once the process conditions reach the steady state, the product is diverted to a second deaerator under vacuum to remove any air or nitrogen entrapped in the product.
10. The product then passes through a jacketed tempering tank, which is maintained at 90°F (32.2°C).
11. The tank has a scrape-wall top-entering agitator that runs at a very slow rpm to keep the tank walls clean of crystals without damaging them.
12. The residence time in the tempering tank is 45–60 min (can be up to 90 min) to ensure complete transformation of all crystals to the beta form.
13. The product is cooled to 70°F (21.1°C) in a cooler using cold water.
14. The product is stored in a jacketed storage tank with a slow agitator to keep the product in motion without breaking the crystals.
15. The product is checked for fluidity before it is either loaded into trucks or sent to packaging.

10.9.6 Critical Control Points

10.9.6.1 Formulation of the Mix

1. The mix formulation must be done accurately.
2. Care must be taken that not even a trace amount of cottonseed or palm hard stock gets blended in the mix.

10.9.6.2 Deaeration

The mix must be deaerated before it enters unit A. In the absence of this step the product fails for the reasons as listed:

1. The product can get thicker (lower fluidity) than desired.
2. The product can separate in the package during storage, showing a floating layer of the solid fraction separated from the product.

10.9.6.3 Freezer (Unit A) Outlet Temperature

1. This temperature will vary depending on the amount of the hard stock in the mix.
2. The freezer outlet temperature (FOT) must be kept as low as possible. Normally it is about 70°F (21.1°C).
3. A lower FOT produces smaller crystals.
4. The FOT should be controlled within ±2°F (1°C).

10.9.6.4 Tempering Temperature

This is maintained at 9°F (32.2°C) to maximize the transformation of the crystals to beta phase.

10.9.6.5 Tempering Time

The residence time in the tempering tank should be at least 45–60 min (can be up to 90 min) to ensure a complete transformation of all beta prime crystals to beta.

10.9.6.6 Agitation in the Tempering Tank

The recommended agitator speed is 5–10 rpm, using a soft scraper to remove the crystals from the wall without breaking them.

10.9.6.7 Hot Water Temperature in the Jacket

The water in the jacket must be maintained at 88–92°F (31–33.3°C).

10.9.6.8 Storage Tank Design

1. It must be a jacketed tank with a cone bottom and heated with warm water.
2. Alternatively, electrical tape on the outside of the tank could be used.
3. The jacket temperature should be maintained at 91–100°F (33–38°C).
4. This will melt the product along the inner wall, producing a very thin layer of liquid oil a few microns thick, which facilitates the movement of the product during a transfer either to a tank truck or to packaging.
5. The tank should preferably have a top-entering scrape-wall agitator running at 5–10 rpm.

10.9.6.9 Storage of Pourable Shortening in the Warehouse

1. The maximum recommended storage temperature is 77°F (25°C).
2. The product can separate into liquid and solids with the liquid floating at the top if it is exposed to a temperature exceeding 86°F (30°C).

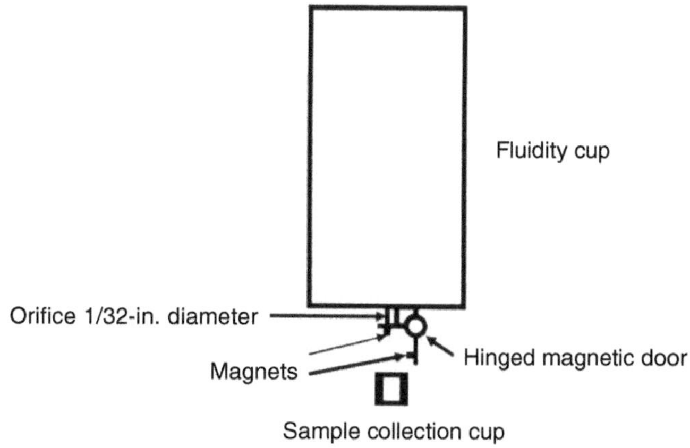

FIGURE 10.18 **Fluidity measuring cup.**

10.9.6.10 Shipping (Transit)

The warehousing temperature conditions also apply for shipping.

10.9.7 Fluidity of the Shortening

This is the most important physical attribute for the pourable liquid shortening. This is measured in terms of grams of shortening passing through a fixed aperture in 30 s.

Fig. 10.18 shows the basic design for a fluidity measuring cup with a sample collection cup.

1. The product, typically at 70°F (21.1°C), is placed in the fluidity measuring cup.
2. The cup has an orifice, which is typically 1/32 in. in diameter.
3. A hinged door is operated with the help of a solenoid valve at the bottom of the orifice.
4. The hinged flap door remains open for 30 s and then automatically closes.
5. The sample is collected in the sample cup.
6. The amount of sample (in grams) collected in 30 s is normally reported at the fluidity of the product.

The actual mode of measurement can vary between companies but the general principle for fluidity of the liquid pourable shortening is the same.

READING REFERENCES

Anderson, K.A., 1995. Margarine processing plants and equipment. In: Bailey's Industrial Oils and Fats, fifth ed. John Wiley Sons, New York.

Baur, F.J., Jackson, F.L., Kolp, D.G., Lutton, E.S., 1949. J. Am. Oil Chem. Soc., 26.

DeMan, J.M., Beers, A.M., 1987. Fat crystal networks: structure and rheological properties. J. Texture Stud. 18 (4), 303–318.
DeMan, J., D'Souza, V.L., deMan, J.M., Blackman, B., 1992. J. Am. Oil Chem. Soc. 69, 246.
Hoerr, C.W., 1960. J. Am. Oil Chem. Soc., 37 (10), 539.
Hoerr, C.W., 1964. J. Am. Oil Chem. Soc., 41.
Hoerr, C.W., Zimba, J.V., 1965. Food Eng.
Hugenberg, F.R., Lutton, E.S., 1963. J. Chem. Eng. Data 8, 606.
Larsson, K., 1966. J. Am. Oil Chem. Soc. 43, 561.
Lutton, E.S., 1946. IBID 68, 676.
Lutton, E.S., 1948. J. Am. Oil Chem. Soc. 70, 248.
Lutton, E.S., 1950. J. Am. Oil Chem. Soc. 27, 276–280.
Lutton, E.S., 1957. J. Am. Oil Chem. Soc. 34, 521.
Lutton, E.S., 1971. IBID 48, 245.
Lutton, E.S., Fehl, A.J., 1970. Lipids 5, 90.
Lutton, E.S., Hugenberg, F.R., 1960. IBID 5, 489.
Malkin, T., 1931. J. Chem. Soc., 2796.
Marangoni, A.G., Narine, S.S., 2002. Physical Properties of Lipids. AVI Publishing.
Müller, A., 1923. J. Chem. Soc. 123, 2043.
Müller, A., 1930. Proc. R. Soc. 127, 417.
Shukla, V.K.S., Nielsen, W.S., Batsberg, W., 1983. Fette. Seifen. Anstrichm. 85, 274.
Von Sydow, E., 1958. Acta Chem. Scand. 12, 777.
Wille, R.L., Lutton, E.S., 1966. J. Am. Oil Chem. Soc. 43, 491.

Chapter 11

Winterization and Fractionation of Selected Vegetable Oils

11.1 INTRODUCTION

Liquid oils are used both as cooking oil and salad oil. Salad oil is defined as a clear oil which must not exhibit cloudiness when stored in the refrigerator at 40°F (5°C) for several hours.

The term winterization implies that the oil is subjected to cold temperature in order to separate some amount of solids and then separate the solids from the liquid fraction. The solids are removed via filtration process. The liquid fraction thus obtained is used as cooking oil, salad oil. This can also be used for formulation of margarine and shortening. Winterization process removes waxes from sunflower oil, safflower oil, canola oil, and corn oil and it removes the small amounts of stearines present in cottonseed oil. Without this process, the oil appears cloudy when stored in the refrigerator. Winterization makes salad oil that remains clear under refrigeration for several hours or even in some cases up to 24 h.

The standard method recommended to check for the cloudiness in the winterized oil by the American Oil Chemists' Society is called "cold" test Method Cc 11-53 (09). This test is conducted by taking filtered and deaerated salad oil in sealed clear glass bottles and holding it at 32°F (0°C) until cloudiness appears in the oil. The typical cold test for sunflower salad oil is 24 h under refrigeration temperature of 32°F (0°C). However, cold test of 5.5 h minimum is common for winterized cottonseed oil in many parts of the world. Solvent winterized, partially hydrogenated soybean salad oil exhibits a cold test of 24 h or longer.

Fractionation is similar to winterization in terms of chilling of the oils in order to create a solid and a liquid phase in the oil. The solids phase in the cottonseed oil is made of saturated triglycerides with high melting point. In case of partially hydrogenated soybean or regular palm oil, the solid fraction is made of saturated triglyceride, disaturated glycerides, and some monosaturated glyceride. This process is commonly performed on the following oils:

1. Partially hydrogenated soybean oil, to make salad oil and soy-stearin fraction which can be used in margarine and shortening formulation.

Practical Guide to Vegetable Oil Processing. http://dx.doi.org/10.1016/B978-1-63067-050-4.00011-8
Copyright © 2017 AOCS Press. Published by Elsevier Inc. All rights reserved.

2. Fractionation of palm oil is done to produce various palm stearin and palm olein fractions. Palm olein is used as cooking and frying oil. Palm stearin is used in formulating palm oil shortening and margarine and to make cocoa butter extenders (CBE).

11.2 WINTERIZATION OF SUNFLOWER SEED OIL

Crude sunflower oil may contain 300–400 ppm of wax in the oil expelled from well-corticated seeds, and this can be much higher and can be up to 2000 ppm if the seeds are not properly decorticated.

The wax content in sunflower oil is reduced through the winterization process. In this process, the oil is heated at first to ensure that all of the waxes are completely melted. The oil is then chilled under controlled cooling rate. This separates the waxes from the oil, which is then separated from the rest of the oil through filtration. Fig. 11.1 shows the schematic diagram for the sunflower oil winterization process.

The typical steps of procedure for the process are described as follows:

1. The feed to the winterization process must be of uniform composition. Therefore, it is advised to use a very large oil supply tank with mechanical agitator to maintain uniform conditions for the feed to the chiller. Sometimes it is even referred to as an ocean of oil for feed.

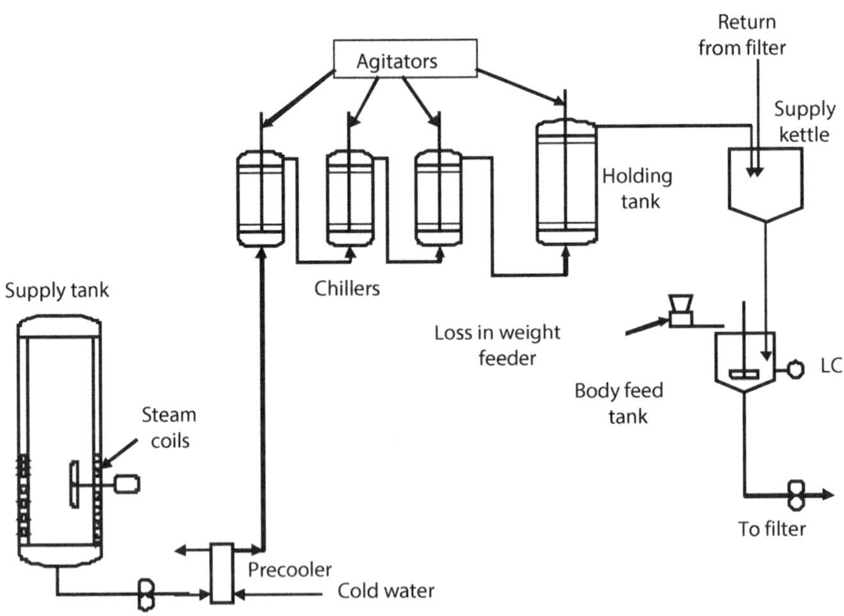

FIGURE 11.1 Schematic diagram for sunflower oil winterization process.

2. The oil from the supply tank is heated to 140°F (60°C)–150°F (66°C) to melt the waxes. The supply tank has a side-entering mechanical agitator, steam coils with temperature controller, and thermodynamic steam trap (not shown in Fig. 11.1).
3. The oil is then pumped through a precooler to cool the oil to 104°F (40.6°C) as it enters the chiller.
4. The chiller is a jacketed stainless steel tank. The oil is cooled at the rate of 5–7°F/h (3–4°C/h).
5. A top-entering scrape-wall agitator running at a speed of 5–10 revolutions per minute (rpm) gently scrapes the inner wall of the chiller. The chilled water in the jacket crystallizes the waxes in the oil. The scrape wall agitator removes the solidified waxes from the walls of the chill tank, keeps them in suspension and, at the same time, keeps the inner wall of the tank free of solids so the heat transfer rate is properly maintained.
6. The oil is cooled down to a final temperature of 42–45°F (6–7°C).
7. The oil is left in one or two holding tanks for 12–24 h.
8. Holding time is critical for the oil for proper growth of the solids.
9. The oil is then heated to 50–54°F (10–12°C) to reduce the viscosity of the oil for better pumping.
10. The winterized oil is then filtered to separate the solid and the liquid fractions under controlled temperature conditions.
11. Wax is removed from the oil in a pressure-leaf filter precoated with diatomaceous earth.
12. The technique for precoating the filter has been discussed in detail in Chapter 6 on bleaching.
13. A constant body feed of diatomaceous earth is maintained in the oil flow into the filter. Without this, the filter screens will clog rapidly, reducing the cycle time for the filter.
14. Filtration is stopped as the differential pressure across the screen reaches 30–35 psi (2.18–2.54 kg/cm^2).
15. The oil inside the filter is blown with air, and the residual oil is blown to the blow down tank. The cake is dried with compressed air. The filter is opened and the cakes are removed from the filter screens.
16. Waxes from sunflower oil are collected, the diatomaceous earth is separated, and the wax is sometimes blended with oils to make finished shortening.
17. The pressure-leaf filter needs periodic hot oil wash to remove any residual wax on or inside the screens.

11.2.1 Cold Test Versus the Wax Content of Sunflower Oil

As discussed earlier, the purpose of winterizing sunflower oil is to reduce the wax content and improve the cold test. Based on the industry experience, the wax content of the oil must be <14 ppm in order to obtain winterized sunflower oil with a cold test of >24 h.

11.3 CRITICAL PROCESS VARIABLES FOR WINTERIZATION OF SUNFLOWER OIL

The winterization process is different from the other processing areas in the sense that here the oil is cooled under controlled conditions to separate the solids from the rest of the oil and then filtered. There are several critical process variables that determine the efficiency of the process and the quality of the winterized oil produced. The following are the important process variables for the process:

1. incoming oil quality,
2. incoming oil temperature,
3. cooling rate for the oil,
4. final oil temperature,
5. agitation during chilling,
6. holding time in the chillers, and
7. filter operation
   - filter precoat and body feed
   - filtration rate
   - pressure differential across the filter screens
   - drying the cake
   - cleaning the filter screens
   - periodic hot oil washing for cloth screens
   - cooling the filter after the hot oil wash
   - maintenance of the filter screens

1. Incoming oil quality

   There are three main impurities in the crude oil that must be reduced with care in the refining and bleaching steps. These are:

Phosphorus	Must be	<1 ppm
Moisture	Must be	< 0.1%
Soap	Must be	0 ppm

   - Any one of the above impurities at higher level can prematurely blind the filter screens, causing slow filtration.
   - This reduces the filter cycle time and, hence, the productivity.
   - High level of phospholipids (phosphorus) can also interfere with the separation of the solids from the oil in the chillers.
2. Incoming oil temperature
   The oil from the supply tank must be well heated and mixed. As discussed earlier, the desired temperature of the oil is 140°F (65°C) to 150°F (66°C) in order to completely melt the waxes in the oil.
   - The temperature of the oil must be well above the complete melt point of the waxes.

- A low temperature may cause prenucleation of the high melting fraction. This can cause the following production difficulties:
  - interference with the growth of proper characteristics for the crystals,
  - poor separation in the chiller, and
  - slowing down of the filtration rate.

A very high temperature will increase the energy cost to precool the oil.

3. Cooling rate for the oil
   The rate of cooling is maintained within a certain temperature range to have controlled formation of the nuclei that would promote the crystal growth.
   a. A rapid cooling rate forms small crystals, which is not desirable in winterization because small crystals can reduce the filtration rate requiring:
      - frequent cleaning of the filter,
      - reduced filter throughput.
   b. Slower cooling rate produces larger crystals. This does not impair the flow of oil through the filter. Therefore improved productivity can be achieved.
   c. The cooling rate for the oil at any plant needs to be established in order to maximize total productivity in winterization.
4. Final oil temperature
   The final oil temperature in the chiller is 42–45°F (6–7°C). A lower temperature is not necessary. It can also increase the oil viscosity and slow down the filtration rate. At higher temperature, there may be some noncrystallized wax left in the oil. This will reduce the cold test of the winterized oil.
5. Agitation
   The oil in the chiller is agitated with a scrape-wall agitator operated at a speed of 5–10 rpm. This agitator removes the solids formed on the wall and keeps the wall surface clean for better heat transfer. A higher speed can agitate the solids too much and disturb the crystal matrix.
6. Holding time
   The chilled oil is held in a holding tank for 12– 24 h for complete crystallization. Sometimes the oil is held for only 4–6 h before filtering. Reduced holding time may reduce the cold test of the winterized oil. This should be tested by the plant. This will also vary with the wax content of the incoming crude oil. Therefore, a longer holding time covers for the higher wax content of the crude oil. Holding tanks are jacketed chilled tanks with temperature control to hold it within ±2°F (±1°C).
7. Filter operation
   Filter precoat and body feed
   The need for filter precoat and the methodology has been described in Chapter 6 under Bleaching. The additional step required in filtering winterized sunflower oil is the body-feed. About 0.25% of diatomaceous earth is added continuously to the oil feed to the filter. This is done with a precision feeder (like a loss-in-weight type feeder) and the diatomaceous

earth is dispersed in the oil in a small feed tank before the filter. Fig. 11.1 shows the body feed tank and the diatomaceous earth feeder. Addition of diatomaceous earth as body feed is vital for the operation of the filter. In the absence of the body feed:

- The screens get plugged up by the waxes.
- The pressure differential increases rapidly across the filter screen.
- The filter loses its throughput.
- Productivity is decreased.

Filtration rate

The recommended filtration rate	0.015 gpm/ft.^{2} of filter area.
	0.125 Imp. gallons/min/m^2.

At higher flow rate:

- The filter screens get prematurely blinded and the filter requires frequent shut downs.
- The screens may get plugged, requiring more frequent hot oil wash.

Proper flow of oil through the filter can be best controlled with the help of a flow controller and a pressure controller. Fig. 11.2 shows the schematic diagram for the pressure leaf filter for filtering waxes from winterized sunflower oil.

- The desired flow rate through the filter is set by the flow indicator controller (FIC). The actual oil flow through the filter is indicated by the flow monitor (FM).

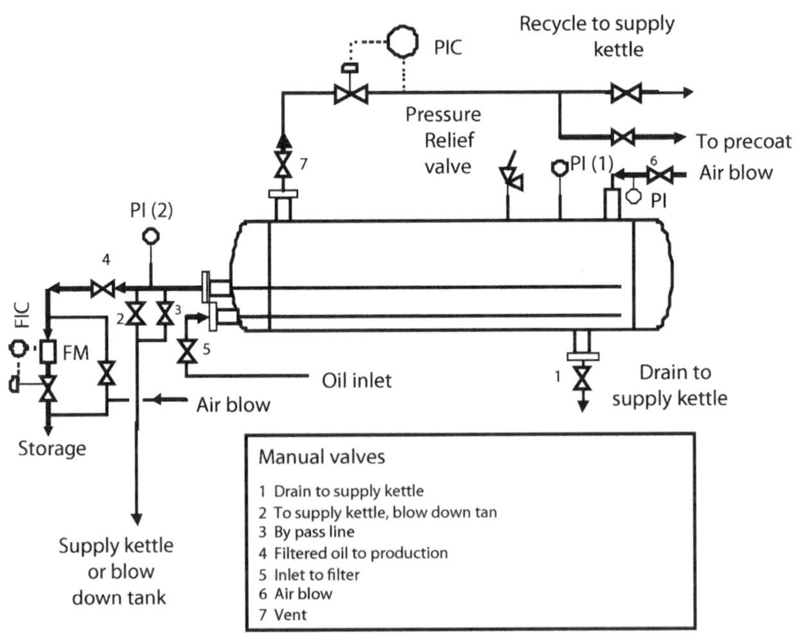

FIGURE 11.2 Schematic diagram for pressure leaf filter.

- Pressure indicator controller (PIC) controls the back pressure in the filter by modulating the pressure control valve (PCV).
- As the pressure inside the filter begins to increase, more oil is returned to the supply kettle (Fig. 11.1).
- The PCV allows the maximum amount of oil bypass to the supply kettle. This minimizes the risk of overpressurizing the screens.
- As the cake on the filter screens builds up, the flow through the filter screens begins to drop.
- The flow control valve (FCV) begins to open to allow more oil flow through the filter.
- At the same time the PCV begins to open in order to maintain a constant back pressure in the filter.
- Ultimately, the FCV will become wide open, indicating maximum output signal from the FIC.
- The PCV will also open to maintain the back pressure in the filter.
- In normal practice, the filtration cycle would be terminated when the PCV valve reaches the wide open position because at this time the filter throughput is at its minimum.
- The actual point of termination of filtration cycle can be determined based on the maximum production per cycle. This can be established through the operating experience at the plant.

Pressure differential across the filter screens

The difference between the pressure in the filter (indicated by the pressure indicators, PI-1) and that in the filtered oil discharge header (indicated by the pressure indicators, PI-2) is known as the pressure differential, P or ΔP, across the filter screens.

The pressure differential across the screen should be zero or near zero at the beginning of the filter cycle. The differential pressure rises as the cake builds up on the screen.

A rapid rise in pressure differential across the screens is caused by one or more of the following factors:

- The screens are either dirty from the previous filter cycle or they are partially blinded.
- Higher than normal oil flow rate through the filter.
- The discharge pressure of the filter feed pump is higher than required.
- The moisture content of the oil is >0.1%.
- Soap content of the oil is >0 ppm.
- The phosphorus content of the oil is >1 ppm.
- The wax content of the sunflower oil is much higher than normal.
- Improper or incomplete precoat of the screens.
- Insufficient amount of body feed or no body feed during filtration.
- Improper chilling of the oil.
- Excessive amount of agitation in the chill tanks.

At the plant, one must investigate the reason for rapid increase of the differential pressure and correct the issue.

The filters must not be operated beyond the safe limit for pressure differential to prevent any damage to the screens. Normally, the safe operating pressure limit is suggested by the filter screen manufacturers.

Drying the cake

The cake on the filter screens must be dried at the end of filtration, before the filter screens are vibrated to release the cake built on the screens. The drying step is very critical in operating the filter because:

- The wet cake does not release itself from the surface of the screens.
- This requires scraping the screen surface. This process increases the risk of damaging the screens if the scraping is done too often or it is done incorrectly.
- The wet cake leaves a residue of the waxes on the filter screens, which can cause premature blinding of the screens during the next filtration cycle.
- The screens require more frequent hot oil wash to remove the waxes from inside of the screens.

The steps for cake drying are listed below:

- At the end of the filtration cycle the oil in the filter is drained back into the filter feed tank by using compressed air through valve #6.
- The cake on the screens is dried by blowing air for 30–40 min through the filter using compressed air at 25–40 psi pressure.
- The volume of air used in this step is typically 0.5 $ft.^3$/min per square foot of filtering area. For example, a 500 $ft.^2$ filter will consume 250 $ft.^3$ of compressed air per minute.

The conventional air blow may not produce the desired results from time to time. This is because the cake formed at the top of the screens may get dry and allow most of the air to pass through, leaving the wet or semi-wet cake at the lower parts of the screens. An alternate method, called "shock-drying," is applied. This procedure is outlined as follows:

1. Referring to Fig. 11.2, close the manual valves, #1, 2, 4, 5, and 7.
2. Open the air inlet manual valve #6.
3. Open bypass valve #3. This is a 1/2 or 3/4 in. valve as compared to valve #2, which is typically 2 or 3 in.
4. Allow the pressure inside the filter to build up to 40–45 psi.
5. Quickly open valve #2 and close valve #3.
6. Let the air flow into any of the tanks chosen (supply kettle, precoat tank, or blow down tank).
7. Repeat steps #1 through #4 several times until the cake is dry.
8. Open the filter and let the screens move over the dump chute for collection of the dry cake.
9. In case of Industrial filter, the screen assembly moves via a hydraulic device. In some other filters, such as Sparkler, and so on, the shell slides, exposing the screens over the discharge chute.
10. The dry cake will fall off the screens into the chute for collection.

Cleaning the filter screens

The screens are best cleaned by vibrating them with a mechanical vibrator, until the cake falls off the screens. In the event that a vibration device is absent one must use a plastic scraper with a smooth, rounded tip so the screen surface is not easily damaged during scraping.

Even with proper drying and using a mechanical vibrator, portions of the cake may remain stuck to the screens. A spatula, as described, should be used with proper care so the screens are not damaged.

Periodic hot oil wash

The filter screens can have a fabric surface or stainless steel mesh surface. Hot oil wash is required for the screens with fabric surface. A good hot oil wash cleans up the pores on the screen surface and improves filtration in the subsequent operation.

The procedure for the hot oil wash is outlined as follows:

- Winterized sunflower oil is heated to 176°F (80°C) in a tank.
- The hot oil is circulated through the filter for sufficient time to dissolve the waxes built up on the screens and some in the interiors of the screens.
- A good indicator for determining the cleaning end point would be zero differential pressure across the filter. PI-1 and PI-2 should read the same pressure, which is more or less the same as the pump discharge pressure minus any line drop.

Cooling the filter after hot oil wash

After the hot oil wash, the filter must be cooled down before the next filtration operation. This is important to prevent melting of any of the waxes in the chilled oil at the start of the next filter cycle. This would lower the chill test (or cold test) of the oil filtered next.

Cooling is done by circulating chilled water through the jacket of the filter until the filter attains ambient temperature.

In many instances the filters do not have any cold water jackets. In such cases, the filters are opened and left open until they reach the ambient temperature.

Maintenance of the filter screens

Proper maintenance of the filter screens is very important. Wax can leak through the damaged screens and produce winterized oil with low cold tests. Therefore, the screens must be handled with care. Some helpful recommendations on proper handling of filter screens are listed as follows:

- The screens, when taken out of the filter, must be handled with extra care.
- The screens must be placed on a rack so they are not in physical contact with one another.
- The screens must be inspected thoroughly for any damage or tear after they are taken out.
- Any damaged or torn screen must be sent to the manufacturer for proper repair, reconstruction, or replacement.

- Any attempt to repair the screens by the plant can be either inadequate, impractical, or produce unsatisfactory performance as compared to those repaired by the manufacturer.
- The "O" rings on the sockets on the oil discharge manifold and the nipples on the screen bottoms must be inspected every time the screens are taken out for any reason.
- With the help of a micrometer, the roundness of the sockets on the discharge header and the nipples on the screens should be checked and replaced if there is any defect.
- The "O" rings should be replaced every 6 months.

Table 11.1 lists the summary of critical control points, their significance, and the consequences for noncompliance.

11.4 TROUBLESHOOTING

The most commonly experienced issues in the winterization process are:

1. Low cold test in the winterized oil.
2. Poor filterability of the winterized oil.

Table 11.2 lists the symptoms, their causes, and suggested corrective actions.

11.5 WINTERIZATION OF SOYBEAN OIL

Although this process is commonly referred to as winterization, in reality it is a dry fractionation process (a process similar to what is used in fractionating palm and palm kernel oils). Partially hydrogenated soybean oil is chilled under controlled conditions, held in a refrigerated cold chamber for at least 24 h, and then filtered in a cold room to avoid remelting of the crystals.

11.5.1 Process Description

1. Soybean oil is partially hydrogenated to an IV of 104–109 under nonselective hydrogenation condition to minimize the formation of *trans* fatty acids.
2. Partially hydrogenated oil is collected in a large tank with heating coils, side-mounted agitator, and temperature control.
3. Low-pressure steam is used on the steam coils with a thermodynamic steam trap.
4. The oil is stored at 120–130°F (49–55°C).
5. The oil is heated to 160–170°F (71–77°C) to completely melt all high melting fractions, which consist of trisaturated glycerides, disaturated glycerides, and monosaturated glycerides.
6. The oil is precooled to 104°F (40°C) and then loaded into chill tanks which are traditionally located in a cold room maintained at 32°F (0°C).

TABLE 11.1 Critical Control Points for the Winterization Process for Sunflower Oil

Process standards	Control limits	Remarks
Moisture in the feed oil.	<0.1%	Higher moisture in the oil slows down the filtration rate.
Phosphorus in the feed oil.	<1 ppm	Higher phosphorus (or phospholipids) interferes with the crystallization process and slows down the filtration rate.
Soap in the feed oil.	0 ppm	Positive soap content in the feed oil slows down the filtration rate.
Feed oil temperature to precooler.	140–150°F (60–66°C)	• The temperature of the oil must be well above the complete melt point of the waxes. • A low temperature may cause prenucleation of the high melting fractions and interfere with the formation of crystal of proper characteristics and cause poor separation in the chiller. • This also slows down the filtration process. • A very high temperature increases the energy cost to precool the oil.
Cooling rate for the oil in the chiller	5–7°F (3–4°C)/h	• Faster cooling may cause supercooling of the solids. • A very slow cooling rate will increase the crystal size and make filtration easier.
Final oil temperature	42–45°F (6–7°C)	• At higher temperature crystallization of wax remains incomplete. • Lower temperature increases the oil viscosity and may slow down filtration.
Agitation	The agitator speed should be low, preferably 5–10 rpm.	• At higher speed the crystals may break down making it difficult to filter the oil.
Holding time in the holding tank	Typically, 12–24 h	• At significantly less than 12 h of holding time the crystal matrix may not be right for filtration. • Filter cycle time may reduce requiring frequent cleaning of the filter. • A longer holding time does not have any negative impact on the process or quality.

(Continued)

TABLE 11.1 Critical Control Points for the Winterization Process for Sunflower Oil (*cont.*)

Process standards	Control limits	Remarks
Filter precoat	Must be done to prevent premature blinding of the screens. Typically 100 lb of diatomaceous earth is required to precoat a filter with 300 ft.^{2} of filtering surface.	• Improper, poor or no precoat will blind the screens fast and reduce filter cycle time reducing productivity.
Body feed	Body feed uses diatomaceous earth @0.25% of the oil flow through the filter.	• At reduced rate of body feed the filter screens plug up reducing the filter cycle time. • A higher rate of body feed is not required but it can also reduce the filter cycle time due to higher dirt load. • At very high dosage of body feed, the pressure drop may not be high even when the cake on the filter screens bridges. This can badly damage the screens.
Filtration rate	Controlled by flow controller FIC. Filter throughput is controlled by the combined action of the flow controller FIC and pressure controller PIC.	A higher flow of oil through the filter can: • Prematurely plug the screens. • Reduce filter cycle time. • Reduce overall productivity. • Improper back-pressure control can have similar effects as above.
Pressure differential across the filter	This must be zero or near zero at start. The goal is to keep it as low as possible without sacrificing production significantly.	• The pressure-differential across the screens is maintained by the pressure controller by modulating the pressure control valve (PCV). • A high pressure-differential can blind the screens rapidly causing loss of production.
Drying the cake	The cake must be dried at the end of every filter cycle	Wet cake needs scraping of the screens. This has the following drawbacks: • The screens can get damaged. • The screens are smeared with the waxes and may require frequent hot oil wash. • This will reduce overall productivity.

Cleaning filter screens	Filter screens need to be cleaned to scrape off the cake adhering to the screen surface.	Scraping must be done with care using a smooth edge plastic scraper without damaging the screen surface.
Hot oil wash	Hot oil temperature: 176°F (80°C). Duration: Until pressure differential = 0. Frequency: As needed	Needed to dissolve and remove the wax build up on the screen and inside to improve filtration rate through the screens.
Maintenance of the filter screens	"O" rings on the screen nipples and the header socket need replacement every 6 months. The sockets and the nipples must be checked for roundness using micrometer or other gauges. The screens, when taken out must be stacked upright on a rack without touching one another. Any tear or damage should be repaired by the screen manufacturer, not by the plant.	Maintenance of the screens are important for obtaining: • Proper filtration. • Proper cold test. • Good productivity.

TABLE 11.2 Troubleshooting Winterization Process

Symptom	Probable cause/causes	Suggested corrective action
Low cold test	• The chilled oil temperature is high. • The holding time is short. • Cooling rate is very rapid, causing supercooling of the solids. • Contamination after filtration • Broken or torn screen allowing unfiltered oil to pass through. • Inadequate precoat application allowing unfiltered oil to pass through. • The "O" ring at the bottom of the screen may be bad passing unfiltered oil forward.	• Check and correct the chilled oil temperature. • Check and increase the holding time. • Check and correct the cooling rate. • Identify the source of contamination and correct it. • Check and replace the broken or torn screen. Send the broken or torn screen to the manufacturer for repair. • Check and correct precoating operation. • Check and replace bad "O" rings.
Filtration rate is slow.	The feed oil may have one or more of the following quality issues: • Moisture content >0.1%. • Phosphorus >1 ppm. • Soap >0 ppm • Wax content is very high. • Insufficient or no body feed.	• Check and correct the oil quality issues. • Use up the high wax oil using slower filtration rate and frequent filter shut down and cleaning. • Blend with low wax containing crude oil if possible to reduce the overall wax feed to the system. • Check and correct the body feed.
Pressure differential is high at start.	• The flow indicator controller (FIC) may have been set for higher flow rate. • The filter screens are not clean.	• Check and correct the flow rate. • Shut down and clean the screens with a hot oil wash.
The cakes on the screens are not dry.	• The drying operation is not being carried out properly. • Air pressure may be too low. • Air-flow rate may be inadequate.	• Check and correct the procedure if needed. • Check and correct the air pressure. • Observe the air pressure during the blowing cycle. The air supply from the reservoir may not be sufficient if it indicates any pressure drop during air blow. • Sometimes, multiple use of compressed air at the plant may result in the short supply of compressed air if the system is not designed for the peak demand.
Rapid rise in pressure differential.	• Inadequate precoating. • Insufficient or no body feed. • Filtration rate is set too high on the controller (FIC). • The PCV may not be opening properly to allow proper amount of oil by-pass.	• Check and correct the precoating operation. • Check and correct the body-feed. • Check and reduce the flow rate. • Check and correct ant issue with either the controller (PIC) or the control valve.

7. The crystals are allowed to grow under this condition for at least 24 h with minimum agitation.
8. A cooling rate of 4–6°F (2.2–3.3°C)/h is recommended to obtain the desired crystal matrix.
9. A more rapid cooling rate forms supercooled solids that take time to crystallize out. It also tends to form small crystals that are difficult to filter.
10. A slower cooling rate produces larger crystals, which is good for filtration.
11. It generally takes about 12 h for the oil to reach the final temperature of 32°F (0°C).
12. The oil must be left in the chiller for a minimum of 12 additional h for crystal matrix to grow. This is when the latent heat of crystallization is removed from the oil and the secondary crystal growth occurs (see Chapter 10).
13. The oil is filtered through a pressure leaf filter with cloth or stainless steel screen surface.

11.5.2 Filtration

A similar setup as shown in Fig. 11.2 is suitable for filtering winterized soybean oil. No precoat or body feed is needed.

For soybean oil, the pressure leaf filter uses a flow rate of:

- 0.04 US gallon/min/ft.^{2} of filter surface area, or
- 0.33 Imperial gallon/min/m^2 of filter surface area.

The filtration rate can be maintained high without building high pressure differential across the screens because of the nature of the crystals obtained from partially hydrogenated soybean oil. Therefore, one must be careful that the cakes formed between the screens do not bridge, for this may seriously damage the screens.

The procedure for drying of cakes by air blow and removal of the cakes from the screens is the same as that described under winterization of sunflower oil.

The solid or the stearin cake is used for formulation of shortening and margarine.

The liquid or the olein fraction is used as cooking and salad oil.

The typical fatty acid composition of the stearin and the olein fractions, obtained in this process, are shown in Table 11.3.

The stearin fraction has lower oxidative stability than the starting oil even with the reduced amount of unsaturated fatty acids. This is because the oil decomposition products, including dimers and trimers are present in the stearin fraction.

Partially hydrogenated soybean oil is also solvent fractionated, which is also called wet fractionation process. The same starting oil is dissolved in normal hexane. The mixture is chilled to allow the crystals to grow and then filtered. The hexane is removed from both fractions by stripping. This process produces higher olein yield and also different fatty acid composition for both olein and stearin fractions.

TABLE 11.3 Fatty Acid Composition of Winterized Soybean Oil Fractions (Unpublished Work of M.K. Gupta)

Analysis	Starting oil	Olein fraction	Stearine fraction
Iodine value (IV)	109	113.1	98.9
Fatty acid Composition			
C-16	10.4	9.2	12.6
C-18	4.8	3.8	6.6
C-18:1	50.7	52.5	47.3
C-18:2	31.8	34.2	31.3
C-18.3	2.1	2.20	1.9
Solid fat content @ 70°F (21.1°C)	1.0	0.0	4.8

TABLE 11.4 Relative Yields of Olein and Stearine Fractions in the Two Winterization Methods

Process	Yield of olein fraction (%)	Yield of stearine fraction (%)
Solvent process	70–80	20–30
Conventional process	60–65	35–40

Yields of olein and stearin fractions are different in the two processes, as shown in Table 11.4. The solvent process provides higher yield of olein compared to the conventional winterization process.

The stearin and the olein fractions have somewhat different composition. Table 11.5 shows the difference in composition between the two stearines obtained from RBHI-109 soybean oil. The stearin fraction from the solvent process has lower IV and contains more palmitic acid and stearic acids compared to the stearin from the conventional winterization process. The olein fraction is somewhat more unsaturated than in the conventional process.

Table 11.6 shows some additional data on the olein and stearin fractions obtained from RBH soybean oil at various IVs.

11.6 FRACTIONATION OF PALM OIL

The natural composition of palm oil is probably one of the most important aspects of the characteristics of the oil that lends itself to fractionation into several useful oil fractions. Fractionation of palm oil has been known in the oil industry but not so much in the United States for various reasons. This is most advanced

TABLE 11.5 Typical Fatty Acid Composition of Olein and Stearine Fractions in the Two Winterization Methods (Starting Oil Soybean RBH I-109) (Unpublished Work of M.K. Gupta)

Analysis	Olein (solvent process)	Stearine (solvent process)	Olein (conventional process)	Stearine (conventional process)
Iodine value	114.5	95	113.1	100
Fatty acid composition (%)				
C-16	7.9	14.9	9.2	12.6
C-18	3.2	9.3	3.8	6.6
C-18:1	53.3	44.7	52.5	47.4
C-18:2	33.2	28.5	34.2	31.3
C-18:3	2.0	2.4	2.20	1.9

TABLE 11.6 Additional Examples of Dry Fractionation of Hydrogenated Soybean Oil

	Olein fraction				Stearine fraction	
Iodine value	Yield (%)	IV	Cloud Pt. (°C)	Cold test (h)	IV	Drop Pt. (°C)
135 (RB)	85–90	119	−11	>24	98	33.5
109	75–80	114	−10	18–24	92	34.5
97	65–70	104	−9	12–18	84	35.5
85	50–55	94	−7	<5	75	36.5
75	40–45	84	−5	<2	68	37.0

Source: From Fractionation and Winterization, Industrial Oils and Fats, 5th ed., John Wiley & Sons, 1996.

in Malaysia, Europe, Canada, and in certain parts of Latin America. Application of palm oil fractionation techniques can be found in Latin America, which can be compared to the advanced techniques used in Europe and Malaysia. A great majority of these plants are using the fundamental technology developed by Desmet Ballestra and Tirtiaux; both from Belgium. Both crude and refined palm oils are fractionated; the latter is being predominant.

Palm oil fractionation, combined with interesterification process, has become the way of *trans* fat reduction in the food industry. In many cases, the companies are combining various palm oil and palm kernel oil fractions or their derivatives to achieve both functional and *trans* fat requirements for the baking industry.

This is a vast and complex area and the author will not be able to cover fractionation of both palm and palm kernel oils in this chapter. Therefore, the focus will be given only to the process conditions and their impacts on palm oil fractionation.

Palm oil can be fractionated into stearin and olein fractions in the dry fractionation process by controlled chilling of palm oil and separating the solids (stearin) from the liquids (olein) through filtration. For a long time, the palm olein received the industry attention as the premium product for its stability and its usefulness as a versatile product. Palm stearin was looked upon as a by-product and had lesser demand. However, the advancement of the fractionation techniques has allowed the oil industry to use various palm stearin fractions to make *trans*-free shortening and margarine. This is very important in the light of the low and no-*trans* movement throughout the world, although there are concerns among the consumer advocates in the United States that the higher saturated fat content of palm oil and also the interesterified shortening and margarine may not be as healthy. This issue is being addressed by the oil industry specialized on palm oil and palm kernel oils. They find that the shortenings that are available today can be used by the US food industry to meet the low *trans* requirement with no significant increase in the saturated fat content in the finished product. These shortenings, using fractions and derivatives of palm kernel oil, are expensive to the food industry in the United States.

11.6.1 Suitability of Palm Oil for Fractionation

The natural fatty acid distribution in palm oil lends itself into fractionation and the production of multiple fractions of palm oil that are naturally stable. Palm oil has the distinct advantage over natural seed oils because of the presence of different saturated triglycerides. Table 11.7 shows the typical triglycerides present in natural palm oil.

TABLE 11.7 Triglyceride Composition in Malaysian Palm Oil

Triglyceride type	Each triglyceride contains	Mole %
No double bonds	MPP, PMP, PPP, PPS, PSS, PSP	7–10
One double bond	MOP, MPO, POP, POS, PMO, PPO, PSO, SOS, SPO	34–36
Two double bonds	MLP, MOO, PLP, PLS, PPL, OSL, SPL, POO, SOO, OPO, OSO, PSL	34–36
Three double bonds	MLO, PLO, POL, SLO, SOL, OOO, OPL, MOL	16–17
Four or more double bonds	PLL, OLO, OOL, OLL, LOL	4–6

Note: M, myristic acid; P, palmitic acid; S, stearic acid; O, oleic acid; L, linoleic acid; palm oil contains negligible amount of linolenic acid.

The presence of trisaturated and disaturated triglycerides in the palm oil facilitates the formation of fat crystals as the oil is chilled. Controlled chilling process makes it possible to obtain various palm oil fractions with the desired physical properties required for the baking industry.

11.6.2 Methods for Fractionation

There are three known methods for fractionating palm oil. They are:

1. dry fractionation (slow cooling and rapid cooling),
2. wet fractionation with detergent, and
3. wet fractionation with solvent.

11.7 DRY FRACTIONATION

Dry fractionation is most commonly used in fractionation of palm oil. This is the simplest, least costly and very effective way of obtaining palm oil fractions of desired attributes. There are fundamentally two steps involved in the dry fractionation process (1) crystallization of the solids and (2) separation of the solid and the liquid fractions.

There are two major companies that offer dry fractionation processes and equipment. They are:

1. Tirtiaux, Belgium.
2. Desmet Ballestra, Belgium.

The fundamental differences between the two processes are that Tartiaux process applies slow cooling process while Desmet Ballestra uses a larger cooling surface that provides more rapid chilling without supercooling the solids. The end products from the two processes are very similar but can be different depending on the source oil and the exact conditions applied.

There are other companies that are also well recognized in this field. They are:

- Crown Iron Works, United States.
- Lipico Technologies, Singapore.
- Oiltek, Malaysia.
- Krupp, Germany.
- Intersonicon, Malaysia.

In the dry fractionation process, the palm oil feed is heated to 155–165°F (68–73°C) to completely melt all solids in the oil. The oil is then cooled very slowly under controlled conditions to obtain the desired crystal structure for the solid fraction. The temperature is almost at the equilibrium condition, so there is little or no supercooling of the solids, and the crystals formed have the proper characteristics for filtration. The oil is filtered to separate the liquid fraction (olein) from the solid fraction (stearin).

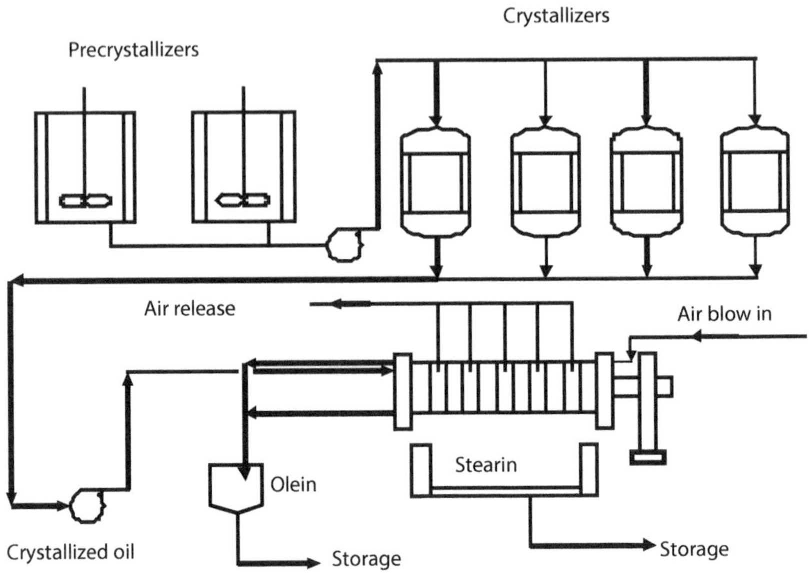

FIGURE 11.3 Schematic diagram for dry fractionation.

Fig. 11.3 shows the schematic diagram for dry fractionation process. The diagram does not show the following items:

- controls,
- utilities, such as steam, compressed air, refrigeration, and electricity.

The oil is precrystallized by chilling the oil to 130°F (50°C) in the precrystallizers, using chilled water, and then the oil is pumped into crystallizers. Generally there are four crystallizers to supply oil to the filter round the clock. The typical operating conditions are listed in subsequent sections.

11.7.1 Precrystallizer

Total precrystallizer capacity	120% of the plant
Oil inlet temperature	155–165°F (68–73°C)
Precrystallizer final temperature	130°F (50°C)
Time for the oil to reach 130°F (50°C)	8 h
Average cooling rate in precrystallizer	3.75°F (2.1°C)/h

11.7.2 Crystallizer

Time to reach 104°F (40°C)	6 h
Average cooling rate in the crystallizer	4.3°F (2.4°C)/h
Holding time at 104°F (40°C) in the crystallizer	4 h
Time to reach 68°F (20°C)	6 h
Average cooling rate from 40 to 20°C	6°F (3.33°C)/h

Note:

There are some variations from the aforementioned conditions which are determined by individual processors to suit their operation to obtain the desired end product.

11.7.3 Filtration

Time to filter each crystallizer: 6 h.

The crystallized oil is filtered through the filter. Fig. 11.3 is showing a plate and frame membrane filter. However, there are rotary vacuum filters and stainless steel continuous belt filters that are used in palm oil fractionation plants. The schematic diagram for the membrane filter is shown in Fig. 11.4. Fig. 11.5 shows the membrane filter in a palm oil fractionation operation.

Crystallized oil enters the frame cavity of the membrane filter. The operating sequences are:

1. Crystallized oil is fed into the filter through port #2.
2. Olein is separated from the stearin in the filter and comes out through port #4.
3. At the end of operation, compressed air is blown on the outside of the membrane in the filter cavity. A pressure of 8–30 bars is applied on the membrane. This compresses the cake and more olein is released.
4. The cake becomes dry.
5. The air is released from the filter.
6. The filter is opened.

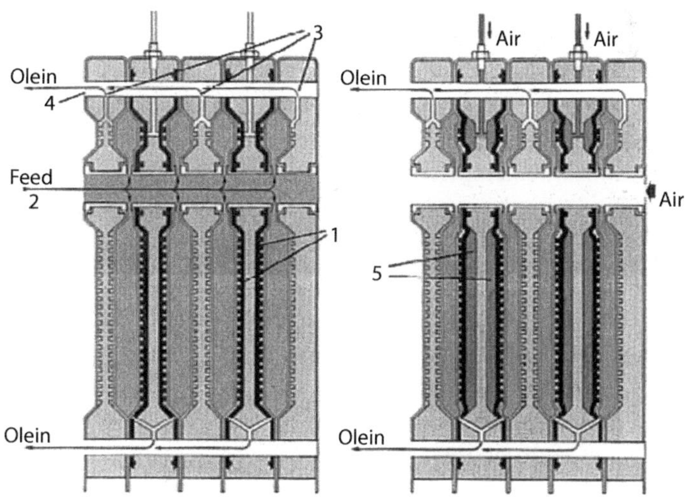

FIGURE 11.4 **Membrane filter.**

FIGURE 11.5 **Membrane filters at Tirtiaux Plant.**

7. The dry cake is collected in a trough below.
8. The cake is melted in the trough, pumped out to a storage tank.
9. Filter throughput is roughly 0.1 Mt/h per square meter of filter surface.

11.7.4 Critical Control Points in Dry Fractionation

The critical control points for dry fractionation of palm oil are listed as follows:

1. Initial oil temperature.
2. Precrystallization.
3. Cooling rate (in the precrystallizer as well as in the crystallizer).
4. Holding time in the crystallizer.
5. Agitation in the crystallizer.
6. Final crystallization temperature.
7. Filtration.

11.7.5 Initial Oil Temperature

As stated earlier, the palm oil feed is heated to 155–165°F (68–73°C) to completely melt all solids in the oil. This destroys any crystal memory from any prior crystallization in the oil.

11.7.6 Precrystallization

This step is necessary to initiate some nuclei formation before the oil is pumped into the crystallizer. This helps the formation and growth of crystals. Sometimes the oil is seeded with crystals to expedite the crystal growth and reduce the residence time in the crystallizer.

11.7.7 Cooling Rate

The cooling rate is controlled with precision to obtain the right crystal type and size. Separation of the stearin from the olein can become difficult if the crystallization process is not properly carried out. The qualities of the two fractions also suffer with poor crystallization.

At the end of the crystallization process, the triglycerides are present in three forms:

1. Solids as cocrystals.
2. As distinct liquid.
3. As liquid trapped inside the crystal matrix.

The aforementioned distribution is linked to the cooling rate, holding time, and agitation. Slow cooling, close to the equilibrium condition, produces the best results. For growth of easily filterable crystals, they should be of uniform dimension and must not settle to the bottom of the crystallizer. It has been suggested by some researchers that for optimum filtration, the palm oil crystals should be of beta prime phase with a crystal size of 0.1 mm.

To achieve this, the cooling water temperature in the crystallizer is reduced at a programmed rate, allowing the oil to cool gradually and maintaining a predetermined temperature differential between the oil and the cooling water.

11.7.8 Holding Time in the Crystallizer

The crystal development requires extraction of the latent heat of crystallization from the crystals as the nuclei begin to coalesce and start forming the crystal matrix. A shorter time may not allow the formation of sufficient fat crystals of appropriate size and matrix needed for good filtration. As mentioned earlier, the goal is to produce hard crystals with uniform size to maximize the filtration efficiency.

11.7.9 Agitation in the Crystallizer

Agitation in the precrystallizer is to move the oil around the cooling surface to reduce the temperature without supercooling the solids. Agitation in the crystallizer is operated at low speed with proper scraping action. The speed is slow enough not to break up the crystals to small size and yet not allow settling of the same to the bottom of the crystallizer. In most cases, the agitators are run at speeds of 5–15 rpm. The internal design of the crystallizer varies greatly between manufacturers. Each design has certain advantages and disadvantages. Choice of a specific design is a matter of the oil processor's preference because they all produce the desired end product, some with more or less operating issues. Fig. 11.6 shows the sectional views of vertical crystallizers with

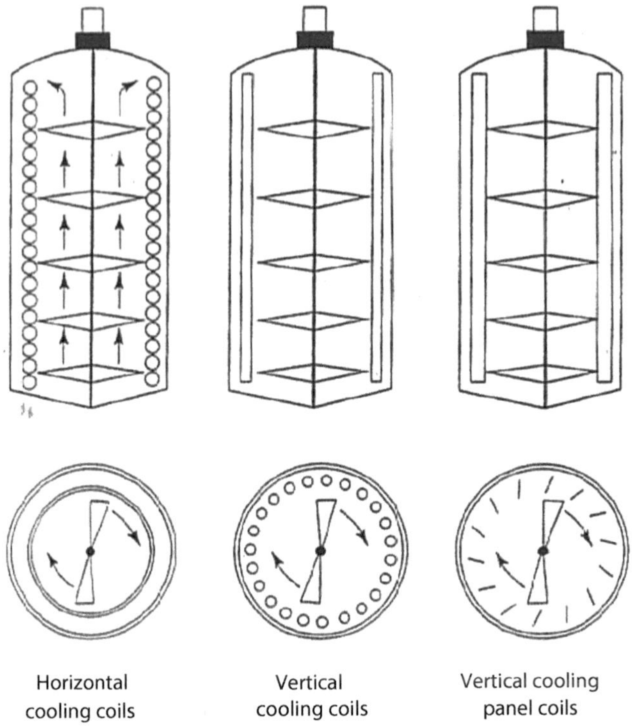

FIGURE 11.6 Crystallizers with propeller-type agitators.

propeller-type agitators with different cooling designs. Fig. 11.7 is showing the cross-sectional views of tubular and concentric design crystallizers.

11.7.10 Final Crystallizer Temperature

The temperature is chosen based on the targeted olein and stearin to be made. The temperature scheme, shown earlier, is used for single fractionation.

11.7.11 Filtration

The efficiency of filtration depends totally on the crystal size and quality. Crystals with softer consistency will hold more olein occluded between the crystals. This alters both yield and the qualities of the two fractions. In addition, the filtration rate is also reduced by the softer crystals.

11.8 TROUBLESHOOTING DRY FRACTIONATION

Although the dry fractionation process has developed to an advanced stage, it is still somewhat of a blend of art, technology, and acquired experience. This is because the palm oil composition (diglycerides, phospholipids, etc.) varies

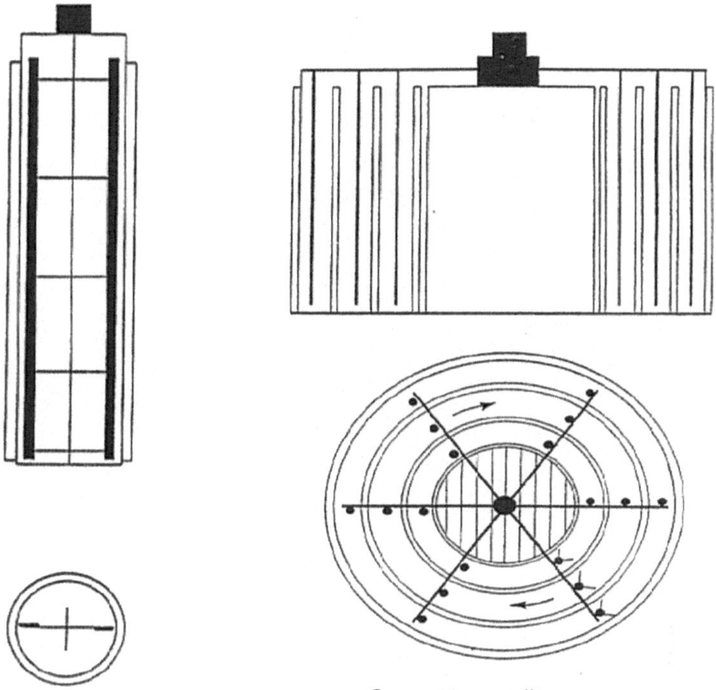

Tubular crystallizer

Concentric crystallizer

FIGURE 11.7 **Tubular and concentric crystallizers.**

and they influence the crystallization process. This is an area that has been controlled and managed by the Malaysian and European palm oil processors like Golden Hope and others by strict control of the feed oil quality to the fractionation process, besides the process control on the fractionation unit.

Table 11.8 lists the most commonly experienced process and quality issues in dry fractionation process, their probable cause or causes, and the suggestions to rectify them.

11.9 MULTIPLE DRY FRACTIONATION

Multiple fractionation of palm oil has become more common to produce various fractions of olein and stearine which are applicable in formulation of shortening and margarine for various food applications. There are two pathways followed in this process after the first fractionation:

1. The olein is fractionated further to make different grades of olein.
2. The stearine is fractionated further to make different grades of stearine and some more olein.

TABLE 11.8 Troubleshooting Single Step Dry Fractionation Process

Symptom	Probable cause/causes	Suggested corrective action
Precrystallizer taking long to cool the oil.	• Incoming oil temperature is higher than 155–165°F (68–73°C). • The coolant temperature may be high. • The cooling surface may be fouled. • The agitator may not be working correctly.	• Reduce the incoming oil temperature if it is high. • Check and correct the temperature. • Shut down and clean the cooling surface. • Check the rpm of the motor. Fix the drive if it is slow. • Check the amperage of the motor. If it is low but the rpm is normal, it may indicate that one or more agitator blades have become loose or have fallen off. Shut down and fix it.
Cooling rate in the crystallizer is slow.	• Incoming oil temperature is higher than 130°F (50°C). • The coolant temperature may be high. • The cooling surface may be fouled. • The agitator may not be cleaning the cooling surface.	• Correct the issue at the precrystallizer. • Check the program for the coolant, determine whether it is equipment related or control program related and correct it. • Shut down and clean it. • Shut down and correct the scraping function of the agitator.
Olein yield is low.	• Inappropriate crystallization process. This can include the operations of: 1. precrystallizer 2. crystallizer 3. agitator • Air pressure on the membrane may not be sufficient.	• Check precrystallizer and crystallizer operating conditions and correct any deviations. Agitator has already been discussed above. • Increase the air pressure on the membrane. This is especially needed when the crystals produced are soft. • This is not possible with the rotary filter or the belt filter. There one needs to increase the vacuum across the filter screen.
Stearin produced is soft.	• Same as above.	• Same as above.

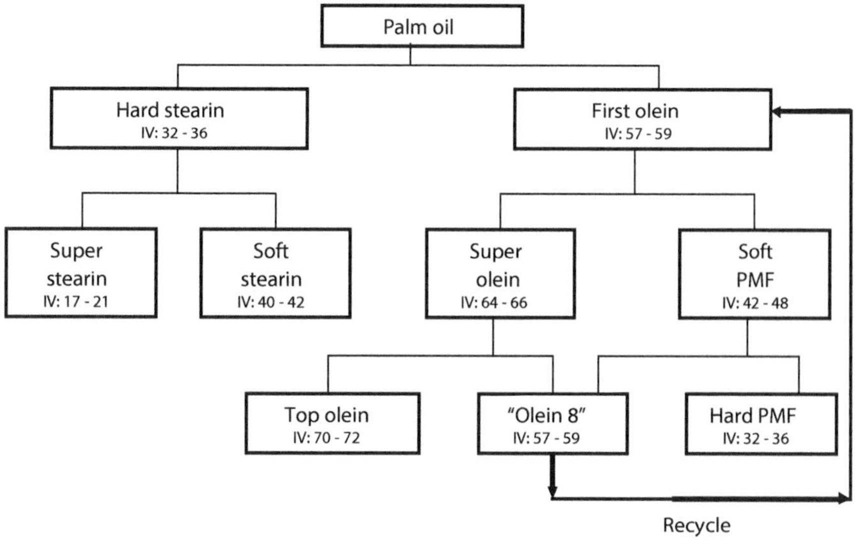

FIGURE 11.8 Multiple dry fractionation.

Fig. 11.8 shows the multiple dry fractionation steps for palm oil. Several benefits are derived from this process.

11.9.1 Benefits of Multiple Dry Fractionation of Palm Oil

Several benefits are derived from the process of multiple dry fractionation of palm oil. For example:

The olein fractions derived from the process can be used as cooking oil and/or salad oil (top olein with cold test >24 h).

The hard and soft palm midfractions can be used to make shortening, margarine, and CBE products. This can be clear from the solid fat content data on hard and soft palm midfractions shown in Fig. 11.9. Table 11.9 lists the various applications of PMFs—the palm oil fractions.

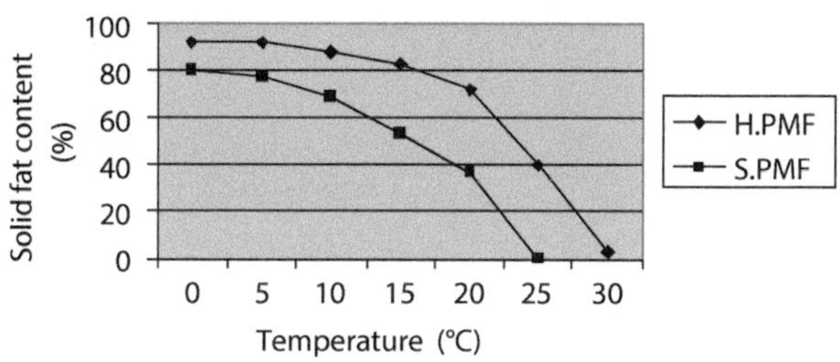

FIGURE 11.9 Solid fat content of palm midfractions.

TABLE 11.9 Applications of Palm Oil Fractions

Product	Palm oil	Olein	Stearin	Super olein	Soft stearin	PMF
Shortening	+++	+++	++	-	+++	+
Margarine	++	+++	+	-	+++	+
Industrial frying	+++	+++	-	+++	++	+
Cooking oil	-	++	-	+++	-	-
Salad oil	-	+	-	+++	-	-
Specialty fat	-	-	-	-	+	++
Cocoa butter extenders (CBE)	-	-	-	-	+	+++
Ice cream	+++	-	-	-	-	-
Icing	++	-	-	-	+	++
Cookies	+++	-	+	-	++	-
Biscuits	+++	+	+	-	++	-
Crackers	+++	+	+	-	++	-
Cakes and pastries	+++	-	+	-	++	-
Noodles	+++	+++	-	-	++	-
Hard coating	-	-	++	-	-	-

+, Limited application; ++, suitable; +++, highly suitable; -, unsuitable.

11.10 WET FRACTIONATION WITH DETERGENT (LANZA PROCESS)

Detergent fractionation is carried out with only crude palm oil (CPO). The process involves cooling, crystallization, and separation of the crystals with the help of a detergent, sodium lauryl sulfate solution, and an electrolyte (magnesium sulfate or sodium sulfate) to aid the separation.

The chilled oil is mixed with a mixture of detergent and electrolyte that has been chilled to the same temperature as the oil. The detergent wets the stearin crystals and releases the olein occluded in the crystals. The crystals and the olein fraction are separated in a centrifuge. The olein fraction is washed to remove the detergent, dried and stored.

The stearin and the detergent with the electrolyte come out together in the heavy phase. The heavy phase is heated to melt the stearin and passed through a second centrifuge where the stearin is separated from the detergent. The stearin is washed, dried, and stored. The detergent is reused.

Until the introduction of the membrane filter, this was the process known to provide higher olein yield than the dry fractionation process. Table 11.10 shows the comparison between the oleins and stearines produced by the dry and

TABLE 11.10 Compositions of Olein and Stearin Fractions in Dry Fractionation and Lanza Fractionation Process

Analysis	Olein fraction			Stearin fraction		
	Dry rapid	Dry slow	Lanza Process	Dry rapid	Dry slow	Lanza Process
Yield (%)	60	68	83	40	32	17
Iodine value	58.4	57.2	57.8	48.7	28.6	28.0
Slip M. Pt. (°C)	23.3	22.2	21.7	45.9	50.7	55.3
Triglyceride (%)						
S_3	0.7	0.4	0.3	13.8	21.9	42.0
S_2 U	47.9	44.1	50.2	46.2	47.0	41.4
SU_2	44.1	44.7	43.2	34.6	27.1	14.2
U_3	7.3	6.8	6.3	5.4	4.0	2.4

the wet fractionation processes. Lanza process produces higher olein yield and stearin fraction with minimum amount of olein. Separation between the two fractions is cleaner. The slow cooling produces better separation of olein and stearin in dry fractionation.

11.11 SOLVENT FRACTIONATION PROCESS

The solvent method is very similar to what was described for fractionation of partially hydrogenated soybean oil, except the solvent used in this process is acetone, while the solvent used in soybean oil fractionation is normal hexane.

The typical ratio between solvent and the oil is 1:3 or 1:4. The oil and solvent mixture is chilled, the crystals are filtered, and the solvent is stripped and recovered from both fractions. As in the case of soybean oil fractionation, the solvent process is not used in the palm oil industry because of the high investment cost, high operating cost, the risk for explosion, and loss of solvent. The membrane press, combined with the multiple dry fractionation technique, has allowed the industry to improve both yield and quality of both olein and stearin fractions. However, solvent fractionation is still very useful in making high value midfractions of palm oil. This process is especially useful where the triglycerides containing long chain fatty acids tend to become very viscous at the very low temperatures required for making certain products.

The fundamental distinction of the solvent fractionation is that it is based on the different solubility of the triglycerides in the solvent rather than their melting points. This point can be illustrated with the following data. Melting points of the following triglycerides are:

POP	37°C
PPO	40°C
PPP	65.7°C

Thus it will be easy to separate PPP from the other two in the dry fractionation process, and most of it would appear in the stearin, but some still may be found in the olein fraction.

The solvent process works on the basis of the relative solubility of the triglyceride molecules. The relative solubility of the three triglyceride molecules in this example is:

$$\mathrm{POP/PPO/PPP} = 1.3/5/0.002$$

Thus, in the solvent process, PPP will remain in the solid state at the low temperature, while the other two will remain predominantly in solution in the solvent and appear in the olein fraction. This also makes it possible to wash the stearin with solvent to produce more pure stearin by removing the olein from the crystal surface.

Fig. 11.10 shows the schematic diagram for the two-stage solvent fractionation process.

1. The oil and the solvent (acetone) are mixed in tank A in the desired ratio.
2. The solution is pumped through the jacketed coolers B and C. The solids are scraped continuously from the walls.
3. The crystals are then separated in a hermetically sealed filter D. This is critical because the plant uses explosion-proof equipment, and no solvent vapor is allowed to be released into the plant.

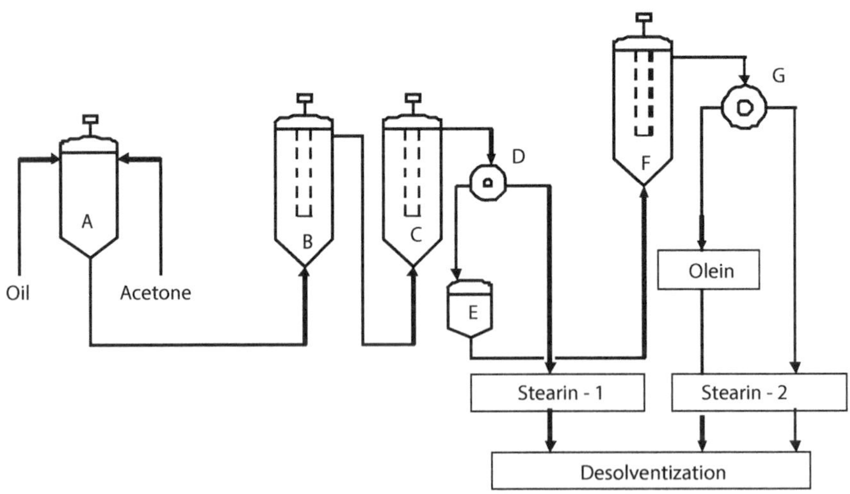

FIGURE 11.10 Two-stage wet fractionation process.

4. The first stearin (stearin-1) is desolventized.
5. The olein + solvent (micelle) fraction is stored in the tank E. The micelle is pumped into the cooler F and cooled to a lower temperature, and the solid crystals formed are separated in a second hermetically sealed filter G.
6. The micella (olein + solvent) is desolventized.
7. Stearin obtained from filter G (stearin-2) is also desolventized.
8. The two stearin fractions have different physical properties and triglyceride compositions.
9. The temperatures in the coolers determine the type of solid fraction that is obtained from the feed.
10. The distilled solvent is condensed, recovered, and reused.

11.11.1 Critical Control Points

The critical control points in the wet fractionation process are:

1. Solvent-to-oil ratio.
   A higher proportion of the solvent helps the separation of the specific triglycerides. On the other hand, a very high solvent–oil ratio makes the process uneconomical and it is not necessary to use higher amount of solvent in the process.
2. Cooling temperatures in the coolers.
   It is important to create the driving force for crystallization of the solids in the feed oil.
3. Cooling rate.
   This is very critical, as it was discussed under dry fractionation. The temperature differential between the coolant and the micelle must be carefully programmed and controlled.
4. Agitation.
   This is important for removal of the crystals from the inner surface of the coolers.
5. Filtration temperature.
   This must be controlled carefully; otherwise, the end products may not meet the target quality.

11.11.2 Comparison Between the Three Methods of Fractionation

The three methods—dry, detergent, and the wet fractionation process—produce both olein and stearin fractions that are distinct from each other. Even the slow versus rapid chilling under dry fractionation produces distinct products.

It must be recognized that these methods cannot be applied in a plant interchangeably. Therefore, careful considerations must be given to the type of product to be made before a specific process can be chosen for the plant.

Owing to the development made in the dry fractionation process, the newer installations mostly use the dry fractionation process. As mentioned earlier, the wet fractionation method is chosen only when very high value specialty products are to be made.

READING REFERENCES

Calliauw, V., Gibon, V., Greyt, W.D., Foubert, I., Dewettinck, K., in press. Research in Production of Soft PMF and Super Olein; Presented at the 4th Euro Fed Lipid Congress, October 1–4, Madrid, Spain.

Davidson, H.F., Campbell, E.J., Bell, R.J., Prichard, R.A., 1996, fifth ed. Bailey's Industrial Oil and Fatsvol. 2John Wiley & Sons.

Deroanne, C., 1976. Societe Belge de Filtration VI, 171–188.

Duns, M.L., 1985. Palm oil in margarine and shortenings. JAOCS 62, 408–410.

Iida, M., Kato, C., 1985. US Patent 4,265,825.

Kellens, M., Hendrix, M., 2000. Fractionation, Introduction to Fats and Oils Technology, second ed. AOCS Press, Champaign, IL.

Krishnamurthy, R., Kellens, M., 1996, fifth ed. Bailey's Industrial Oil and Fatsvol. 4John Wiley & Sons.

Michael, B., 1998. Modification of fats and oils. Fats and Oils HandbookAOCS Press, Champaign, IL, (Chapter 6).

Skau, E.L., Dopp, W.N., Burleigh, E.G., Banowertz, L.F., 1950. JAOCS 27, 556–564.

Thomas, III, A.E., 1985, fourth ed. Bailey's Industrial Oils and Fatsvol. 1John Wiley & Sons.

Van den Berg, H.J., 1979. U.S. Patent 4,161,484.

Veronique, G., Tirtiaux, A., 2000. Fractionation Combined With Interesterification: A Tool Towards the Formulation of Zero-Trans New Products. AOCS-OTAI Publication, New Delhi, India.

Yosof, B., 1996, fifth ed. Bailey's Industrial Oil and Fatsvol. 2John Wiley & Sons.

Chapter 12

Insight to Oil Quality Management

12.1 INTRODUCTION

The primary purpose of this chapter is to provide the readers with some insight to oil quality management and discuss the tools that can be used to achieve the objective.

It is important for the oil processors to recognize the fact that it is the primary responsibility of the oil processor to deliver acceptable quality oil or formulated margarine or shortening products either to the consumers or to the end users who use these ingredients to formulate finished products. This would require the oil processor to maintain appropriate operating standards throughout the operation, including: (1) setting the raw material specifications, (2) raw material receiving and storage, (3) processing, (4) in-process oil storage, (5) finished product storage and handling, and (6) warehousing and distribution of the bulk and packaged products.

There are certain areas that are beyond the control of the standalone oil refiners. There are some areas that are also beyond the control of the oilseed crushers. These areas will be discussed along with the critical control points for oil processing.

12.2 MANAGING OIL QUALITY

In today's business environment, where the use of shelf-stable product is increasing in every country, whether it is a developed or a developing country, managing oil quality has become a bigger challenge than ever before. High-quality oils are in increasing demand because of the following reasons:

1. High-quality oil *produces* high-quality finished product.
2. High-quality oil *leads to longer shelf life* for the packaged products. This allows the food manufacturer the advantage in extending code date, production scheduling, warehousing, product distribution, and achieving lower cost.
3. Most food manufacturers have started using "on-time" delivery policy for receiving their raw materials. This means the oil quality must be satisfactory at all times as it arrives at the end user's plant, otherwise the food plant may have to be shut down for lack of oil supply.

Practical Guide to Vegetable Oil Processing. http://dx.doi.org/10.1016/B978-1-63067-050-4.00012-X
Copyright © 2017 AOCS Press. Published by Elsevier Inc. All rights reserved.

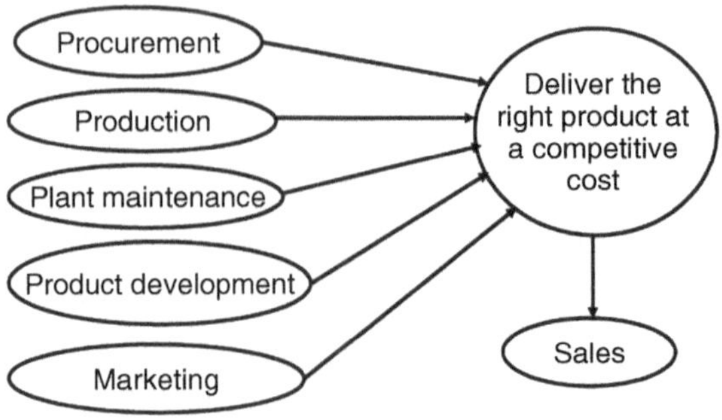

FIGURE 12.1 Alignment of various functions in a company to produce and deliver right quality product.

Therefore, it has become necessary for the oil processors to deliver oil to the end users with zero deviations. To meet this challenge, the oil processors need to take certain steps as outlined in following sections.

12.2.1 Step #1: Have a Clear Product Objective

This requires the company to look into the following areas:

1. Product definition—a clear definition of all key product attributes.
2. Proper raw material quality—it is necessary to look for the best quality crude oil in order to reduce plant losses and improve cost.
3. Full knowledge about storage stability of the product.

The first step in any venture is to have a clear vision about the product to be made based either on the market survey for the bulk products or consumer survey for the new or innovative consumer products.

Physical and chemical behavior of the product must also be understood clearly to make proper selection of raw materials, packaging material, storage, and distribution system so the product can be delivered to the point of use without defects.

All functional groups, such as product development, purchasing, marketing, manufacturing, accounting, sales, distribution, and so on, must be fully aligned. These groups should have one common objective, which is to deliver the right product to sales on time (Fig. 12.1).

12.2.2 Step #2: Have the Right Capability in Place

The plant must have trained personnel who are capable of performing the following functions:

1. Assessing the existing process capability—it is important for the plant personnel to be fully aware of the capability of the available processing system in terms of capacity, delivered quality, energy requirements, etc.

2. Setting appropriate process operating standards—this specifies the temperature, pressure (if applicable), flow, heating, and cooling requirements for any process system to deliver product of the right quality and at the desired capacity.
3. Setting appropriate storage conditions—this applies for the in-process oils, as well as the finished products.
4. Establishing proper means of delivery of product either to the industrial end users or to the consumers.
5. Assessing whether or not any new equipment is needed to achieve higher throughput or for manufacturing any new product that marketing and sales may want to introduce to the market.

12.2.3 Step#3: Measurements of Quality and Setting Standards

There must be a quality assurance function in place which can perform the following functions:

1. Set quality standards and the criteria for accepting the incoming raw and packaging materials.
2. Set in-process oil quality and process standards.
3. Set and monitor the finished product standards.
4. Set and monitor the standards for product in storage and shipping.
5. Set product safety standards for the organization.

12.2.4 Step #4: Measurement of Performance

There must be an infrastructure to monitor the compliance and provide feedback to the appropriate management in the company. The specific tasks are as follows:

1. Evaluate the degree of compliance in accepting all incoming materials.
2. Monitor performance and finished product quality.
3. Upgrade raw material and product quality as needed.
4. Offer methods for reducing product defects.
5. Report the findings periodically to the appropriate management.
6. Follow up on the compliance by the plant.

12.2.5 Step #5: Understand the Behavior of the Oil and Learn How to Protect It From Degradation

Oil is a chemical product that can undergo decomposition starting from the time the seeds are harvested. This continues throughout the processing, storage, transport, and at the end user's plant. However, there are a number of steps that the oil processor can take which can provide protection to the oil from undue degradation that can provide longer storage stability for the oil.

It must be pointed out here that oil stability is normally measured by PV or OSI. However, these measurements do not necessarily provide any information regarding the flavor stability of the oil in storage or in finished products.

To appreciate this, one needs to refer back to the Chapter 1 on requirements for successful production and delivery of vegetable oils and Chapter 2 on basic oil chemistry and also the processing steps from crude oil storage through deodorizing and product distribution.

12.3 MODES OF OIL DECOMPOSITION

Oil decomposes primarily in three ways:

1. oxidation
2. hydrolysis
3. thermal degradation

1. Oxidation

   When a molecule of oxygen reacts with the unsaturated fatty acid in the oil, it produces numerous oil breakdown products. Some of these act as catalysts for further oxidation of the oil. This reaction is accelerated by higher temperature. Therefore, controlling the temperature of the oil and minimizing exposure to air (oxygen) throughout oil processing is of utmost importance. There are three ways the oil can undergo oxidation, namely:

   - autoxidation
   - enzymatic oxidation
   - photooxidation

     Autoxidation

     Autoxidation is initiated by the reaction between a molecule of unsaturated fatty acid and a trace metal "initiator," such as iron, nickel, copper, and so on. This produces a very powerful oxidation catalyst called "free radicals." These compounds have free electrons to donate. This makes these compounds strong oxidizing agents. These compounds can oxidize more unsaturated fatty acids in the oil, produce more free radicals, and can carry on this reaction until there are no more unsaturated fatty acids left in the oil or the supply of oxygen is depleted (Fig. 12.2). This is a complex reaction process. This is also the primary mechanism for oil oxidation in refining, handling, storage, and distribution. Free radicals are formed in the oil whenever the oil is heated. Therefore, it is necessary to maintain proper temperature control in oil processing.

     Autoxidation can produce free radicals that can react with each other, forming dimers or polymers. These compounds are characterized as oxidative polymers. These compounds are strong prooxidants. Some of these compounds contain as much as 9 times more oxygen than an oil molecule. These compounds are unstable. They can decompose, form more free radicals, and continue the process of oil decomposition even under nitrogen

(RH) + Initiator = R^{+}		Initiation (1)
Unsat. FA Metal ion	Free alkyl radical	
R^{+} + O_2	= $ROO\cdot^{+}$	Autoxidation (2)
Free radical Oxygen = Peroxy radical (alkoxyl radical)		
ROO^{+} + (RH) Peroxy radical	= ROOH + R^{+} Hydroperoxide Alkyl radical	Propagation (3)
R^{+} + R^{+}	= RR	Termination (4)
ROO^{+} + ROO^{+}	= ROOOOR	Termination (4)
R^{+} + ROO^{+}	= ROOR	Termination (4)

FIGURE 12.2 Autoxidation of unsaturated fatty acid.

protection and also at freezing temperature. Therefore, the oil must always be protected from undue exposure to oxygen (air) and the trace metals are reduced to very low levels as discussed in Chapter 6 on oil bleaching.

Autoxidation in the oil cannot be stopped but the reaction can be controlled by using proper temperature of the oil, storing under nitrogen, and sometimes by using antioxidants which react with the free radicals and protect the oil up to a certain point.

Enzymatic oxidation

The oilseeds and fruit palm contain a group of enzymes called lipoxygenase. These enzymes can oxidize the oil in the oilseeds or fruit palm under suitable conditions.

Lipoxygenase activity can be reduced by minimizing the damage to the seeds, not exposing the seeds to higher temperature, or by thermal deactivation of the enzyme as it is routinely done in crude palm oil (CPO) production.

Photooxidation

The unsaturated fatty acids in the oil can oxidize rapidly if the processed oil contains chlorophylls and the breakdown products of chlorophylls and is exposed to visible light. The oil, under this condition, oxidizes nearly 1500 times faster than in the autoxidation process. The oil in processed foods can undergo similar oxidation if there is riboflavin in the product (a common constituent of dairy products).

The oil can also oxidize when exposed to ultraviolet light in the presence of a metal initiator as in autoxidation. This reaction is as fast as the autoxidation process.

Photooxidation is seen primarily in finished products packaged in clear packaging material or oils packaged in clear bottles.

Photooxidation can be greatly reduced by using opaque packaging for products, as well as for the oil. The refined oil must also be bleached to very low chlorophyll content as discussed in Chapter 6.

2. Hydrolysis
   Hydrolysis occurs when a molecule of water reacts with an oil molecule, releasing a molecule of fatty acid. These acids are then called free fatty acids (FFA) because they are no longer attached to the glycerol backbone of the triglyceride molecule (see Chapter 2).
   Hydrolysis can proceed only when the oil and water are in intimate solution. This can happen only if there is a surfactant present in the oil. Natural surfactant, such as phospholipids (see Chapters 2 and 4), can promote hydrolysis. Some additional surfactants, such as monoglycerides and diglycerides, can be produced in the refining step (Chapter 5). Some FFAs are also produced in hydrogenation (Chapter 7) and deodorization (Chapter 8).
   Enzymatic hydrolysis
   Enzymatic hydrolysis of the oil in oilseeds and also in fruit palm is caused by the enzyme called lipase. These enzymes can cleave the fatty acid from the triglyceride molecule under suitable conditions. This process can be controlled in oilseeds by maintaining proper storage temperature and proper handling of the seeds. Lipase activity is high when the seed storage temperature is 45–60°C at high humidity. In fruit palm the enzymes are deactivated by treating the fruits to steam in a giant autoclave or in a microwave chamber.
   Palm fruits are more fragile and are prone to have ruptured skins during harvest and transport. This causes very rapid hydrolysis of the oil in the fruit. The author has witnessed an FFA content of 50% in the oil in palm fruit when it is damaged badly. Normal FFA in soybean crude oil is 0.6%–1.0%. However, through lipase action this value can be considerably higher. Careful handling of oilseeds, as well as fruit palm can reduce the lipase activity and thermal deactivation can help eliminate lipase in both oilseeds and fruit palm.
3. Thermal decomposition
   Thermal decomposition of oil occurs when it is subjected to high heat for an extended period or if this is done repeatedly over a period. In this process, the fatty acid moiety is cleaved from the glycerol backbone. Two or more of these compounds can bond and form dimers and polymers. These polymers have different properties than the oxidative polymers.

12.4 AREAS IN OIL QUALITY MANAGEMENT

The quality management of the oil begins with the oilseeds. The specific areas are listed as follows:

1. quality of seeds (fruit palm in the case of palm oil)
2. seed drying, transport, and storage
3. seed crushing and crude oil handling
4. crude oil receiving and storage
5. degumming/acid pretreatment

6. refining
7. bleaching
8. hydrogenation
9. deodorization
10. handling of deodorized oil
11. loading oil and shipping of deodorized oil

1. Quality of seeds (fruit palm)
   High-quality oilseeds produce better-quality oil. Crude oil obtained from damaged seeds produces crude oil of poor quality. The seeds can be damaged before harvest, during harvest, and postharvest. A number of preharvest conditions can significantly reduce the quality of the oil in the seeds.
   Preharvest
   1. Dry weather conditions (drought) before harvest can cause insect infestation. This results in oxidation and hydrolysis of the oil in the seeds.
   2. Wet weather before harvest makes seeds too moist and moldy. This also increases the chlorophyll content in the crude oil. Chlorophylls are hard to remove without overbleaching the oil (Chapter 6). This reduces the RBD oil quality.
   3. Physical damage may occur to the oilseeds during harvest. This exposes the interior of the oilseeds to the atmosphere and increases both lipoxygenase and lipase activity in the oilseeds, causing oxidation and hydrolysis to the oil.

   During harvest
   Oilseeds can be damaged during harvest and storage. Field-damaged seeds tend to produce crude oil with darker color and poor flavor stability. The crude oil is difficult to refine and bleach and also is subject to higher refining losses.
   Postharvest
   The seeds can be damaged in handling and delivery, causing hydrolysis and autoxidation of the oils in the seeds.
   Fruit palm
   Fruit palm is very prone to physical damage even under the most careful harvesting procedure. The skin of the fruit is ruptured during harvest, exposing the oil to oxygen and moisture in the environment in the presence of the enzymes lipoxygenase and lipase in the fruit. This causes oxidation and hydrolysis to the oil in the fruit. This is why CPO can contain as much as 5% diglyceride or higher. With careful harvesting and selected crushing, it is possible to have 3% or less diglyceride in the CPO.
   The diglyceride content in the seed oil is considerably lower because the seeds are not as vulnerable to handling as fruit palm.
2. Seed drying and storage
   Oilseeds must be dried before storage. High moisture in the seeds in storage can be harmful to the oil in the seeds. Following are the necessary steps in drying and storage of oilseeds.

1. Dry the oilseeds to 10%–12% moisture content before storage. Incomplete drying can increase lipase activity.
2. Too rapid drying can generate cracks on the seed surface, increasing lipase and lipoxygenase activity.
3. The seeds must be stored at a temperature <45°C (114°F).
4. At temperature >45°C (114°F) and moisture content of 14% or higher, enzyme phospholipase-D is activated. This transforms hydratable phospholipids to nonhydratable phospholipids, making the crude oil more difficult to refine (Chapter 5).
5. This also causes color fixation, requiring additional bleaching treatment, which removes more tocopherols from the oil and can reduce its oxidative stability.

3. Seed crushing and crude oil handling

The last step in the solvent extraction process is the stripper where the solvent is removed from the oil. The temperature of the oil is about 220°F (104.4°C). At most seed crushing operations, the oil is stored at this temperature. The oil is not filtered or cooled before storage.

When crude oil is stored at this elevated temperature, free radicals are formed at a rapid rate and the primary oxidation proceeds. The peroxides in the oil rise and some break down into secondary oxidation products, which are constituted primarily of aldehydes, ketones, and partially decomposed triglyceride molecules. Some of these compounds are captured in the analysis called anisidine value (AV). These reactions proceed rapidly at high temperature.

The recommended storage conditions for the crude oil at the extraction plant are as follows:

1. Crude oil must be filtered and cooled to <140°F (60°C) before storage, where it is directly sent for degumming.
2. If the crushing is done solely for selling crude oil to other refiners, the crude oil must be filtered and cooled down to <104°F (40°C).
3. The storage tank must have a side-mounted agitator with internal baffle arrangements, low-level cut-off for the agitator, and temperature indicator controller.
4. The tank must be bottom filled or the feed line should be extended to the floor of the tank.
5. The lines, if needed to be cleared, must be blown with nitrogen and not with steam.

4. Crude oil receiving and storage

The crude oil receiving standards have been discussed in Chapter 3. The oil must be unloaded and stored with care. The crude oil must be refined as soon as possible. Prolonged storage can increase the FFA and also the oxidation level in the crude oil. Higher FFA increases the refining loss. Highly oxidized crude oil cannot produce good-quality refined oil.

Following are the recommendations related to crude oil receiving and storage:

1. Crude oil must be stored in the same fashion as described under Crushing.
2. The tank must have a side-mounted mechanical agitator with low-level cut-off for the agitator and temperature indicator controller. With no agitation in the tank, the gum rapidly settles to the bottom of the tank. This can cause heavy oil loss in refining. It can also cause some flavor reversion in soybean oil.
3. The lines should never be blown with steam. Steam can hydrate the gum (phospholipids) and make it settle to the bottom of the tank. This can increase the refining loss.
4. Recommends use of nitrogen, or air if nitrogen is not available.
5. The tank must be cleaned every 3–6 months (preferably every 3 months).
6. Crude oil temperature must be maintained at $<120°F$ (49°C), and preferably $<105°F$ (40.6°C).

Consequences of higher storage temperature or longer storage time for the crude oil

As mentioned earlier, preoxidized crude oil does not produce good quality finished oil. The freshly deodorized oil may have a fresh PV value of zero, even when it is obtained from crude oil with high PV. However, the oil in storage loses flavor rapidly with the initial high PV in the crude oil, especially when the PV is >8. The experimenters refined, bleached, and deodorized soybean oil that had following PVs:

Sample #	PV of the crude oil
1	8.3
2	22.1
3	56.5
4	73.0

The deodorized oil samples were stored at 140°F (60°C). An expert panel evaluated the flavor of the oil samples. The following results were observed:

1. All four samples had acceptable initial flavor with the actual flavor grade ranging from 7.0 to 8.0.
2. The oil produced from the crude oil with a PV of 8.3 had the highest initial flavor grade.
3. The oil made from the crude oil with a PV of 73 had the lowest flavor grade.
4. After 3 days of storage, the flavor grade of Sample #1 was 7.3 while the rest of the samples had flavor grades of 6 or less.
5. After 5 days of storage, the flavor grades of all four samples were poor.

Fig. 12.3 shows the progression of the flavor grades in the four oil samples stored at 140°F (60°C).

Based on these results, the experimenters recommended that the PV of the crude oil in storage must be as low as possible and should not be

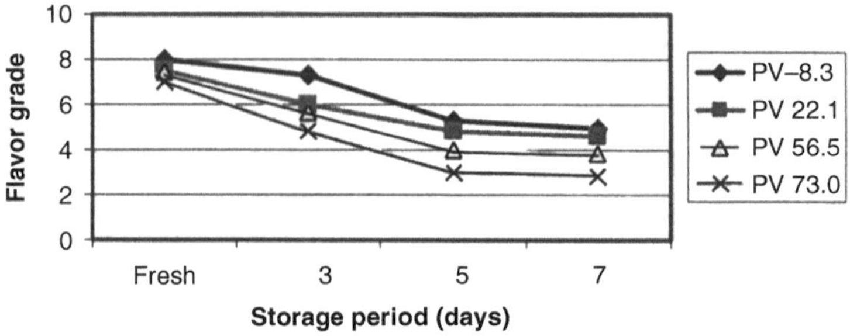

FIGURE 12.3 **Flavor stability of preoxidized crude soybean oil deodorized and stored at 140°F (60°C).**

allowed to rise above 8. The author recommends the crude oil PV to be maintained at <4. The oil must be refined immediately or as quickly as possible if the PV begins to rise in the crude oil storage tank.

The experimenters also stated that when the PV breaks down, only 10% of the original triglyceride molecule is distilled in the deodorizer at elevated temperature under vacuum. The remaining 90% of the molecule remains in the oil. These partially decomposed oil molecules are nonvolatile, highly reactive, and produce numerous oil decomposition compounds, such as oxidative dimers and polymers. Oxidative polymers are formed whenever the oil is heated to high temperature for an extended period. These compounds, as stated earlier, are strong oxidizing catalysts for the oil and degrade the oil flavor in storage, as well as in the products made from the preoxidized crude oil. The oil flavor degrades even when it is stored under nitrogen, frozen, or stored in the dark. This phenomenon cannot be explained by the standard oil oxidation measurements, such as PV or OSI. This phenomenon is termed hidden oxidation.

As stated earlier, high peroxide in the oil produces oxidative dimers and polymers. These compounds can produce poor flavor in the fresh oil even at a fairly low concentration. The same experimenters deodorized the oxidative dimers from the oil and blended it back into the oil with good flavor at a level of 0.5, 1.0, and 2.0%. The results showed that:

1. the oil flavor became progressively less acceptable as the dimer content in the oil increased, and
2. the oils also showed poor flavor stability in storage.

Fig. 12.4 shows the progression of the flavor grade in the oil containing the various levels of oxidative dimers and stored at 140°F (60°C).

The previous data indicate that proper handling and storage of the crude oil and in-process oil during the subsequent processing steps are very critical to the stability of the oil. The key is to maintain the oil at low PV. Bleaching (under vacuum) or deodorization reduces the PV to zero, but that does not guarantee good flavor stability for the deodorized oil in

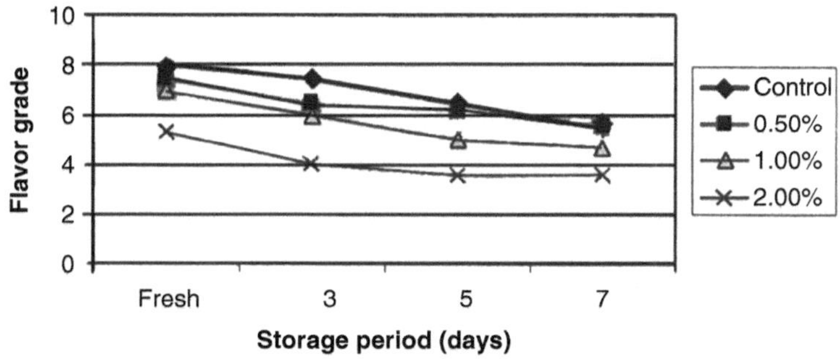

FIGURE 12.4 Effect of dimers in deodorized oil stored at 140°F (60°C).

TABLE 12.1 Recommended PV and Anisidine Values (AV) for Crude Oil

Analysis	Maximum Value
PV	<8 preferably <4
AV	<4 preferably <2

storage. When the PV is reduced, the AV and sometimes the oxidative polymers in the oil go up. High AV as well as the oxidative polymers can adversely affect the oil flavor stability. Therefore, both PV and AV should be considered as critical control points for the crude oil, as well as later in the process. Table 12.1 shows the recommended values for PV and AV for crude oil to obtain finished oil with good flavor stability.

Case study

Cottonseed oil was received from a supplier that met all fresh oil quality specifications, such as FFA, PV, and flavor. The oil was used for industrial frying in a continuous fryer. The oil in the fryer started to get heavily oxidized very rapidly.

The oil supplier was contacted and they were asked about the history of the oil. The following information was provided by the supplier:

1. The crude oil was purchased by the standalone refiner from another source 3–4 months before it was processed. The actual age of the crude oil was unknown.
2. The crude oil samples from the storage tanks were analyzed for PV and AV.
3. The PV ranged from 18–38.
4. The AV ranged from 10–22.
5. The PV in the fresh oil receipt ranged from 1–2.
6. The FFA in the fresh oil was 0.04%.
7. The AV in the fresh oil ranged from 9–18.

Normally, the AV in the fresh oil is <4.

This is a classic example of poor-quality crude oil and its impact on the quality of the finished oil obtained from it.

5. Degumming/acid pretreatment

The primary objective of water degumming soybean oil is to recover the phospholipids to make lecithin. The purpose of acid degumming, ultra degumming, and acid pretreatment of the crude oil is to further reduce the phospholipids content in the crude oil before refining. All of these treatments reduce the nonhydratable phospholipids in the crude oil.

German scientists have shown that phospholipids, if left in the refined oil, can cause flavor reversion. The researcher suggested repeated water washing of the crude oil using water containing sodium silicate and citric acid. Citric acid was used for removing trace metals from the crude oil.

Scientists in the United States showed that oil containing linolenic acid (soybean oil) developed fishy flavor in the deodorized oil after storage. This is more pronounced when the crude oil is oxidized.

A similar phenomenon was observed by these researchers when soybean oil was left in contact with trimethylamine oxide.

It was reported that out of the three unsaturated fatty acids in soybean oil, linolenic acid produced the highest level of di- and trimethylamines related to the fishy odor in soybean oil.

Flavor-reverted soybean oil was found to contain 0.67%–2.72% nitrogen. There was a strong correlation between flavor-reverted soybean oil and nitrogen-bearing compounds in the oil, such as amines.

The previous discussions lead us to conclude that the phospholipids in the crude oil must be reduced to a low level before refining for producing improved flavor in the refined oil.

USDA research showed that it is better to degum the soybean oil before storage and also the phosphorus content should be <50 ppm, although the trading rule allows up to 200 ppm of phosphorus in the degummed soybean oil. This research also suggested long mixing times for refining soybean oil. This also applies to cottonseed and corn oil. Degummed soybean oil was less preferred over the crude oil for controlling the primary separator in the oil refinery. At some point, refiners used a blend of degummed and nondegummed crude soybean oil to obtain better control in separation of soap and refined oil in the primary separator. However, improved equipment and operating techniques have allowed oil processors to handle degummed soybean oil effectively.

Reducing the phosphorus content to <10 ppm has been found to improve the oil flavor. Ultra degumming can reduce the phosphorus content of good-quality crude soybean oil to <5 ppm. This oil can be successfully refined by the physical refining process (see Chapter 5).

Case study

Cottonseed oil is considered the "gold standard" for frying potato chips. In a controlled test, 90% cottonseed oil and 10% lightly hydrogenated commercial garden variety soybean oil were blended. Potato chips were fried in the control oil and the test blend. The following results were obtained:

1. The experts identified the test product fresh and after aging as being different from the gold standard (control) and less acceptable.
2. The consumers could not identify the difference in the fresh product but identified the aged product to be different from the control and also less acceptable.

In order to test the hypothesis on the impact of phospholipids in the crude soybean oil, the following experiment was conducted:

1. Ultra degummed soybean oil from high-quality crude was lightly hydrogenated and added to cottonseed oil at the 50% level.
2. Potato chips were fried in both oils.
3. Neither the experts nor the consumers could tell the difference between the two in fresh, as well as in the aged products.

This is a case that highlights the benefit of good degumming and preferably, ultra degumming of crude soybean oil to obtain better flavor stability.

Incidentally, the cost of the ultra degummed soybean oil was higher than that of the cottonseed oil. Thus, this could not be a commercial option but it proved the hypothesis that the crude soybean oil with the lowest level of phosphorus (<5 ppm) can produce finished oil with superior flavor stability.

6. Refining

Refining techniques and the critical control points for vegetable oil refining have been discussed in Chapter 5. It has been discussed in this chapter that:

1. Poor-quality crude oil with high FFA requires a higher dosage of caustic.
2. Crude oil with high phospholipids content generally requires stronger caustic or longer mixing time.
3. Crude oil with darker color requires stronger caustic treatment.
4. Sometimes, extremely poor-quality crude oil may require double refining.
5. Chapter 5 shows that the phosphorus level in the refined oil must be 3 ppm or less. At higher phosphorus content in the refined oil, the bleaching process may not reduce it sufficiently (<1 ppm) as specified in Chapter 6 for the bleached oil.

Comments

1. It has been mentioned that the water-washed oil should have a maximum soap content of 100 ppm.
2. In each and every one of the previously mentioned situations, there is always some neutral oil that reacts with the caustic. This produces a higher amount of soap in the refined oil, making it harder for the water-washing centrifuge to remove the soap from the refined oil.
3. These conditions also produce diglycerides and monoglycerides in the refined oil.
4. These compounds are emulsifiers. They make separation of the soap and oil more difficult, resulting in more neutral oil in soap and a higher than normal amount of soap in the refined oil. This can have the following impacts on the process:

a. High soap in the refined oil requires a higher dosage of bleaching clay for its removal, causing an increase in the FFA level in the bleached oil.
b. A higher level of FFA in the bleached oil can increase catalyst usage in hydrogenation by forming nickel soap.
c. A higher level of FFA will increase the oil loss. The additional amount of FFA will also carry out some additional amount of neutral oil in the distillate from the deodorizer.
d. With the use of the standard amount of bleaching clay, the soap may not be removed from the refined oil. This can impact the hydrogenation reaction.
e. If the bleached oil has positive soap content, it can make the deodorized oil darker. It can also either prolong the deodorization time to reduce the FFA or lead to higher than normal FFA and darker color in the deodorized oil.
f. With excessive or repeated refining, the oil loses some additional amount of tocopherols. This reduces the stability of the finished oil.
g. High phosphorus left in the bleached oil can poison the hydrogenation catalyst.

7. Bleaching
The principles and practices for the bleaching process have been described in detail in Chapter 6. The objective of bleaching is to reduce the levels of all trace impurities in the oil. This includes trace metals, color compounds, such as chlorophylls, some carotenes, and phosphorus. In addition, bleaching must reduce the soap in the oil to zero ppm.
Normally, the crude soybean oil is bleached to remove the chlorophyll and not to reduce the red color as much. The red color is removed from the oil during deodorization by heat bleaching. Bleaching the oil to a very low red color may cause color fixation in the oil. Therefore, the refining color control is focused on the reduction of the chlorophylls.
There are two types of chlorophylls present in seed oils, namely chlorophyll A and chlorophyll B.
Soybean oil and canola oil generally contain the highest levels of chlorophylls. The other oils, such as sunflower oil, cottonseed oil, and so on, can also have chlorophyll. Wet harvest condition increases the chlorophyll content in the oilseeds. Since chlorophylls are soluble in the oil, the wet harvest condition produces greener crude oils. High chlorophyll in the crude oil requires a higher than normal dosage of bleaching clay. This can have several negative impacts on the bleached oil.
   1. Higher dosage of the bleaching clay increases the FFA content in the bleached oil by hydrolyzing some of the neutral oil in addition to hydrolyzing the soap in the refined oil.
   2. Higher FFA in the bleached oil can become a catalyst poison as discussed in Chapter 7 by forming nickel soap of the fatty acids.

3. Higher FFA in the bleached oil will increase the neutral oil loss in the process as discussed earlier.
4. Overbleaching will reduce the amount of tocopherols in the bleached oil, reducing the stability of the finished oil.
5. Overbleaching of the refined oil can decompose the chlorophylls in the oil, making the oil more susceptible to photooxidation.
   Breakdown of chlorophylls
   Chlorophylls break down under acidic pH and in the presence of oxygen (air). Chlorophylls break down into three products:
   - pheophytins
   - pheophorbides
   - pyropheophorbides

   These products are 10 times stronger photosensitizers than the parent chlorophylls.
   Undue exposure to air under an acidic environment can take place in the following processing steps:
   1. Acid degumming or acid pretreatment
   2. Bleaching (uses citric or phosphoric acid and acid-activated bleaching clay)
   3. Under poor vacuum conditions in the vacuum bleacher
   4. Under atmospheric bleaching conditions

   These breakdown products are not detectable at the same absorption wavelength on the spectrophotometer as chlorophyll. Therefore, one may not know about the presence of these breakdown compounds under the standard operating tests.
   It is recommended that:
   1. The phosphorus content of the bleached oil must be <1 ppm
   2. The soap in the bleached oil must be 0 ppm
   3. Atmospheric bleaching must be avoided
   4. Vacuum must be maintained at an operating pressure of 100 mm of mercury column max., and preferably <50 mm of the same

8. Hydrogenation
   Linolenic acid loses one double bond as the first step in the hydrogenation process and produces isolinoleic acid. Natural linoleic acid has the double bonds at the 9–10 and 12–13 positions. Isolinoleic acids produced from hydrogenation of linolenic acid are of two types. They are:
   - *Type-1*: The double bond at the 12–13 position is hydrogenated but the double bonds at the 9–10 and 15–16 positions remain in place and not hydrogenated.
   - *Type-2*: The double bond at the 12–13 position is hydrogenated but the double bonds from position 9–10 and 15–16 shift to positions 10–11 and 14–15, respectively.

   Type-1 isolinoleic acid was identified to be a contributor to the fishy flavor in the soybean oil. Additional research indicates that at least one or both of the double bonds in the *Type-1* isolinoleic acid are in *trans* configuration. This leads one to consider the fact that the following hydrogenation conditions could lead to the formation of *Type-1* isolinoleic acid in the hydrogenated oil.

1. More selective hydrogenation conditions
2. Poor catalyst activity
3. Poor hydrogen gas quality
4. High phosphorus in the bleached oil
5. Presence of positive soap in the bleached oil
6. High FFA in the bleached oil
7. Catalyst poisons like sulfur (which may come from not removing the mercaptans present in the natural gas used for making hydrogen gas)

In an essence, poor-quality bleached oil can produce *Type-1* isolinoleic acid in the hydrogenated oil, which may produce fishy flavor in the finished soybean oil in storage.

It has been discussed in Chapter 7 that a slower hydrogenation rate can cause hydrolysis of the neutral oil, elevating the FFA content in the hydrogenated oil beyond the normal rise of up to 0.05%. This extra FFA means lower overall oil yield and higher neutral oil loss in the deodorizer.

9. Deodorization

Deodorizers are designed to reduce the FFA and the red color and also to remove the odor-bearing compounds from the bleached oil. The critical operating parameters have been discussed in detail in Chapter 8. The primary requisites for obtaining good deodorized oil are:

1. Good-quality bleached oil
2. Good vacuum
3. Proper deodorizer temperature
4. Cooling of the deodorized oil
5. Appropriate team stripping rate

Good-quality bleached oil

Chapter 6 indicates that the bleached oil must meet the following analytical standards:

1. PV leaving the bleacher	0 mEq/kg
2. Phosphorus, P	< 1 ppm
3. Soap	0 ppm
4. Chlorophyll (soybean and canola oils)	< 30 ppb

With good-quality refined oil and proper bleaching conditions, the PV in the bleached oil must be zero. If it is not zero, then there is poor vacuum. The consequence of poor vacuum is higher levels of oxidative dimers and polymers in the bleached oil going to the deodorizer. Deodorization cannot remove these oxidative polymers from the oil and the oil will have poor flavor stability.

Phosphorus, chlorophyll, and soap cannot be removed in the deodorizer.

High phosphorus and soap content in the bleached oil produces darker color and also potentially higher FFA in the deodorized oil as discussed in Chapter 8.

Poor vacuum, high deodorizer temperature, and lack of cooling of the deodorized oil lead to a high level of oxidative polymers in the finished oil. Therefore, the oil would be expected to have poor storage stability.

Handling and storage, as well as shipping and distribution have been discussed in detail in Chapter 9.

12.5 SUMMARY OF OIL QUALITY STANDARDS

Crude oil quality	
PV	<8 mEq/kg
(Preferably)	<4 mEq/kg
pAV	<4 AVU
(Preferably)	<2 mEq/kg
Bleached oil quality	
PV leaving the bleacher	0 mEq/kg
Phosphorus, P	<1 ppm
(Preferably)	<0.5 ppm
Soap	0 ppm
Chlorophyll (soybean and canola oils)	<30 ppb
Iron	<0.5 ppm
(Preferably)	<0.3 ppm
Calcium	<0.5 ppm
(Preferably)	<0.2 ppm
Magnesium	<0.5 ppm
(Preferably)	<0.2 ppm
Refined, bleached, and deodorized oil must be low in:	
Phosphorus	<1 ppm,
(Preferably)	<0.5 ppm
PV (fresh out of the deodorizer)	<0 mEq/kg
PV (as delivered to customer)	<1 mEq/kg
(Preferably)	<0.5 mEq/kg
pAV	<6
(Preferably)	<4
Conjugated dienes	<0.5%
(Preferably)	Trace
Dimers	<0.2%
(Preferably)	Trace
Polar compounds	<4%
(Preferably)	<2%
Monoglycerides	<0.5%
Diglycerides	<1%
Chlorophyll	<30 ppb
Iron	<0.5 ppm
(Preferably)	<0.3 ppm
Calcium	<0.5 ppm
(Preferably)	<0.2 ppm
Magnesium	<0.5 ppm
(Preferably)	<0.2 ppm

READING REFERENCES

Chang, S.S., Brobst, K.M., Tai, H., 1961. Characterization of flavor reversion of soybean oil. JAOCS 38, 671.

Chang, S.S., Kummerow, F.A., 1954. The relationship between the oxidative polymers of soy beanoil and flavor reversion. JAOCS 31, 324.

Evans, C.D., List, G.R., Beals, R.E., 1974. Iron and phosphorus contents of soybean oil from normal and damaged beans. JAOCS 51, 444.

Evans, C.D., Frankel, E.N., Cooney, P.M., Helen, A., 1960. Thermal dimerization of fatty acid hydroperoxide. JAOCS 37, 452.

Frankel, E.N., 1985. Flavor Chemistry of Fats and Oils. AOCS Press, Champaign, IL.

Frenkel, E.N., 1987. Handbook of Soy Oil Processing and Utilization. AOCS Press, Champaign, IL, p. 229.

Goss, W., 1946. Solvent extraction of oilseeds. Oil Soap 23, 241.

List, G.R., Mounts, T.L., Lancer, A.C., 1992. Factors affecting the formation of non hydratable phospholipids in soybean. JAOCS 69, 443.

Evans, C.D., List, G.R., Moser, H.A., 1973. Long term storage of soybean and cottonseed salad oils. JAOCS 50, 218.

Mounts, T.L., List, G.R., Heakin, A.J., 1979. Post harvest handling of soybeans. JAOCS 56, 883.

Mounts, T.L., Nash, A.M., 1990. HPLC analysis of phospholipids in crude oil for evaluation of soybean deterioration. JAOCS 67, 757.

Rickford, W.G., 1941. Oil Soap 18, 05.

Perkins, E.G., Wantland, L.R., 1973. Characterization of nonvolatile compounds formedduring thermal oxidation of 1-linoleyl-2,3 distearin III. Evidence for presence of dimeric fatty acids. JAOCS 50, 459.

Sifensieder, L.L., 1937. Zig 64, 122.

Thompson, S.W., Taylor, W.G., Gudheim, A.R., 1946. Second conf. on flavor stability of soybean oil, Chicago.

Usui, R., Endo, Y., Kanada, T., 1984. Agric. Biol. Chem. 48, 99.

Usuki, R., Endo, Y., Suzuki, T., Kanada, T., 1983. Proceedings of 16th ISF Congres, Budapest, Hungary.

Chapter 13

Trans Fat Alternatives and Challenges

13.1 INTRODUCTION

13.1.1 Pioneering by Europe

Trans fatty acids in edible oils have been under scrutiny in Europe for decades. In the United States, the health effects of *trans* fatty acids in hydrogenated fats have been questioned since the 1960s. Europe went ahead of the rest of the world to put regulatory standards in place on *trans* fat, limiting its amount in edible fats and formulated foods. Denmark took the pioneering action on *trans* fat in fats and foods containing fats. (Excerpt from the website www.tfx.org.uk shows that Executive Order No. 160 of 11 March 2003 on the Content of *Trans* Fatty Acids in Oils and Fats.) It also lists the following information regarding the regulation:

- From 1 June 2003, the content of *trans* fatty acids in the oils and fats covered by this Executive Order shall not exceed 2 g/100 g of oil or fat.

13.1.2 *Trans* Fat Regulation in the United States

As of January 2006, the Food and Drug Administration (FDA) regulation declared that the food industry must comply with the following requirements:

- The food label must indicate the amount of *trans* fat content per serving.
- The *trans* fat content can be declared as 0 g per serving if the *trans* fat content in the product is <0.5 g per serving.

Sample of the US Food Label is shown in Fig. 13.1. It shows saturated fat with the recommended daily value (DV) but none for the *trans* fat.

13.1.3 *Trans* Fat in the United States Diet and the Sources

1. FDA estimates that the average daily intake of *trans* fat in the United States population in 2006 was about 5.8 g or 2.6% of calories per day for individuals 20 years of age and older.

Practical Guide to Vegetable Oil Processing. http://dx.doi.org/10.1016/B978-1-63067-050-4.00013-1
Copyright © 2017 AOCS Press. Published by Elsevier Inc. All rights reserved.

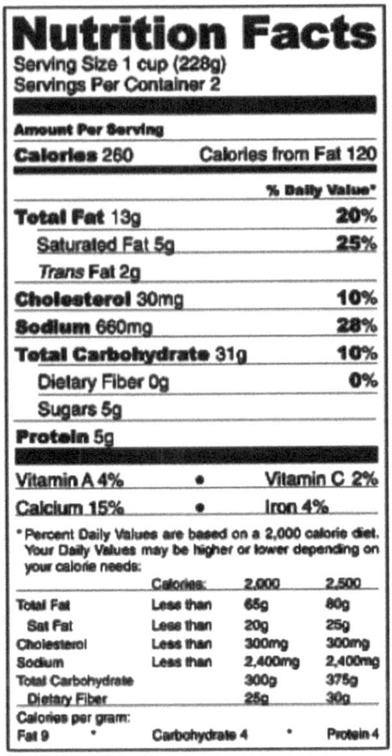

FIGURE 13.1 **Sample of food label in United States showing *trans* fat content per serving.** The amount of *trans* fat present per serving of the standard fats in the United States are listed in Table 13.1. *(This information was taken from the FDA CFSAN document published on July 9, 2003)*

2. On average, Americans consumed approximately 4 to 5 times more saturated fat than *trans* fat in their diet.
3. *Trans* fat could be found in vegetable shortenings, some margarines, crackers, candies, cookies, snack foods, fried foods, baked goods, and other processed foods made with partially hydrogenated vegetable oils.
4. Small amounts of naturally occurring *trans* fat can be found in some animal products, such as butter, milk products, cheese, beef, and lamb.

The amount of *trans* fat present per serving of the standard fats as required by the FDA in 2006 in the United States is listed in Table 13.1. This information was taken from the FDA CFSAN document published on July 9, 2003.

13.1.4 Subsequent Developments in FDA Regulations on *Trans* Fat

In 2013, the FDA determined that there is "no longer a scientific consensus" that partially hydrogenated oils (PHOs) are safe for their intended use in food, made a tentative determination that they are no longer GRAS "under any condition of use in food," and issued a request for comments on its proposal.

TABLE 13.1 Saturated and *Trans* Fat per Serving of Various Fats

Product	Total fat (g)	Saturated fat (g)	*Trans* Fat (g)	Combined saturated and *trans* fat (g)	Cholesterol (mg)
Butter[a]	10.8	7.2	0.3	7.5	31.1
Margarine, stick[b]	11	2.1	2.8	4.9	0
Margarine, spread[b]	9.7	1.8	2.7	4.5	0
Margarine, tub[b]	6.7	1.2	0.6	1.8	0.1
Margarine, bottle[c]	0.4	0.1	0	0.1	0.2

Serving size: one tablespoon.
[a]*Butter values from FDA Table of* Trans *Values, dated 1/30/95.*
[b]*Values derived from 2002 USDA National Nutrient Database for Standard Reference, Release 15.*
[c]*Prerelease values derived from 2003 USDA National Nutrient Database for Standard Reference, Release 16.*

On June 16, 2015 FDA finalized its determination that PHOs, the primary dietary source of artificial *trans* fat in processed foods, are not "generally recognized as safe" or GRAS for use in human food. Food manufacturers will have 3 years to remove PHOs from products. The FDA made this determination based on extensive research into the effects of PHOs, as well as input from all stakeholders received during the public comment period since 2013, when the FDA first announced the agency's position on artificial trans fat and PHO.

The FDA has set a compliance period of 3 years. This will allow companies to either reformulate products without PHOs and/or petition the FDA to permit specific uses of PHOs. Following the compliance period, no PHOs can be added to human food unless the FDA approves their use through petition or special approval process.

The FDA data indicate that the consumption of *trans* fat in the United States has been declining, mostly because of the effort of the food manufacturers who have made a positive impact on the *trans* fat reduction in foods.

The new food nutrition label for packaged foods and restaurants were released by the FDA in May, 2015. The food companies will have to comply by January 2018. This does not include any Daily Value (DV) for *trans* fat intake. Fig. 13.2 shows the FDA data on the *trans* fat contents of US food products.

New York City banned PHO in restaurant foods in 2006, and the state of California did the same in 2008.

In the United States, the states can restrict the use of artificial *trans* fat in foods served at restaurants and food services or use of reduced salt in their cooking but they cannot regulate on the GRAS status for any food material or additives.

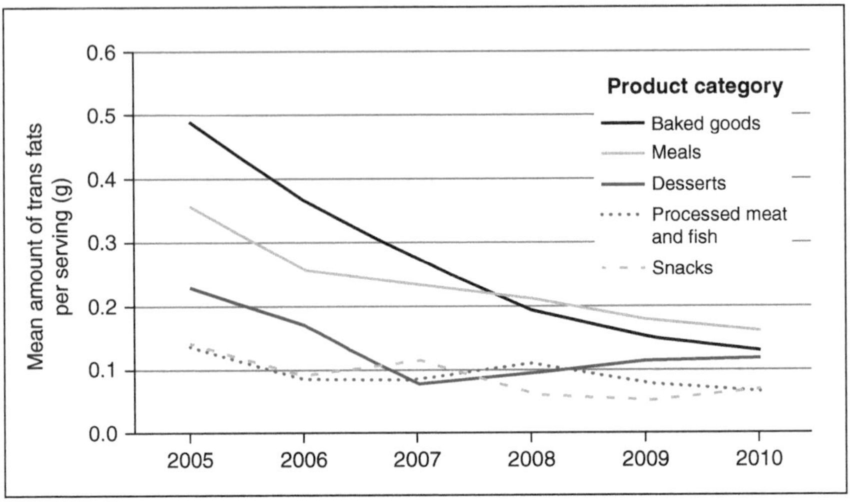

FIGURE 13.2 **Average *trans* fat in US food products.**

13.1.5 *Trans* Fat Regulation in Canada

Health Canada, the equivalent of the US FDA has not declared the artificial *trans* fatty acid as unsafe (or not GRAS as by the FDA). However, British Columbia went its own way in 2009, when it restricted artificial *trans* fats in all foods prepared and served in the province, including in restaurants, bakeries, schools, and healthcare institutions.

Canadian mandatory *trans* fat regulation is different from that of the United States, as shown further. It was more rigorous than the US regulation at that time.

On June 20, 2007, the Minister of Health announced that Health Canada adopted the recommendations of the *Trans* Fat Task Force with respect to the amount of *trans* fat in foods. These recommendations from the *Trans* Fat Task Force were twofold:

1. Limit the *trans* fat content of vegetable oils and soft, spreadable margarines to 2% of the total fat content.
2. Limit the *trans* fat content for all other foods to 5% of the total fat content, including ingredients sold to restaurants.

The Minister called on the food industry to achieve these limits within 2 years. The Minister also announced that if significant progress has not been made in the next 2 years, Health Canada will develop regulations to ensure that the recommended levels are met.

In doing so, companies and food manufacturers are encouraged to replace *trans* fats with healthier alternatives, such as monounsaturated and polyunsaturated fats and to not replace *trans* fats with saturated fats.

To ensure that the industry is making progress in meeting the 2 and 5% limits of the total fat, Health Canada will closely monitor the actions of the industry via the *Trans* Fat Monitoring Program.

Canada is the first country to publish this type of monitoring data.

A *Trans* fat Task Force was created, which was responsible to monitor the *trans* fat in the food products every 6 months, compile, and report the information to the industry and to the Ministry.

There were four monitoring data accumulated from 2007 to 2009. The products from the industry, restaurants, and food services were monitored for *trans* fat contents. It was found that the industry had significantly lowered the *trans* fat level in the products of several areas.

In 2011 the third set of data report stated that there were reduction in the *trans* fat in many areas but the Canadians were consuming high amounts of fat and *trans* fats.

In 2012, the fourth set of data report stated:

1. The fourth set of monitoring data continues to indicate that the nutrition labeling regulations are an effective motivator for industry to reformulate their products, as many food manufacturers have reduced the *trans* fat content of their products to meet the 5% *trans* fat of total fat content limit.
2. The results also indicate that some of the small- and medium-sized family and quick service restaurants have been successful in reducing the *trans* fat levels of their products to meet the limits.
3. There has also been progress, although slightly slower than other areas of the food service industry, in foods collected from cafeterias located in institutions.
4. Finally, the results continue to show that there are some sectors that face challenges in reducing the *trans* fat content of their products. For example, some bakery products, desserts, and cookies remain high in *trans* fat. The hurdles they face include maintaining the functional properties of their products; however, alternatives are now available for all applications. In these sectors the level of success has also been lower.

The ministry published a report on the "Risk Assessment of Exposure to *Trans* Fat in Canada" in October 2012.

13.2 NUTRITIONAL LABELING REGULATION

13.2.1 *Trans* Fat Claims

To bc labeled "Reduced in *Trans* Fatty Acids," a food must be

- Compared to the original food.
- Contain at least 25% less *trans* fatty acid per serving than the original food.
- Contain no more saturated fat than the original food.
- Unable to qualify for "Free of *Trans* Fatty Acids" claim.

To be labeled "Lower in *Trans* Fatty Acids," a food must be:

- Compared to another food.
- Contain at least 25% less *trans* fatty acid per serving than the food compared to.
- Contain no more saturated fat than the food compared to.
- Unable to qualify for "Free of *Trans* Fatty Acids" claim.

13.2.2 Nutrition Labeling Regulation

13.2.2.1 Trans *Fat Claims*

To be labeled "Free of *Trans* Fatty Acids," a food must be:

- <0.2 g *trans* per serving
- <2 g saturates + *trans* per serving
- <15% of energy from saturates + *trans*

Food industry prefers "Zero *Trans* Fats"

To be labeled "0.0 g of *Trans* Fatty Acids" on Nutrition Facts Panel, a food must be:

- <0.1 g *trans* per serving
- *trans* level not dependent on saturates level

13.2.3 For 30-g Serving

- 15% fat in food: 1.1% *trans* in fat
- 20% fat in food: 0.8% *trans* in fat
- 25% fat in food: 0.7% *trans* in fat
- 30% fat in food: 0.6% *trans* in fat

13.2.4 For 10-mL (9.2-g) Serving

- canola, sunflower, corn oil: qualify for "Zero *Trans*" claim
- soya, cottonseed, peanut oil: qualify for "0.0 g *Trans*" claim if <0.54% *trans*

The impact of *trans* fat on human health has been discussed in Chapter 7 on hydrogenation. In addition, the methods for obtaining low-*trans* hydrogenated fats have been discussed in Chapter 7.

The United States and some other countries are facing the following challenges in finding the ultimate solution for the *trans* fat alternatives for:

1. functionality
2. technology
3. availability
4. economics

Functionality is especially important because the United States and several other countries whose vegetable oil industry is based on seed oils are at a distinct disadvantage. Palm oil and palm stearin (PS) contain naturally occurring saturated triglyceride molecules that allow one to use some of the solid fractions from the fractionation process of palm and palm kernel oil in making shortening and margarine products with zero *trans* fat.

The consumer advocate group the Center for Science in Public Interest (CSPI, Washington, DC, United States), has been actively objecting to the use of palm oil in the United States because of its high-saturated fat content. However, palm oil and its fractions have been in use practically all over the world. In the United States, the food industry is using palm oil, palm kernel oil, their fractions, and their derivatives successfully for the past 10 years and meeting the saturated fat content requirements in finished products.

Seed oils do not have naturally occurring saturated triglyceride molecules. Therefore, the seed oils have been hydrogenated for nearly a century to provide a stable fat system for shortening and margarine products. Therefore, these oils have to be either partially hydrogenated to create solid fats, or the liquid oil has to be interesterified with fully hydrogenated fat in to obtain proper functionality of these fats.

13.2.5 Influence of *Trans* Fats

As discussed in Chapter 7, *trans* fatty acids provide certain unique traits in the hydrogenated fat systems, such as:

1. Higher-melt point.
2. Higher-oxidative stability.
3. Rapid melting characteristics for the shortening and margarine by providing the desired solids profile over a range of temperature.
4. Solid content in fat is measured by two methods and are expressed as solid fat content (SFC) and as solid fat index (SFI). The recent trend is to use SFC for measuring and expressing the solid fat content in a fat system. The difference between the two methods is that the latter one (SFI) produces a slightly flatter solid curve than the former (SFC).
5. Provides other functional properties in baking fats and margarine.

These properties of *trans* fat have enabled the food scientists and the oil technologists to achieve the following important functionalities in food products:

1. Obtain and maintain proper batter consistency.
2. Maintain a sharp melting curve for margarine.

Trans fatty acids also improve the beta prime stability in the fat crystals, which improves the performance of the shortening in reducing batter density and increasing batter viscosity and even structural stability in baked cakes, pastries, and icing.

Technology to meet this challenge in the United States and the countries using primarily seed oils are available today to the oil processing industry.

However, availability of high-stability/high-solids seed oil is a major challenge, and because of the short supply of these modified composition oils, the cost of *trans*-free product is more challenging if one depends only on seed oils. However, using palm oil fractions and in some cases palm kernel oil fractions has been done successfully.

13.3 SOURCE OF *TRANS* FATTY ACIDS

As discussed in Chapter 2, *trans* fatty acids are actually unsaturated fatty acids. During hydrogenation, the hydrogen atoms on the same side of the double bonds (*cis*-position) rotate and place themselves on the opposite side of the double bonds. Chapter 7 discusses the conditions that promote formation of higher *trans* fatty acids and also the techniques that could lower their levels in hydrogenated oils.

Animal fat, such as dairy fat, lard, and tallow contain small amounts of naturally occurring *trans* fats.

Chapter 7 shows how the *trans* fat content rises in the oil during hydrogenation. It reaches a maximum and then declines. *Trans* fat content in fully hydrogenated fat is essentially zero.

However, if fully hydrogenated fat and liquid oil are blended together and the solids content is determined at various temperatures, one would find that the SFI curve would appear as a straight line over the typical temperature range of 50–104°F (10–40°C). It would appear essentially as a horizontal straight line over the range of temperature, as shown in Fig. 13.3. The SFC curve also would have a slight slope.

This type of zero *trans* fat blend could be used for industrial frying but will not have any functional property in formulating baked products.

The edible oil industry worked in this area, looking for alternatives to hydrogenated fats.

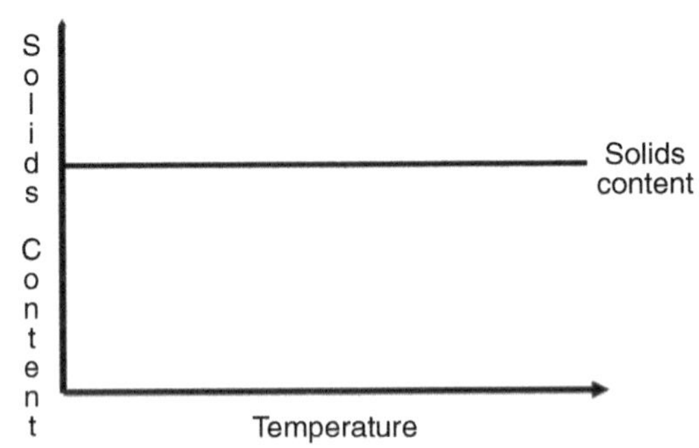

FIGURE 13.3 Flat solids curve.

13.4 TECHNICAL ALTERNATIVES AVAILABLE TODAY

The oil industry can meet the challenge of making no-*trans* shortening for the baking industry. There are a few ways the *trans* fat can be reduced in shortening and margarine. These perform well in the finished products. The biggest challenge lies in availability of raw material and economics.

13.4.1 Technical Solutions for *Trans* Fat Reduction

The following technical methods are available for *trans* fat reduction in the fats:

1. Hydrogenation of oil under special conditions.
2. Use of a noble metal, such as platinum, for hydrogenation instead of a nickel catalyst.
3. Interesterification (transesterification) of a mixture of fully hydrogenated fat and liquid oil under specific reaction conditions or interesterification of a liquid oil with palm oil fraction with higher solids.
4. Use of modified composition seed oils for frying and baking applications.
5. Use of pourable shortening formulated with fully hydrogenated fat and naturally stable or modified composition liquid oils, using a special crystallization process (see Chapter 7). This can be used for frying and some baking applications.
6. Use of various fractions of palm oil and palm kernel oil to make either blended or interesterified products that meet the specifications for the baking industry. In some products a small amount of fully hydrogenated palm oil, cottonseed oil, or soybean oil is used to meet the finished product standard. Fully hydrogenated fractions do not increase the *trans* fat in the shortening, it provides the crystal structure and the required consistency for the shortening.

13.4.2 Hydrogenation Under Special Conditions

Use of higher-than-standard reaction pressure, using lower temperature and less-selective conditions, produces lower-*trans* fatty acids in the hydrogenated fat. Various examples have been illustrated in Chapter 7. The resultant oil is higher in stearic acid content compared to the standard hydrogenated fat. The higher-pressure reaction requires a completely redesigned reactor. This would require every oil manufacturer to buy new reactors and other accessories, making the existing hydrogenation system obsolete. This is a tremendous economic burden that the consumers will have to bear.

13.4.3 Use of Platinum Catalyst

Platinum catalyst allows the reaction to be carried out at much lower temperature and produces lower-*trans* fatty acids in the process. The process also produces higher stearic acid. Additionally, the entire process economics is based on some intelligent assumptions regarding the recovery of the precious metal and

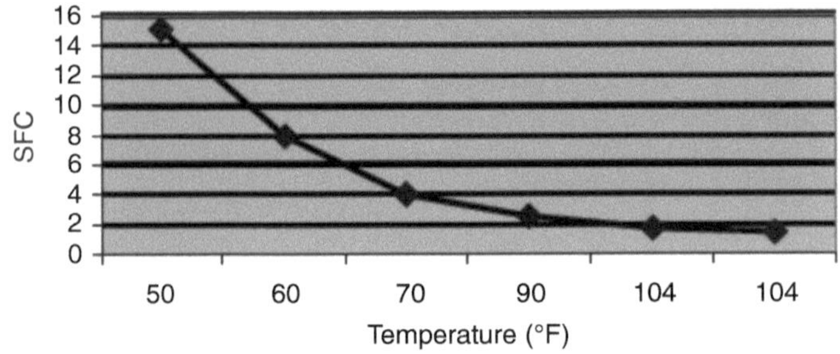

FIGURE 13.4 Solids curve after interesterification. *SFC*, Solid fat content.

the number of uses one can get out of the catalyst without altering the product characteristics. These are unknown factors because no vegetable oil plant has been able to control the loss of platinum, and, hence, could not establish a long-term history for an operational process.

13.4.4 Interesterification

Interesterification has also been referred to as transesterification in the vegetable oil process industry for decades. The common term used for this reaction in earlier days was "ester interchange in fats." This is a true representation of what this reaction is all about. This meant that in this process, two dissimilar oils are reacted where the fatty acids from the triglycerides are exchanged between their parent glycerol backbones, producing a mixture of triglycerides that are different from either of the reacting triglycerides. The ester interchange is also called by many oil processors and researchers as ester–ester interchange. The commonly used term used today is interesterification, implying that the ester linkage is taking place between two different triglyceride molecules under appropriate reaction conditions. The fatty acids can be rearranged within the same triglyceride molecule. This process is called the random molecular rearrangement.

The blend with a straight horizontal melting curve, as shown in Fig. 13.3, can be converted to transesterified product that would have a solids profile somewhat like one shown in Fig. 13.4.

Interesterification reaction is mostly carried out under random conditions. This means the fatty acid from one triglyceride molecule moves to occupy any position on the other glycerol back bone or may exchange positions with another fatty acid moiety in the same triglyceride molecule. This is why there are very little trisaturated triglycerides left after the reaction between a fully hydrogenated stock with liquid oil. With a mixture of 50% fully hydrogenated triglyceride (hard stock) and 50% liquid oil, one can find around 4% or less saturated triglycerides left in the equilibrium product after the interesterification reaction. This significantly lowers the melt point of the product compared to the starting blend of hard stock and liquid

TABLE 13.2 Melt Points of Blends Before and After Interesterification

Liquid oil (%)	Hard stock (%)	Melt point *before* interesterification (°F/°C)	Melt point *after* interesterification (°F/°C)
Soybean (75)	Cottonseed (25)	140/60	90/32.2
Coconut (90)	Cottonseed (10)	136/57.8	106/41.1

oil. Table 13.2 lists examples of some of those where the melt point after interesterification was considerably lower than that of the oil blend before interesterification.

13.4.4.1 Benefits of Interesterification

Interesterified shortening can provide the melting characteristics and the performance needed in the baking shortening and margarine. The following is an example of all-purpose shortening made by interesterification of fully hydrogenated cottonseed oil and modified composition soybean oil. The shortening produced satisfactory results compared to commercial all-purpose shortening made via the hydrogenation process, in cookies and crackers. Table 13.3 lists some of the results of this test.

The data presented in Table 13.3 show three distinct differences between the standard hydrogenated shortening and the interesterified shortening:

1. The standard hydrogenated shortening had significantly higher-*trans* fat content (31.2% vs. 2% in the interesterified product).
2. The interesterified shortening had significantly higher-saturated fat content (51.4% vs. 31% in the hydrogenated product).
3. Cholesterol-promoting *trans* fatty acids were significantly lower in the interesterified product.

In this test, both shortenings produced vanilla cookies and crackers that were identical in all attributes and shelf life stability.

13.4.5 Modified Composition Oils

Modified composition seed oils have been in existence for over four decades in the United States. However, the demand for these oils was not significant because of their high cost. As a result, the seed companies were not heavily engaged in the expansion of these seeds.

Since the declaration of the FDA decision on *trans* fat labeling on food packaging, the food industry has been pushing very hard to change to low and no *trans* fats in their food formulation, and the seed companies also increased their activities in this area.

TABLE 13.3 Comparison of Transesterified Shortening Against Standard Shortening

Analysis	Standard shortening	Test product
FFA (%)	0.03	0.03
PV (mEq/kg)	1.0	0.8
Mettler drop point (°F/°C)	116.2/46.8	116.6/47.0
Fatty acid composition (%)		
C-16	**15.3**	**17.4**
C-18	**14.7**	**34.0**
C-18:1	58.1	15.9
C-18:2	11.4	31.7
C-18:3	0.1	0.6
C-20	0.4	0.4
Trans fat (%)		
C-18:1	27.3	0.0
C-18:2	3.9	2.0
Total *trans* fat (%)	**31.2**	**2.0**
C-16 + *trans* (%)	56.5	19.4
Solid fat content at (°F/°C)		
50/10	27.5	28.9
70/21.1	19.4	18.2
104/40	9.4	9.9

FFA, Free fatty acid.
Bold emphasizes saturated fat.

Modified composition oils offer definite advantages in frying and baking applications for the following reasons:

1. Being low in linolenic acid, these oils do not have to be hydrogenated for frying, baking, or other food applications where lightly hydrogenated soybean oil was used for required oxidative stability in the oil.
2. In margarine and shortening, these oils can replace the lightly hydrogenated canola or soybean oil.
3. These oils can be used as the source for the liquid oil that can produce stable transesterified products.
4. These oils can replace lightly hydrogenated soybean or canola oil from the pourable shortening to provide zero *trans* product with no loss of functionality.

TABLE 13.4 Fatty Acid Composition of Modified Composition Seed Oils and Other Stable Oils

Modified composition oils	C-16	C-18	C-18:1	C-18:2	C-18:3
Low-linolenic soy	11	5	27	54	<3
USB soybean	3	3	37	54	3
Regular soy	10	5	23	55	8
High-oleic sunflower	3.6	4.3	82.2	9.9	Trace
Mid-oleic sunflower	4.6	4.2	61.3	27.3	Trace
Regular sunflower	7	4.5	18.7	67.5	Trace
High-oleic canola	3.3	2.3	78.8	5.1	5.2
High-oleic/low-linolenic canola	4	2	74	14	<3
Mid-oleic canola	4.6	4.2	61.3	27.3	Trace
Low-linolenic canola	4	3	65	18	4
Regular canola	4	2	58	20	9
Corn	11	2	25	60	1
Cottonseed	24	2	20	55	1
Palm	43	5	39	11	Trace
Palm olein	40	4	43	11	Trace

Table 13.4 shows the fatty acid compositions of various modified composition seed oils and also the currently available stable liquid oils.

Oils are used in applications listed:

1. salad oil
2. frying
   a. salty snack food
   b. food services
   c. making pan-fried products
3. baking
   a. in cookie dough
   b. spraying on crackers
4. margarine
5. candies
6. breakfast cereals, bars, etc.
7. drinks

The oils that are currently used for these applications are:

Application	Type of oil used
1. Salad oil	Liquid nonhydrogenated
2. Frying	
Salty snacks	Partially hydrogenated
Food services	Mostly hydrogenated
Pan-fried products	Mostly hydrogenated
Salty snacks	Some products are fried in nonhydrogenated oils
3. Baking	Hydrogenated
4. Margarine	Liquid and partially hydrogenated
5. Candies	Partially hydrogenated
6. Breakfast cereals	Liquid and partially hydrogenated
7. Drinks	Liquid and partially hydrogenated

Modified composition seed oils can be used in all of these applications, replacing the liquid and partially hydrogenated oils.

13.4.6 Use of Pourable Shortening

Chapter 10 on the fundamentals of fat crystallization illustrates the process of making pourable shortening. This is not a new concept but the only difference is that:

1. The traditional pourable shortening uses partially hydrogenated soybean oil or partially hydrogenated canola oil as the major component in the formula.
2. Partially hydrogenated oil provides oxidative stability to the pourable shortening, but at the same time it contributes a significant level of *trans* fat in the shortening.
3. The modified composition seed oils can be used to replace the partially hydrogenated oil, obtain the same functionality, and oxidative stability without the *trans* fat in the shortening.

13.5 CHALLENGES

The existing practice has been to use partially hydrogenated soybean or canola oil as the liquid fraction in the hydrogenated shortening.

To make *trans* free shortening, one must use a nonhydrogenated liquid oil fraction in the formula.

The liquid oil used in the transesterified product must have a good oxidative stability without hydrogenation.

The oil industry is facing several challenges today in making and delivering stable shortening by the transesterification process. This is because transesterified products do not exhibit good shelf life when regular soybean or canola oil is used in the formula. Soybean and canola oil contain 8% linolenic acid, which oxidizes rapidly and reduces the shelf life of the shortening and the products formulated with it. To improve the shelf life of these products, one needs to use liquid oil that has high-oxidative stability.

13.5.1 Challenge #1: Getting Stable Liquid Oil in an Adequate Supply

The liquid oils with high-oxidative stability that are suitable for producing stable transesterified products are:

1. palm olein
2. cottonseed oil
3. corn oil
4. modified composition seed oils

Unfortunately, the existing environment in the United States does not accept a widespread use of palm olein because of its high-saturated fat content and the emotional atmosphere lingering in the food industry in the country since mid-1980s. Some companies are using palm oil, palm kernel oil, and their fractions in making baking shortening and are able to meet the requirements for the total saturated fat, *trans* fat, and shelf life requirements for the baked products. However, in certain instances the serving size had to be reduced to meet the regulatory requirements on both total saturated and *trans* fat in the products, while using some of these newer shortenings.

Cottonseed contains almost 26% total saturated fat. Therefore, meeting the saturated fat content in the food product is somewhat of a challenge with the cottonseed oil. Corn oil, although it contains only 13% saturated fat, has never been popular in baking because of its inherent strong characteristic flavor.

13.5.2 Challenge #2: Supplies of Modified Composition Seed Oils

Although the modified composition seed oils with very low linolenic acid, or essentially no linolenic acid, can deliver good stable transesterified shortening, these oils remain to be in short supply. Oilseed companies, such as Dow Agrotec, DuPont, and Monsanto have ramped up their production of modified composition oilseeds. Table 13.5 lists the potential supplies of some of these oils.

The demand for hydrogenated shortenings and baking oils for industrial and restaurant frying is about 15 billion pounds. Not enough modified composition oil is available to meet the total demand. Some of the frying needs could be satisfied with corn oil and the cottonseed oil. Modified composition seed oils remain in short supply in the United States. This brings up the question of using palm oil and palm oil fractions to fill the gap?

13.5.3 Challenge #3: Consumer Advocates in the United States

Consumer advocates in the United States have been expressing their concern about the higher-saturated fat content in the transesterified shortenings made

TABLE 13.5 Projected Productions of Some Selected Modified Composition Seed Oils

Oil type	Projected production (million pounds)
Low-linolenic soybean	500
Mid-oleic soybean	75
High-oleic canola	>1200
High-oleic sunflower	70
Mid-oleic sunflower (Nu Sun)	642

Crop year 2006–07.

from the modified composition oils and fully hydrogenated cottonseed or palm oil. However, if one combines all cholesterol-promoting components in the hydrogenated shortening (*trans* fat and palmitic acid in Table 13.3) and considering stearic acid to be neutral cholesterolemic; the interesterified product contains about one-third of the total cholesterol-promoting fat as compared to the standard hydrogenated shortening. The consumer product advocates are not satisfied with the higher saturated fat in the transesterified product and they have serious concerns about using palm oil or palm oil fractions in the formula because of the high-saturated fat in this oil. It might be wiser for the food industry in the United States to use more palm oil and palm oil fractions in the products to eliminate the artificial *trans* fat or PHO.

13.5.4 Challenge #4: Use of Regular Soybean Oil is Reducing Shelf Life Stability of the Transesterified Shortening in Some Applications

A number of US oil processors have been formulating transesterified baking shortening using liquid soybean oil as one of the components fortified with TBHQ as antioxidant at the maximum allowable (200 ppm) level. The food companies, in many instances, have found that products made with this shortening exhibited shorter shelf life than the hydrogenated shortening.

13.5.5 Challenge #5: Economic Challenge

This consists of several components:

1. Cost of interesterification.
2. Cost of modified composition oils.
3. Additional cost for the crystallizers.
4. Some degree of product reformulation to accommodate the transesterified shortening to match the current product attributes.

Interesterification process, whether it is conventional (chemical process) or an enzymatic process, adds \$0.05–\$0.07 per pound for transesterification per pound of shortening, over and above the cost of hydrogenation.

Cost of modified composition oils is anywhere from \$0.15–\$0.20 per pound over the garden variety counterparts. Using 50% modified composition seed oil in the formula, the total cost of the transesterified shortening is increased by \$0.14–\$0.18 over a pound of conventionally made hydrogenated shortening.

The interesterified shortening contains much lower amounts of trisaturated triglycerides in the end product. Unless some hard stock is added to the interesterified shortening, it crystallizes at a slower rate. This requires additional crystallization and work unit capacity (see Chapter 10).

In some cases, the processed product made with the transesterified shortening may appear somewhat different in certain attributes compared to the product made with the standard hydrogenated shortening. This requires some reformulation of the product by the food companies to attain the desired finished product attributes. It also requires additional formulation, plant trials, storage stability tests, etc. The food industry has come a long way in the past decade in adjusting their product formulas to make the finished product taste the same as the control product made with hydrogenated shortening.

13.6 INTERESTERIFICATION PROCESS

Today, the oil industry produces transesterified products by:

1. chemical process
2. enzymatic process

The fundamental differences between the two processes are described:

13.6.1 Chemical Process

It is a transesterification process where the fatty acid distribution is strictly random and is governed by the law of probability. This means that the fatty acid from one triglyceride can exchange places within the same triglyceride molecule, as well as those in the other triglyceride molecule. Fig. 13.5 shows a simple illustration where two triglyceride molecules can have six different triglycerides formed after transesterification.

In this example, each of the two triglyceride molecules has been shown to have only one type of fatty acid, A or B in all three positions. In natural oil, the fatty acids are randomly distributed with the unsaturated fatty acids being located in the 2-position. If one of the triglycerides is fully saturated and the other one is a liquid oil with different degree of saturation, the fatty acid profile of the product after transesterification will be more complex.

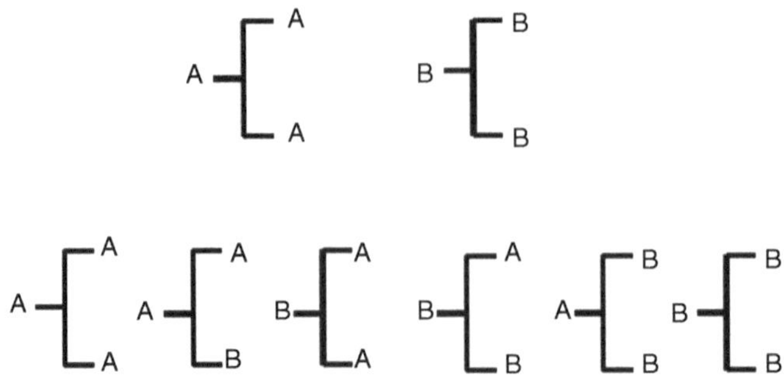

FIGURE 13.5 Distribution of fatty acids in random interesterification. Six possible triglycerides after interesterification.

13.6.2 Enzymatic Process

In this process, the enzyme reacts with the fatty acids in 1,3-positions without affecting the fatty acid in the 2-position. Therefore, the final product is not totally randomized and the triglyceride molecules still retain part of the original structure and the nutritional traits.

The remainder of this chapter will be devoted to:

1. chemical random interesterification process
2. enzymatic interesterification process
3. the critical process controls and their significance for both processes
4. consequences of deviations from the standard operating procedures
5. troubleshooting

13.7 CHEMICAL INTERESTERIFICATION PROCESS

13.7.1 Description of a Chemical Interesterification Process

This process can be carried out either in a batch or in a continuous system. To make shortening or margarine, a mixture of a hard triglyceride and a soft triglyceride are reacted. The reaction can be carried out without any catalyst at high temperature, such as 482–572°F (250–300°C). However, at this high temperature, the oil oxidizes and polymerizes rapidly. Therefore, a catalyst is needed so the reaction can be carried out at a much lower temperature, protecting the oil from oxidation and polymerization.

13.7.2 Reaction Mixture

The reaction mixture consists of the following oil blend:

1. fully hydrogenated oil (seed oils, palm oil, or palm kernel oil) or a hard fraction of palm oil or palm kernel oil,
2. liquid oil (palm olein, modified composition seed oils), and
3. a catalyst (most commonly used catalyst), sodium methoxide.

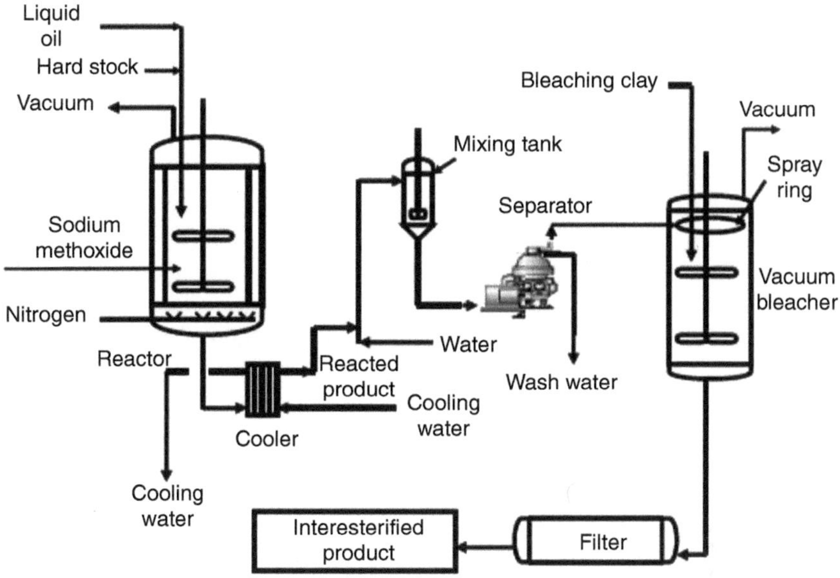

FIGURE 13.6 Schematic diagram for a batch chemical transesterification reaction.

13.7.3 Reaction Steps

Fig. 13.6 shows the schematic diagram for a batch interesterification process.

1. The oil blend is pumped into the batch reactor, which is operated under an absolute pressure of <20 mbar.
2. The reactor is a pressure vessel made of 304 or 316 stainless steel.
3. The oil blend is heated to approximately 260°F (126°C) and held at this temperature for approximately 60 min to remove the moisture and dissolved air from the oil. It can help the deaeration process. Nitrogen gas is bubbled at a very slow rate through the bed of oil without affecting the vacuum in the reactor.
4. The oil blend is cooled to 160°F (71°C) and the vacuum is broken with nitrogen sparge through the bottom of the oil bed in the reactor.
5. Catalyst (sodium methoxide) is added to the oil (not at the surface). The sodium methoxide addition system is operated with a feeder valve with an air-lock device to prevent the air from entering the reactor. However, even with the best precautions, a small amount of air can come into the reactor.
6. Catalyst dosage is 0.05%–0.1% of the oil in the reactor.
7. Reaction is carried out with agitation.
8. A reddish brown color appears in the reaction mixture as soon as the catalyst is added.
9. The mixture is heated and the reaction is carried out at 176–212°F (80–100°C).

10. Although most of the reaction occurs quite fast (10–15 min), typically the reaction is carried out for 30–60 min.
11. The reaction mixture is cooled down to 190–195°F (88–91°C) either in the reactor or through an external cooler before deactivation of the catalyst, as shown in Fig. 13.6. This is done to prevent any flashing of the water that follows the acid addition.
12. The catalyst is deactivated by introducing citric acid, phosphoric acid, or water into the reaction mixture in the reactor.
13. Acid deactivation is a preferred way because the water addition results in some side reactions, such as the formation of diglycerides.
14. A certain amount of wash water is added after neutralization of the reaction product.
15. The oil and the aqueous phases are separated in a centrifuge.
16. The reaction product from the centrifuge is then bleached under vacuum with activated bleaching clay to remove the residual impurities from the product. It also hydrolyzes the residual soap (see Chapter 6).
17. The bleaching vessel is a pressure vessel, made of 304 stainless steel, and with mechanical agitation.
18. Bleaching clay dosage is typically 0.2% and should be no more than 0.5% in a well-controlled process. A higher-bleaching clay dosage damages the product.
19. This step takes 30–45 min.
20. The bleached product is filtered to remove the bleaching clay.
21. The filtered product is then deodorized. (see Chapter 8 for procedure.)
22. Citric acid is used at the end of deodorization (see Chapter 8) to chelate any metal ions.
23. The product is cooled to appropriate temperature and stored, preferably under nitrogen protection (see Chapter 8) in 304 or 316 stainless steel tank.

13.7.4 Critical Control Points in the Chemical Interesterification Process

The critical control points in the chemical interesterification process are listed:

1. incoming oil quality
2. drying and deaeration of the oil before catalyst addition
3. amount of catalyst added
4. agitation during drying and reaction
5. reaction time
6. neutralization of the catalyst at the end of the reaction
7. separation of the reacted product and the aqueous phases after neutralization
8. bleaching
9. deodorization and storage

13.7.4.1 Oil Quality

The oil used for this process must be of very good quality. The following impurities and their levels are critical for the overall reaction, the quality of the finished product, and the yield in the process:

• FFA (%)	Preferably <0.03%; not >0.05%
• PV (mEq/kg)	Not >0.5
• Moisture (%)	Preferably <0.01%; not >0.03
• Soap (ppm)	0 (zero)
• Phosphorus (ppm)	Preferably <1 ppm and not >2.0 ppm
• Aldehydes and ketones	Trace
• *p*-Anisidine value	Preferably <6; not >10

13.7.4.2 FFA

The FFA content of the refined oil must be low because it reacts with the catalyst and produces the following by-products that affect both quality and yield.

Sodium methoxide + FFA = Sodium hydroxide + Fatty acid ester of methanol (FAME).

Sodium hydroxide + Neutral oil = Sodium soap + Diglyceride (possibly monoglyceride).

FAME, monoglyceride, and diglyceride contribute directly to the losses in the process.

13.7.4.3 PV

The PV of the starting oil must be <1. At higher PV, the following side reactions occur, affecting both quality and yield of the final product.

Sodium methoxide + PV = Methanol + Peroxy radicals.

Methanol + Neutral oil = FAME + Diglyceride (possibly monoglyceride).

The same impurities are produced by high PV as FFA.

13.7.4.4 Moisture

The following reactions occur in the reaction mixture when the moisture in the incoming oil is high.

Sodium methoxide + Water molecule = Methanol + Sodium hydroxide.

Methanol + Neutral oil = FAME + Diglyceride.

Sodium hydroxide + Neutral oil = Soap + Diglyceride + Monoglyceride.

As one can see, each of the three impurities produces FAME, which is a direct loss of the neutral oil. Formation of diglyceride and monoglyceride also represents a loss in the total neutral oil and also in terms of the overall performance of the transesterified product.

13.7.4.5 Drying/Deaeration of the Oil Before Reaction

From the earlier discussions, one could conclude that the drying step is critical for the process because:

1. It reduces the moisture content of the refined oil. Chemically refined and water washed oil contains 0.4%–0.5% moisture. At 25 mbar and 258°F (125°C), the equilibrium moisture in oil is 53 ppm. However, it does not reach that low in the vacuum dryer at 60 min. In reality, at <20 mbar and 258°F (125°C), it reduces the moisture content to <100 ppm (<0.01%).
2. The dissolved air is also removed from the oil in the drying/deaeration step. This reduces the oxygen content of the oil that reduces the peroxide formation at the elevated temperature.

13.7.4.6 Amount of Catalyst

Typical dosage for the sodium methylate catalyst is 0.05%–0.1% of the oil. At a lower-catalyst dosage, the reaction can be slow and will take longer to reach the equilibrium or it may not reach the desired reaction end point even with the longer-reaction time.

At higher-catalyst usage, the reaction is faster but there is also an increased level of side reactions.

At the end of the reaction, more acid is needed to neutralize the excess catalyst, producing a higher amount of sodium soap of the acid. This requires extra water in the washing step, causing higher-product loss during separation.

Additionally, higher-soap content in the feed to the bleacher will generate higher FFA from the reaction between the soap and the acid-activated clay. This increases the neutral oil loss in the deodorizer (see Chapter 8).

13.7.4.7 Agitation During Drying and the Reaction

Vaporization of the moisture from the oil and deaeration takes place at the surface of the liquid in the reactor during the drying and deaeration step.

Therefore, it is important for the oil from the interior of the reactor to reach to the surface as many times as possible to release the moisture and the dissolved air.

The catalyst is not soluble in the oil until it forms the intermediate by replacing one fatty acid moiety from the triglyceride molecule. This is called sodium glycerate, which is the true catalyst for the reaction. Therefore, intimate mixing between the oil and the catalyst is critical, and good mechanical agitation is essential for the reactor.

Good agitation is also beneficial to bring the catalyst and the oil in close contact for the reaction.

13.7.4.8 Reaction Time

The reaction equilibrium is generally reached in 10–15 min. However, a reaction time of 30–60 min is common in the industrial operation to ensure complete reaction. A shorter-reaction time might leave more nonreacted trisaturated glycerides.

A longer time is generally not needed to achieve the reaction equilibrium but it does not do any harm. An extremely long reaction time will darken the product and may cause color fixation.

13.7.4.9 Neutralization of the Catalyst

At the end of the reaction the catalyst must be neutralized. This is done by adding citric acid or phosphoric acid to the reaction product.

Although water alone can be used to deactivate the catalyst, it is better to use acid deactivation because plain water can produce high levels of diglycerides and FAME in the product, as it was discussed under Sections 13.7 and 4.1.

The added acid forms sodium salt of the acid. This reduces the chance of forming diglycerides in the product. In actual practice, it is necessary to use acid followed by some amount of water to create enough density difference between the oil phase and the aqueous phase for better separation in the centrifuge and also to dissolve the sodium salts of the acids formed in the neutralization process.

Neutralization can be performed by using silica hydrogel, which removes the soap. The product can then be bleached without water washing. This approach may increase the operating cost because of the high cost of hydrated silica, but would reduce the overall capital cost for the process by eliminating the need for the water-wash centrifuge. The justification for the use of silica hydrogel depends on the amount of soap in the product and the amount of silica hydrogel needed for the removal of soap. The advantages of using silica hydrogel are listed:

1. Eliminates the capital requirement for installing the centrifuge and the required accessories.
2. By removing the soap, the FFA in the product does not increase during bleaching, which is caused by the hydrolysis of soap by the acid-activated clay.
3. Hydrated silica is also a good adsorbent for the removal of the oil decomposition products and the polar compounds formed in the oil during the process.
4. It substantially reduces the requirement of bleaching clay, which helps maintain good quality in the product, because a higher level of the bleaching clay tends to reduce the tocopherols in the oil.

13.7.4.10 Separation of the Aqueous and the Oil Phase

The neutralized product is then mixed in with wash water and centrifuged.

The amount of moisture left in the oil phase after centrifuging is <0.5%. This leaves very little amount of the salts in the product.

13.7.4.11 Bleaching

The water-washed and centrifuged product is now bleached with acid-activated bleaching clay to remove all impurities in the finished transesterified product. The process is the same as described in Chapter 6.

The bleaching clay dosage must be controlled carefully without excessive use. The most common range for the clay usage is 0.2%–0.5%. A higher use of bleaching clay can cause the following undesirable reactions in the product:

- The acid activated clay at the higher dosage can hydrolyze some neutral oil-forming FFA and diglyceride. This reduces the overall yield plus leads to some loss of functionality of the product.
- Higher use of bleaching clay oxidizes some of the natural tocopherols in the oil, reducing its oxidative stability.
- Increased loss because the bleaching oil absorbs about 33% of its own weight of neutral oil in the process.

The bleaching step also requires good mechanical agitation for obtaining an intimate contact between the oil and the bleaching clay.

13.7.4.12 Deodorization and Storage of the Final Product

The bleached product is deodorized to remove the FFA in the feed and also the volatile impurities formed during the process. It has been mentioned in the earlier discussion that because of the side reactions, a number of compounds are formed in the reaction product. These are:

- FAME
- diglycerides
- monoglycerides
- soap

Out of these three impurities, FAME and at least 25% of the monoglycerides are removed in deodorization. Diglycerides are not removed under the deodorization condition because of their higher-boiling point.

The peroxy free radicals can produce some oil oxidation compounds, such as dimers and trimers. Some of these compounds remain in the oil after deodorization and can reduce the flavor stability of the product.

13.7.5 Questions Related to Chemical Interesterification

13.7.5.1 How to Determine the Reaction End Point

There are no real-time analytical methods to determine the reaction end point. Mettler drop point [AOCS Method: Cc 18-80 (09)] is normally used to establish the reaction end point. It is necessary to conduct several plant runs to establish the time required to reach the specific end point. The progress of the random interesterification reaction can be monitored by ultraviolet (UV)/visible spectrophotometry. It was done at least on the bench scale test by Lampert and Liu (2001) in their US patent. They demonstrated that the reaction could be monitored up to the point of completion. This method has never been explored for its applicability in actual operation.

13.7.5.2 Stability of the Chemically Interesterified Product

The oxidative and flavor stability of the randomly interesterified product has been questioned by several investigators. It is claimed that the interesterified product has lower-oxidative and flavor stability compared to the original oil mix (in deodorized condition). This could be explained with the following comments:

- In nature, the unsaturated fatty acids are present mostly in the 2-position on the triglyceride molecule.
- This position is more stable against oxidative damage.
- In the random interesterification process, positions of these fatty acids change, and they can be found on position 1 or 3 on the triglyceride molecule.
- It has been found that this positional shift makes the end product more vulnerable for oxidation. This is true for both random rearrangement, as well as interesterified products.
- There is also some loss of tocopherols in the final product, which makes it prone to oxidation.

These aspects will be discussed further in the later Section 13.9 comparing the chemical and enzymatic interesterification processes.

13.7.5.3 Losses in the Process

Other than any spills or degradation of product, it has been reported by the industry that the typical neutral oil loss in the reaction is 1.5%–2% with dry inactivation of the catalyst. This is based on the catalyst concentration of 0.1%. A loss of 5% has been reported with wet catalyst deactivation in the process.

The losses occur at several areas:

- Formation of FAME, diglyceride, soap, and some monoglyceride in the drying and deaeration step due to high FFA, PV, and moisture.
- Oil absorbed by the bleaching clay.
- Deodorization loss (mostly volatiles and some neutral oil in the distillate).
- Additional neutral oil loss due to the excess FFA produced in bleaching.

13.7.5.4 Troubleshooting Random Interesterification Process

Table 13.6 lists the most common issues faced in the interesterification process, their causes, and suggested solutions.

Following are the typical issues faced by the plant personnel:

- Slow reaction.

TABLE 13.6 Troubleshooting Transesterification Process

Symptoms	Probable cause(s)	Suggested action steps for correction
Slow reaction:		
The end product did not reach equilibrium in the predetermined reaction time	• Incoming oil quality is poor; it is possible that the oil contains: FFA >0.05% PV >1 Moisture >0.01%	• Check for FFA, PV, and moisture in the feed oil and correct the issue
	• The reactor temperature may be low	• Check and increase the oil temperature if it is low
	• Vacuum in the drying/deaeration step is poor; this will have insufficient drying and deaeration of the oil blend resulting in catalyst loss	• Check and correct the vacuum; this can be due to one or more of the following: • The vacuum system is not working properly (see Chapter 14) • There may be air leak at the agitator shaft gland or seal at the top
	• Agitator in the reactor is not functioning properly	• Make sure it is running at the designed rpm • If the speed is correct, check the amperage on the motor • Lower than the rated amperage will imply that one or more agitator blades are either loose or have fallen off Shut down and fix the agitator
	• Catalyst dosage maybe low	• Increase the amount of catalyst if it is found to be low
Deodorized product is dark	• Incomplete neutralization • Incomplete water washing	• The impurity, such as soap, left in the bleached oil can darken the color of the product
	• Poor bleacher vacuum	• Poor vacuum can darken the oil color via oxidation and color fixation
	• Reaction time is too long	• Find the cause and correct it
	• Air leak into the bleacher vessel	• Same as aforementioned
	• Bleacher temperature is high	• Same as aforementioned
	• Air leak in the deodorizer system	• Same as aforementioned
	• Reflux of the distillate into the deodorizer; this can happen if the catchall is not properly drained	• Clean the catchall more frequently

- Reaction equilibrium is not reached at the end of the period normally allocated for the reaction based on prior experience.
- The deodorized product is darker than normal.

13.8 ENZYMATIC INTERESTERIFICATION PROCESS

13.8.1 Introduction

Enzyme hydrolysis has been a known phenomenon for centuries. The pancreas in the human body releases the enzyme that helps hydrolyze the fat (oil) we consume. This principle was studied closely by scientists, and over the past decades on the bench-scale treatment of fat, using pancreatic lipase has come to the commercial scale using lipase produced from microorganisms.

13.8.2 Catalyst

The catalyst in this case is triacylglycerol hydrolase, a biologically derived lipase from heat-loving microorganisms called *Thermomyces lanuginosus*. This enzyme is *sn*-1,3 specific, which means that the enzyme hydrolyses the triglyceride molecule at the 1,3-positions and does not affect the fatty acid in the 2-position. Thus, being so specific, the interesterified fat retains some of its original natural fatty acid configurations.

The enzyme is immobilized on the internal surface of macroporous support particles, such as silica granulates. This means the enzyme is fixed onto the carrier for better operation and economy.

The catalyst in the process to be described is made by Novozyme of Denmark. The immobilized catalyst in this case is called Lipozyme TLIM.

There are other enzymes with very specific selectivity and applications. Those will not be discussed in this chapter.

13.8.3 Purpose of Immobilization of the Enzyme

Immobilization of the enzyme brings the following benefits to the process:

- The enzyme is retained in the carrier and is not carried into the product.
- The immobilization allows one to use the enzyme for multiple production runs.
- The reaction rate is faster because this process increases the specific surface area for the catalyst (enzyme).

Fig. 13.7 shows the immobilized Lipozyme TLIM from Novozyme (Denmark), viewed through a lighted microscope.

13.8.4 Reaction Steps in Enzymatic Interesterification Process

The process can be carried out either in batch or continuous reactor. The sequence of the reaction steps are shown schematically in Fig. 13.8. This consists of three steps, as outlined in the figure.

FIGURE 13.7 **Immobilized lipozyme TLIM from Novozyme.** Courtesy of Novozyme.

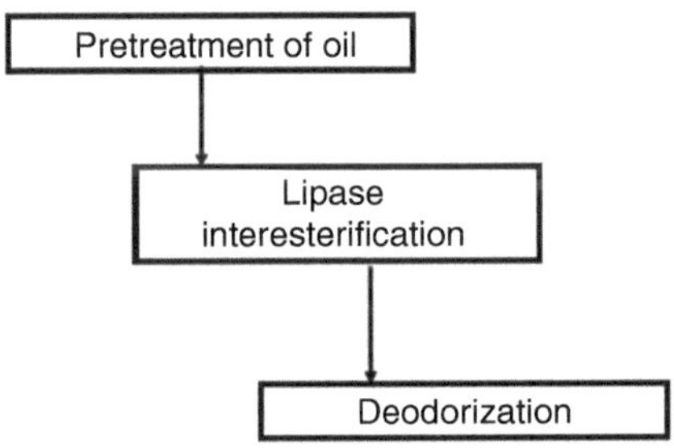

FIGURE 13.8 **Schematic diagram for enzymatic interesterification process.**

13.8.5 Pretreatment

Pretreatment consists of refining, bleaching, and deodorizing the oil before it is used. The oil must have very low levels of impurities, as shown in Table 13.7.

The aforementioned impurities inactivate the catalyst and reduce productivity. Therefore, their levels must be as low as possible in the feed oil.

13.8.6 Lipase Interesterification

This can be done in one of two ways:

1. batch process
2. continuous fixed bed process

The continuous fixed process can be of two configurations:

- fixed bed multiple reactor process
- single fixed bed continuous process

TABLE 13.7 Maximum Recommended Impurities in the Oil

Attribute	Recommended maximum level
PV (mEq/kg)	<1
Aldehydes and ketones (ppm)	Trace <0.5
Phosphorus, PPM	<0.5, No higher than 1.0
Soap, PPM	0 (Zero)
Nickel, PPM	<0.2
Iron, PPM	<0.2

13.8.7 Batch Process

1. In this process, the enzyme is dispersed in the oil. The reaction is carried out at a temperature <158°F (70°C). At higher temperature the enzyme is inactivated.
2. The enzyme is separated from the reaction oil at the end of the reaction through the filtration process.
3. The process is simple but the cost of production is higher because the catalyst cannot be recovered for reuse and also the productivity is low.

13.8.8 Continuous Multiple Fixed Bed Process

In this process, several fixed bed reactors are used in series. Fig. 13.9 shows the schematic diagram for the process.

1. Feed oil is introduced from the top of the reactors and passes in series through the reactors.
2. Catalyst is packed in the reactors either as wet slurry in oil or dry pack followed by oil injection into the reactor under vacuum to remove all trapped air from the reactor and the catalyst bed.

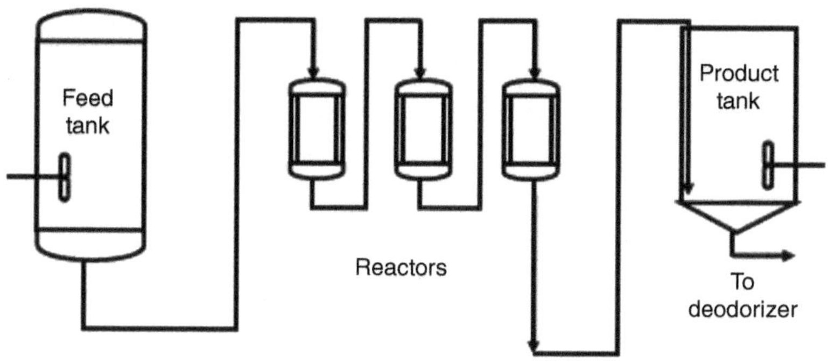

FIGURE 13.9 Schematic diagram for multiple fixed bed process.

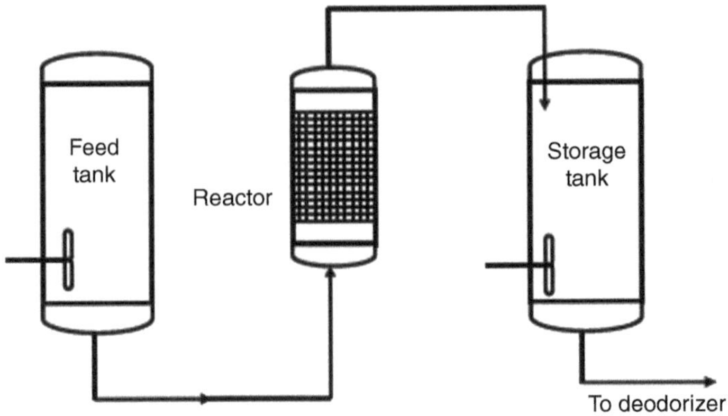

FIGURE 13.10 **Single fixed bed continuous process.**

3. The reactivity of the enzyme in the reactors begins to drop off as more product is made.
4. The average production in this design is 1–2 kg of interesterified oil/kg of enzyme/h.

13.8.9 Single Fixed Bed Continuous Process

1. The most commonly used design is where the catalyst is packed in a single column.
2. The catalyst is loaded in the same manner as described earlier for multiple fixed bed reactors.
3. The reaction rate depends on the rate of oil flow through the catalytic enzyme grid.

Fig. 13.10 shows the schematic diagram for a single reactor continuous process. The oil enters the reactor at the bottom and leaves the reactor from the top. The oil goes to the storage tank, from where it is sent to the deodorizer.

13.8.10 Enzyme Activity

- The reaction rate starts at 10 kg of interesterified product/kg of catalyst/h. The reactivity drops with time.
- The enzyme remains active for up to 100 days (2400 h).
- The flow of oil through the reactor needs to be adjusted downward to get the required conversion as the catalyst is used longer and the activity drops.

13.8.11 Productivity

Productivity depends greatly on the incoming oil quality. With good quality feed oil one could obtain the following production:

- A minimum of 2500 kg of interesterified product/kg of enzyme in an operating plant.

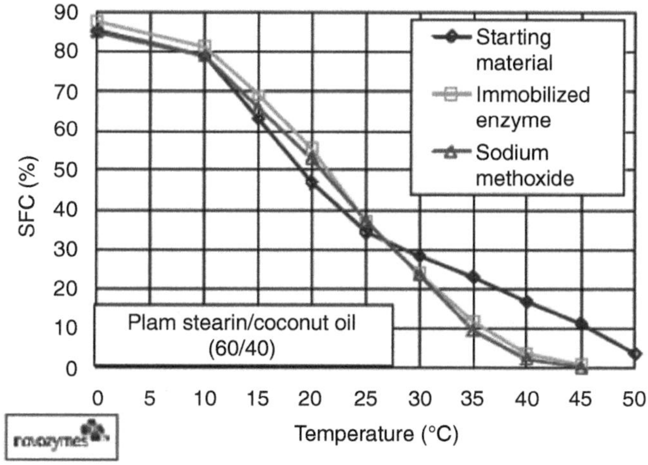

FIGURE 13.11 Solids curves obtained by the two interesterification processes. Courtesy of Novozymes.

- Higher productivity has been reported in pilot plants.
- The productivity is high initially, and then it tapers off as the catalyst remains in the reactor longer.

13.8.12 Deodorization

The finished product is deodorized and stored under nitrogen protection, as described in Chapter 8 and Chapter 9.

13.9 COMPARISON BETWEEN THE CHEMICAL AND THE ENZYMATIC INTERESTERIFICATION PROCESSES

Both chemical and enzymatic processes produce the end products that are very similar in their physical attributes, such as melting characteristics. They have similar functional properties in product applications. Fig. 13.11 shows the solids curves for a 60/40 blend of PS and coconut oil interesterified by both chemical and enzymatic processes, published by Novozyme.

It was mentioned earlier that more natural tocopherols are lost in the chemical process, compared to the enzymatic process. Besides higher tocopherols, the enzymatic process produces lighter-colored product and contains lower levels of diglyceride. Table 13.8 shows some data published by Desmet Ballestra on various interesterified products made from PS/sunflower oil (SFO) blends.

Chemical interesterification is truly a random reaction process because it alters the fatty acids in all three positions in a triglyceride molecule. Enzymatic process randomizes the fatty acids in the 1 and 3 positions in the triglyceride molecules. Thus some of the original traits of the oils are retained in the enzymatically interesterified product. Table 13.9 lists the various areas where the two processes differ and the relative merits and drawbacks of the two processes.

TABLE 13.8 Product Analysis in Interesterified Products from Chemical (Chem) And Enzymatic (Enz) Processes

PS/SFO	10/90			20/80			30/70			40/60			50/50		
	Feed	Chem	Enz	Feed	Chem	Enz	Feed	Chem	Enz	Feed	Chem	Enz	Feed	Chem	Enz
Lov ibond color															
Yellow	11	15	**10**	15	10	**9**	16	11	**8**	19	16	**8**	19	18	**10**
Red	1	2	**1**	1.2	2.3	**0.9**	1.8	2.2	**1.2**	2	3	**1.4**	2.1	3.4	**1**
Tocopherols (ppm)	701	252	**505**	639	197	**412**	581	281	**426**	546	185	**425**	463	182	**366**
Tocopherols retained															
% of original	—	35.9	**72.0**	—	30.8	**64.5**	—	48.4	**73.3**	—	33.9	**77.8**	—	39.3	**79.0**
DAG (%)	1.5	3.9	**2.0**	1.7	3.7	**3.0**	1.9	4.5	**3.5**	2.1	4.2	**3.0**	2.4	4.9	**3.5**

DAG, Diglyceride; PS, palm stearin; SFO, sunflower oil blend.
Bold emphasizes the enzymatic process.

TABLE 13.9 Comparison Between the Chemical and the Enzymatic Processes

Basis	Chemical process	Enzymatic process
Type of catalyst	Sodium methoxide or other chemicals	Immobilized lipase lipozyme *TLIM*
Initial cost of catalyst	Low	High
No. of uses for the catalyst	Once	Multiple
Initial capital cost for installation	Higher	Lower
Operating cost	Lower	Higher
Net cost (capital + operating)	Same as the enzymatic process	Same as the chemical process
Side reactions	High; produces FAME, monoglycerides, diglycerides, soap, and methanol	No side reactions
Process loss	High; due to several side reactions; oil absorption by the bleaching clay in postbleaching	Low
Environmental impact	Less friendly	Enviromnetally friendly
Feed oil	RB	RBD
Feed oil quality required	Very high	Very high
Reaction temperature °F (°C)	176–212 (80–100)	<158 (70)
Final product attributes Physical (solids content) Functionality	The physical properties and the functionality are identical to the product from the enzymatic process	The physical properties and the functionality are identical to the product from the chemical process
Stability of the transesterified product: 1. Oxidative 2. Flavor	Both oxidative and flavor stability tend to be lower than the original oil blend (in the deodorized state); this is partly because the oil blend is bleached before and after the interesterification process (referred to as postbleaching), causing higher loss of tocopherols	Both oxidative and flavor stability are closer to those of the original oil blend (in the deodorized state); the oil blend is bleached only once before the interesterification reaction
Retained tocopherols (measured in ppm)	30%–39% of the original feed	64%–79% of the original feed

FAME, Fatty acid ester of methanol; RB, refined and bleached; RBD, refined, bleached, and deodorized.

READING REFERENCES

De Greyt, W., 2004. Chemical vs. enzymatic interesterification. In: IUPAC-AOCS Workshop on Oils and Oilseeds Analysis and Production; Tunis, Tunisisa, December 6–8, 2004.

Eckay, E.W., 1945. US Patent 2,378,005.

Gooding, C.M.,1943. US Patent 2,309,949.

Kellens, I., Mark, J., 1996. Workshop on Oils Ineresterification, Texas A&M University, College Station, TX, United States.

Lau, F., Hammond, E., Ross, P., 1982. Effect of randomization on the oxidation of corn oil. JAOCS 59, 407–409.

Lampert, D., 2000. Processes and products of interesterification. Introduction to Fats and Oils Technology, second ed. AOCS Press: Champaign, IL.

Lampert, D., Liu, L., 2001. US Patent 6,238,26B1.

List, G., Emken, E., Kwolok, W., Simpson, T., 1977. Zero *trans* margarins; preparation, structure and properties of interesterified soyean oil-soy trisaturate blends. JAOCS 54, 408–413.

Liu, L., Lampert, D., 1999. Monitoring chemical interesterification. JAOCS 76, 783–787.

Norris, F.A., Mattil, K.F., 1946. Oil Soap 23, 289–291.

van Loon, C., 1927. Dutch Patent 16,703,1927; US Patent 1,873,513.

Veronique, G., Tirtiaux, A., 2000. Fractionation Combined With Interesterification: A Tool Towards the Formulation of Zero-Trans New Products. AOCS-OTAI Publication, New Delhi, India.

Chapter 14

Familiarization With Process Equipment

14.1 INTRODUCTION

Oil process plant is designed to deliver certain volume of products with very specific quality standards set by the company to meet customer needs. This requires the knowledge of the process, product traits and the properties of the raw materials as has been described in the previous chapters in this book. The current chapter addresses another aspect of the plant operation that includes understanding the plant equipment, their capabilities, maintenance requirement, etc.

Each production or process unit at the plant is designed to perform in a particular way. The plant can deliver the finished product with the required quality standards as long as the feed conditions and operating conditions are maintained according to the raw material and process standards as discussed in the previous chapters. The plant's goal should be to deliver the best quality product at the maximum production capability of the plant. To achieve this objective, the process supervisor must have a thorough familiarity with the following areas:

1. Plant safety procedure.
2. Quality standards for the in-process oil and the finished products.
3. Plant sanitation procedure.
4. Capacity of every process equipment, such as
   - Pounds/kg per hour.
   - Pounds/kg per batch in batch process.
   - Cycle time of production/batch of product, etc.
5. The operating conditions, such as temperature, pressure, flow rate, bleaching clay dosage, catalyst requirement, and so on.
6. Supply of oil, power, process hot and cold water, steam, nitrogen, process air and instrument air, etc., to the process.
7. Locations for the electrical cut-offs, shut-off valves for all utilities such as steam, water, air, and nitrogen.
8. Required maintenance program for the process equipment and the frequency of maintenance.

Practical Guide to Vegetable Oil Processing. http://dx.doi.org/10.1016/B978-1-63067-050-4.00014-3
Copyright © 2017 AOCS Press. Published by Elsevier Inc. All rights reserved.

9. Functions and operating conditions for all auxiliary equipment, such as ejectors, compressors, pressure controllers, temperature controllers, PLC System (if exists).
10. Troubleshooting of any main process equipment or the auxiliary equipment.
11. Troubleshooting product quality issues.
12. Appropriate utilization of the company resources, such as the quality control (QC), engineering, maintenance, and so on.
13. All emergency procedure related to operating equipment, evacuation procedure, seeking medical assistance in case of injury, etc.
14. Procedures for start up, shutdown including emergency shutdowns.
15. Cost of process materials, utility, such as power, steam, water, labor, and so on.
16. Complete knowledge of manufactured cost of the product and be able to compare this with the standard target costs set by the company.
17. Automation of the various parts of the process.

14.2 PROCESSING EQUIPMENT AND ACCESSORIES

14.2.1 Process Equipment

The typical process equipments used in an oil processing plant are listed as follows:

1. tanks
2. centrifuges
3. converters
4. deodorizers
5. vacuum dryers
6. vacuum bleachers
7. filters
8. slurry tanks
9. heat exchangers
10. piping

14.2.2 Process Accessories

The typical process accessories used in an oil processing plant are listed as follows:

1. vacuum ejectors
2. agitators (mixers)
3. pumps
4. valves (manual, control)
5. cooling towers
6. motors
7. electrical starters, switches, cut-offs, etc.
8. fans/blowers

9. compressors
10. air dryers
11. steam tracing
12. steam traps
13. steam purifiers
14. seals
15. gaskets

14.2.3 Process Instruments

The most common process instruments used in an oil processing plant are:

1. temperature indicators/controllers,
2. pressure indicators/controllers,
3. flow indicators/controllers,
4. level indicators/controllers,
5. automatic control valves,
6. gas purity indicators,
7. oxygen monitors in the deodorized oil storage tanks, and
8. process alarms.

14.2.4 Process Equipment

14.2.4.1 Tanks

Storage tanks are used at various stages of oil process, such as:

1. crude oil storage
2. in-process oil storage
3. deodorized oil storage

Seed oil processing plant uses carbon steel tanks for all stages of the process. Some oil processing plants have deodorized oil storage tanks made of stainless steel (Grade 304 or 316). However, this is not absolutely necessary in a seed oil operation.

Palm oil, palm kernel oil, or coconut oil contains significantly higher levels of FFA in the crude and in-process oils. Therefore, the plant needs to use stainless steel (Grade 304 and 316) tanks and accessories throughout the process.

There are certain requirements for storing oils at various stages of the process, such as:

14.2.4.2 Crude Oil Storage Tanks

1. The tank must have side mounted agitator with baffle arrangement to prevent the meals from settling to the bottom of the tank. The agitator must have a low oil-level cut-off switch to protect the agitator shaft against bending as well as from undue whipping of air into the crude oil.

2. The tank must have a temperature indicator, preferably a temperature control and temperature monitoring device.
3. It is recommended that these tanks be heated with hot water and not with steam. This prevents undue overheating of the oil and rise of FFA due to high temperature or possible leak in the steam coil.
4. If hot water system is unavailable, one could use low pressure steam 10–25 psig (1.79–2.89 kg/cm^2). There should be a thermodynamic trap on the steam coil to maintain proper condensate discharge temperature to control the oil temperature.
5. The tank must have a man head for tank entry for the purpose of repair, cleaning, etc.
6. Atmospheric vent at the top of the tank for the passage of air in and out of the tank as the tank is filled or emptied. The size of the vent must be adequate to allow sufficient air flow matching the larger pumping rate of either the tank loading or unloading pump.
7. A hatch on the top of the tank for manual taping for checking the oil inventory or collecting oil samples with a zone sampler (if the automatic level gauge is not installed on the tank).
8. Alternatively, a level gauge [differential pressure gauge (DP)] for indicating the level of oil in the tank.
9. Besides the local display, all of the measurements, such as oil temperature, oil height, and so on can also be monitored remotely with the help of a programmable control system.
10. A sampling port with a valve located within 3–4 ft. (91.44–121.92 cm) from the bottom of the tank. A ball valve is recommended as a low cost sampling device, otherwise a Strahman valve is preferred where oil samples can be obtained without having to discard any oil from the sampling line.

14.2.4.3 Tanks for Hydrogenated Stocks

1. Hydrogenated oils require higher storage temperatures. Therefore, low pressure steam is preferred for heating the oil instead of hot water.
2. The steam pressure must be 10–25 psig (1.79–2.89 kg/cm^2).
3. It is recommended to use a thermodynamic traps for the steam coils used for heating the oil.
4. The tank must have side mounted agitator with baffle arrangement to maintain a uniform temperature and composition throughout the tank and not to allow any separation of the solids or overheating of the oil on the steam coil surface.
5. The agitator must have a low liquid-level cut-off switch to protect the agitator shaft against bending, damaging the agitator shaft as well as from undue whipping of air into the oil.
6. The temperature of the hydrogenated oil should be no higher than the melt point of the stock by 10°F (5°C).
7. The tank must have a temperature indicator, preferably a temperature monitoring device.

8. The tank must have a man head for tank entry for the purpose of repair, cleaning, etc.
9. A level gauge (DP gauge) for indicating the level of oil in the tank.
10. Atmospheric vent at the top of the tank for the passage of air in and out of the tank as the tank is filled or emptied. The vent diameter must be sized as described earlier for the crude oil storage tank.
11. A hatch on the top of the tank for manual taping for checking tank inventory or collecting oil samples with a zone sampler.
12. Besides the local display, all of the measurements, such as oil temperature, oil height, and so on can also be monitored remotely with the help of a programmable control system.
13. A sample port with a valve within 3–4 ft. (91.44–121.92 cm) from the bottom of the tank. A ball valve is recommended.
14. The tank does not have to be nitrogen blanketed if the hydrogenated oil is blended shortly after hydrogenation and deodorized immediately.

14.2.5 Comments on the Atmospheric Vent

This serves as a safety component to protect the tank from exploding or collapsing. The vent is located at the top of the tank and is designed to allow sufficient flow of air exceeding the maximum oil loading or unloading rate (in terms of volume). The storage tanks are built with weak tops. Hydrogenated oil always contains some dissolved hydrogen gas. Therefore, in the event of any explosion the weak tank top blows out.

In the event the vent is too small, the tank can be over pressurized during oil loading. If the vent is plugged, the pressure in the tank will eventually match the discharge pressure of the oil loading pump or suction of the unloading pump. In that case, the tank top would yield and preserve the rest of the tank from serious damage.

If the vent is plugged for any reason, the oil discharge from the tank creates partial vacuum in the tank. The roof of the tank can collapse under this condition. Therefore, the vent must be sufficiently large and must be maintained clean.

14.2.5.1 Tanks for Deodorized Stocks

The requirements are similar to those for hydrogenated stocks. In addition, the deodorized oil storage tanks must have nitrogen protection as described under storage of deodorized oil Chapter 8.

14.2.6 Designs for Common Oil Storage Tanks

The oil storage tanks can have the following configurations:

1. flat bottom (sloped bottom)
2. cone bottom
3. dish bottom
4. jacketed

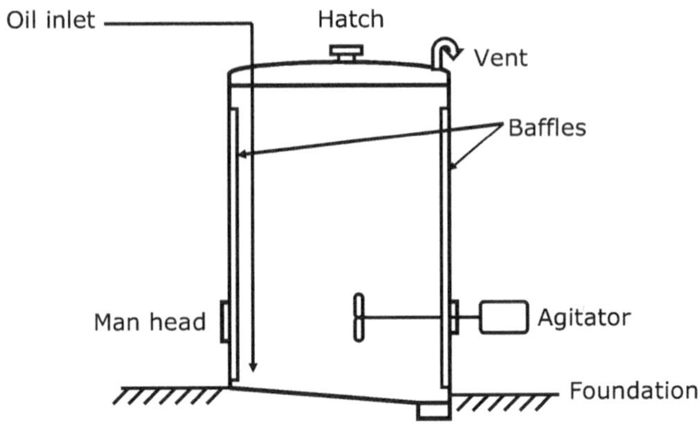

FIGURE 14.1 Flat bottom tank.

14.2.6.1 Flat Bottom Tanks

Flat bottom tanks are most commonly used in storing large volumes of oil such as millions of pounds of crude oil. These tanks are also used in storing in-process oils and deodorized oils at the plants that are much smaller than the larger crude oil tanks.

14.2.6.1.1 Advantage

These tanks are relatively easy to construct and are less costly compared to the three other types listed previously. Fig. 14.1 shows the typical schematic diagram of a flat bottom tank.

14.2.6.1.2 Disadvantage

It is not possible to remove all of the oil from the tank, especially when a centrifugal pump is used. This increases the potential for cross-contamination between different oil stocks.

14.2.6.2 Large Sloped Bottom Tanks for Crude Oil Storage

Fig. 14.2 shows the typical cross-sectional view of a large tank. This type of tank is used for storing millions of gallons of oil. The tank is built on concrete foundation with sand filled inside a circular reinforced concrete foundation. These tanks are provided with a certain amount of flexibility to float. The oil outlet line is normally connected with a flexible connection to the suction of the transfer pump.

The tank normally requires two side mounted mechanical mixers because of the large diameter. The sump is located at the center of the floor with the suction pipe dipped into the sump.

The tank is also equipped with a level gauge and a temperature indicator. It is better to use a positive displacement pump for transferring oil out of the tank.

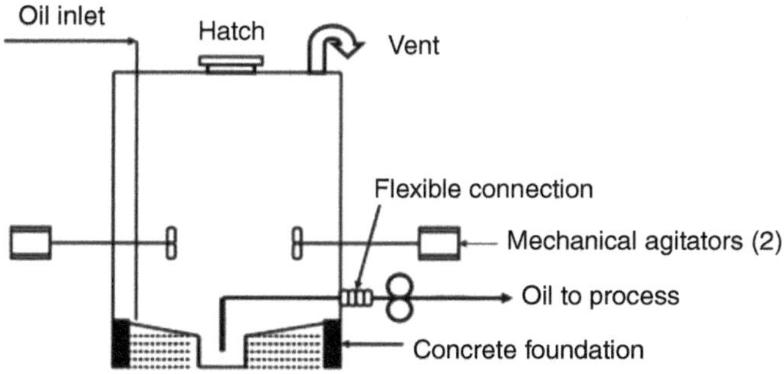

FIGURE 14.2 Large oil storage tank.

14.2.6.3 Cone Bottom Tanks

These tanks have similar accessories as the flat bottom tanks. These tanks are most commonly used in the packaging area where a discrete batch or type of product is delivered to the packaging line. Fig. 14.3 shows the schematic diagram for the cone bottom tank.

14.2.6.3.1 Advantage

There is little or no possibility for cross-contamination of incompatible products unless it occurs due to human error.

14.2.6.3.2 Disadvantage

There are no specific disadvantages of this design.

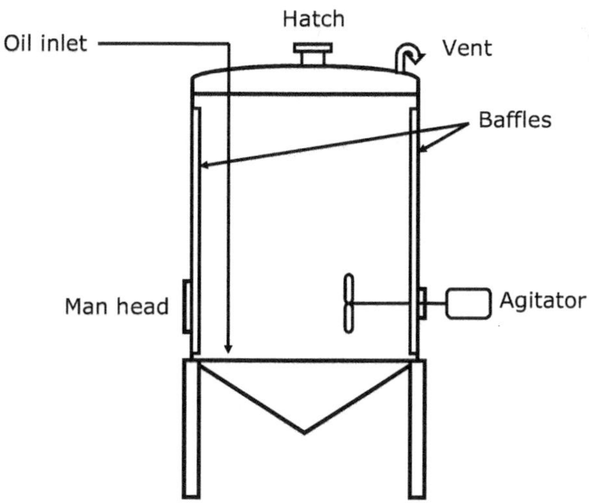

FIGURE 14.3 Cone bottom tank.

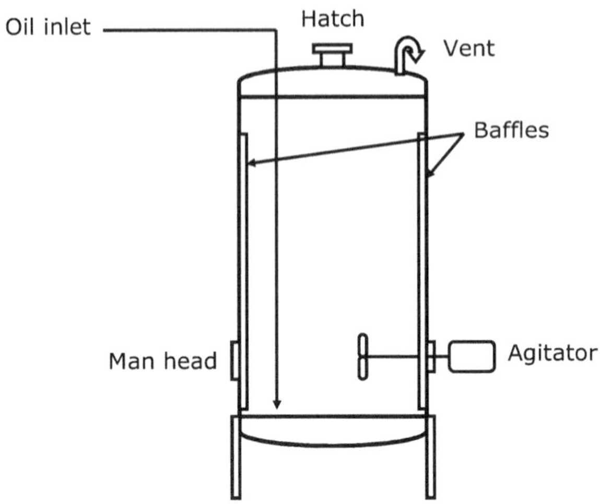

FIGURE 14.4 Dish bottom tank.

14.2.6.4 Dish Bottom Tanks

Dish bottom tanks have similar uses as the cone bottom tanks. Fig. 14.4 shows the schematic diagram for a dish bottom tank.

Advantages of the dish bottom tanks are the same as for cone bottom tanks simply because of the elevation of the tank bottom (Fig. 14.4).

14.2.6.5 Jacketed Tanks

These are special tanks and are relatively small in size. Hot or refrigerated water passes through the outer jacket to maintain a certain temperature without disturbing the fat crystals in the product in the tank.

14.2.7 Process Supervisor's Responsibility Regarding the Tanks

For proper operation of the plant a process supervisor needs to be familiar with the following:

1. Capacities of each tank in the department.
2. Type of oil stored in each tank.
3. All accessories associated with the tank, that is,
   - type of agitator, motor horse power (HP), impeller design, etc.,
   - type of the oil transfer pump at each tank, its capacity, motor HP, etc., and
   - type of heating, steam, hot water; steam pressure, hot water temperature, source of steam and hot water, etc.

Note

The tanks should be dedicated for the same stock or compatible stocks with proper care that no significant cross-contamination occurs in the tanks. Any change in the service for the tank must be well documented to avoid miss-pumping of different oils.

14.2.7.1 Centrifuges

Centrifuges for degumming, refining, and water washing have been discussed in the chapters on degumming (Chapter 4) and refining (Chapter 5). The primary centrifuge in refining is hermetically sealed to prevent oil from coming in contact with the air. Decanters, such as old Sharples separators are not hermetically sealed. Alfa Laval centrifuge, model B214C is not hermetically sealed but the oil discharge is under some back pressure. Therefore, the oil is protected from coming into contact with air.

The separator (centrifuge) can be self-cleaning, where the centrifuge bowl does not require frequent cleaning.

The residual soap left in the refined oil from the primary separator is reduced in a water wash centrifuge. The process supervisor should be familiar with the following:

1. The type of centrifuge being used in the process.
2. Type and production capacity of each centrifuge.
3. All components of the centrifuge.
4. Procedure for cleaning the centrifuge.
5. Routine maintenance requirements and their frequency.
6. Start-up shut down, operation, and troubleshooting procedures for all centrifuges.
7. Be able to detect and take action when a centrifuge appears to have high vibration. This will require balancing the bowl and/or checking the shock absorbing pads at the bottom of the centrifuge.

Note

The plant is advised not to attempt to balance the centrifuge bowls. They must be sent to the manufacturer of the machine to get it balanced.

14.2.7.2 Converters

There are two types of converters that are commonly used. The types and the features of these converters and their operation have been discussed in the chapter on hydrogenation (Chapter 7).

Converters are pressure vessels. Therefore, any welding of the nozzles or on the vessel must follow the ASME code and should be inspected by a certified engineer familiar with the ASME code and testing for satisfactory welding.

There are several critical areas that a process supervisor must be familiar with, such as:

1. Type and the capacity of each converter in the department.
2. Maximum allowable operating pressure for each converter including the flanges.
3. Typical cycle time/batch for each type of product made.
4. Impact of brand mix on the overall production in the converter.
5. Type of agitator, RPM (revolutions per minute), type of seal, type of drive, type of lubricant for the seal, etc.
6. Horse power of each agitator motor.
7. Type of drive for the agitator, such as belt drive or direct drive.
8. Vacuum system in use (steam ejector or vacuum pump).
9. Type and the amount of catalyst to be used per batch of each type of product (catalyst loading).
10. Start up and the operating procedure for the converter.
11. Safety devices on the process.
12. Emergency shut down procedure.
13. Process troubleshooting.
14. Troubleshooting hydrogenated oil quality issue.

14.2.7.3 Deodorizers

Chapter 8 describes the different types of deodorizers that are used in the oil refineries. These are also pressure vessels like the converters, vacuum dryer, and vacuum bleacher. It is the process supervisor's responsibility to be familiar with the following:

1. Types of deodorizers in the department.
2. Production capacity of each deodorizer with the product mixes for the company.
3. Pressure rating of each deodorizer including flanges.
4. Design information on:
    - the deodorizer internals,
    - pumps (type, flow rate, discharge head, etc.),
    - heat exchangers (type, temperature at the inlet and exit of each stream, pressure at the inlet and exit of each stream),
    - vacuum ejector (type, capacity, steam usage, water usage absolute pressure produced), and
    - type of stripping steam distributor.
5. Number of stock changes per day.
6. Safety equipment and their functions.

7. Steam consumption for heating during start up, normal production and also for the ejector system.
8. Steam pressure for heating and the vacuum system.
9. Condenser water flow rate, condenser water discharge temperature at each stage of the vacuum system.
10. Cleaning procedure for the deodorizer and the vacuum system.
11. Frequency of cleaning.
12. Troubleshooting vacuum system.
13. Troubleshooting oil quality issues.

14.2.7.4 Vacuum Dryer

Vacuum dryer is used to reduce moisture in the water washed oil as described in Chapter 4 on refining. The vacuum dryer is also used in the two-step bleaching process described in Chapter 6 on bleaching.

Vacuum bleacher is a pressure vessel which normally operates at a maximum pressure of 100 mm of mercury (vacuum of 660 mm of mercury). The process supervisor must be familiar with the following:

1. The production capacity of the unit.
2. Maximum operating pressure.
3. Minimum operating temperature.
4. The expected moisture content of the oil at the exit of the unit.
5. The operating conditions for the vacuum system, such as steam pressure, steam volume, condenser water temperature, and volume.
6. Troubleshooting vacuum, oil quality, and production volume issues.

14.2.7.5 Vacuum Bleacher

Vacuum bleacher is also a pressure vessel with a top entering agitator as described in Chapter 6 on bleaching. The impeller design along with the baffles keeps the bleaching agent in suspension. The vessel is operated under a maximum operating pressure of 50 mm of mercury (vacuum of 710 mm of mercury).

The process supervisor must be familiar with the same items listed previously under vacuum dryer.

14.2.7.6 Filters

Filters are used to remove the bleaching agents in the bleaching process (Chapter 6) and to remove the catalyst in the hydrogenation process (Chapter 6).

As discussed in Chapters 5 and 6, there are different types of filters. In most refineries pressure leaf filters are used. These filters require precoating of the screens with diatomaceous earth that creates a thin porous bed on the screen. This allows the oil to be filtered without clogging the screens. This has been described in the Chapter 6 on bleaching.

The process supervisor must be familiar with the following and train the operators for efficient operation:

1. The filter precoating technique as described in Chapter 6.
2. The filter size, capacity, and the typical filter cycle time between cleaning.
3. Proper cleaning procedure for the filter screens as described in Chapter 6 on bleaching.

> **Note**
>
> The damaged filter screens must be sent back to the filter manufacturer for repair.

14.2.7.7 Heat Exchangers

Heat exchangers are used either to heat or cool the oil. Heating or cooling of the oil is accomplished in following ways:

1. Heating the oil with hot process water, steam, or thermal fluid.
2. Cooling the oil with process cold water or city water.
3. Interchange heat between the hot and cold oil streams to conserve heat energy. This application of heat exchangers is also referred to as heat economizing.

The heat exchanger is designed to provide the required amount of thermal energy needed for either heating or cooling a fluid stream in the process. The rate of heat transfer is important for the selection of a particular type of heat exchanger. Rate of heat transfer is expressed as BTU/h/ft.^{2}/°F, or CHU/h/cm^2/°C. The type and the actual size of the heat exchanger depend on the required heat load and the final temperature requirement for the fluid in any application.

It is important for a process supervisor to be familiar with the following:

1. Design data on the heat exchangers in every application in the department.
2. Sources and supply for the heating and the cooling media and the locations of their shut-off valves.
3. The designed performance of each heat exchanger and the reasons for their nonperformance.
4. The required cleaning frequency for the heat exchangers. This can vary; depending on the type of heat exchanger, type of heating media and its service, such as deodorized or undeodorized oil.
5. Methods for cleaning the heat exchangers.

14.2.7.8 Types of Heat Exchangers

Various types of heat exchangers are used in an oil processing plant. Today's modern plants use coaxial heat exchangers with or without fins, plate and frame, spiral, and shell and tube heat exchangers. Table 14.1 lists the various types of heat exchangers and their applications.

TABLE 14.1 Types of Heat Exchangers and Their Applications in Oil Processing

Type of heat exchanger	Typical application	Suitability	Remarks
Coaxial heat exchangers with or without fin-tubes.	These are used for heat economizing as well as for direct heating.	• Both low and high temperature applications. • The temperature can range from 100 to 1000°F (38–538°C).	• Relatively low construction cost. • Poor heat transfer efficiency. • Low cross-contamination of products. • Easy to clean when fouled.
Shell and tube heat exchangers.	For direct heating, cooling, and heat economizing.	Suitable for both low and high temperature range as listed for the coaxial heat exchangers.	• Construction cost is high. • Heat transfer efficiency is higher than the coaxial heat exchangers but lower than that of plate and frame or spiral heat exchangers. • Difficult to clean. • Higher potential for cross-contamination.
Plate and frame heat exchanger.	For heat economizing or direct heat with steam.	• Low and intermediate temperature range (up to 285°F (140.5°C). • With Viton Gasket the temperature can be up to 450°F (232°C). • Gasket-less units can go up to 600°F (371°C).	• Lower cost. • Very efficient heat transfer. • Low product cross-contamination. • Easy to clean. • Easy to replace the plates (when needed).
Spiral heat exchangers.	For heat economizing.	Suitable for temperature up to 500°F (260°C).	• Heat transfer efficiency is better than the shell and tube heat exchanger but lower than that for plate and frame heat exchanger. • Cross-contamination is higher than in the plate and frame heat exchanger unless the installation is made correctly for complete drainage of the liquid during cleaning. • Very difficult to clean.
Scraped wall heat exchanger.	Used to crystallize fat for making: 1. margarine 2. shortening	Low temperature for crystal generation.	• Low heat transfer due to the nature of the application. • High operating and maintenance cost. • Low cross-contamination. • Easy to clean.

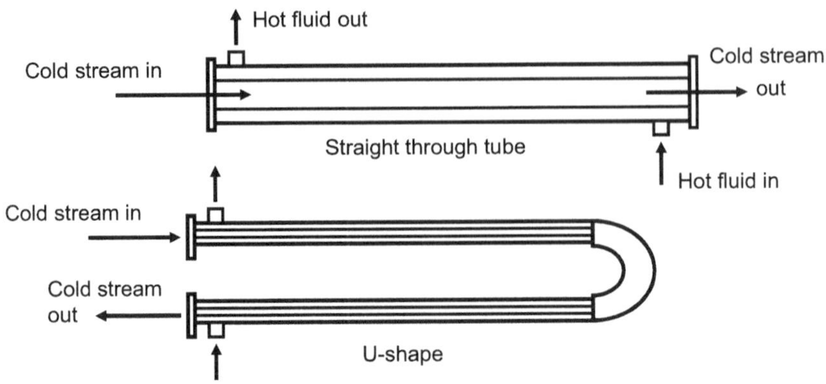

FIGURE 14.5 Coaxial heat exchangers.

14.2.7.9 Coaxial Heat Exchangers

Fig. 14.5 shows the schematic view of coaxial heat exchangers. These heat exchangers have an outer jacket through which the hot fluid is pumped. The cold stream passes through the inner tube. The inner tube can have fins on the outside to increase heat transfer. These heat exchangers are used mostly in the margarine and shortening chilling process. The heat exchangers can withstand high pressure (Table 14.1).

14.2.7.10 Shell and Tube Heat Exchangers

These are used extensively throughout the oil refinery. A single-pass heat exchanger is shown in Fig. 14.6. These are used for:

1. Heat economizing where the heat from the hot stream of oil is utilized to heat the incoming cold stream.
2. Direct heating with steam.

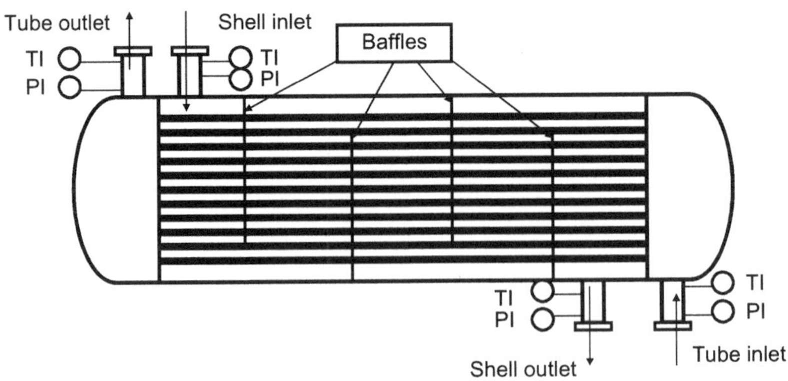

FIGURE 14.6 Shell and tube heat exchanger.

FIGURE 14.7 **Plate and frame heat exchanger.**

3. Direct cooling hot oil with cold process water.
4. The cold stream can pass through a single-pass or a multipass tube bundle.
5. These heat exchangers can be designed to work with high pressure steam.

14.2.7.11 Plate and Frame Heat Exchangers

Fig. 14.7 shows the outer view of a plate and frame heat exchangers. These heat exchangers are used in oil processing for:

1. Heat economizing where the heat from the hot stream of oil is utilized to heat the incoming cold stream.
2. Direct heating with low pressure steam.
3. Cooling oil with cold process water.

Heat transfer rate in these heat exchangers is much higher than that for all other types of heat exchangers. The limitations are that the plates cannot tolerate high pressure steam and the gasket limits the upper temperature limit. APV Paraflow unit is designed to prevent direct physical contact between the gasket and the steam (or other fluid) and with Viton gasket the temperature of heating fluid can be much higher (Table 14.1).The gasket-less design from Alfa Laval (Figs. 14.8 and 14.9) makes the unit suitable for high temperature.

FIGURE 14.8 **Gaskets-less (welded) plate and frame heat exchanger from Alfa Laval.**

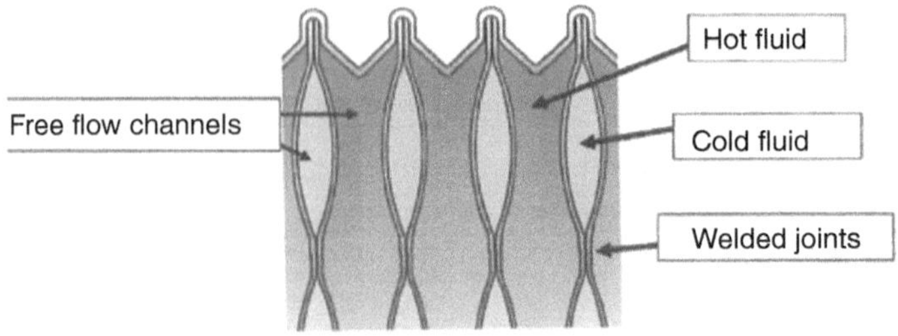

FIGURE 14.9 Free flow channels in gasket-less (welded) plate and frame heat exchanger from Alfa Laval.

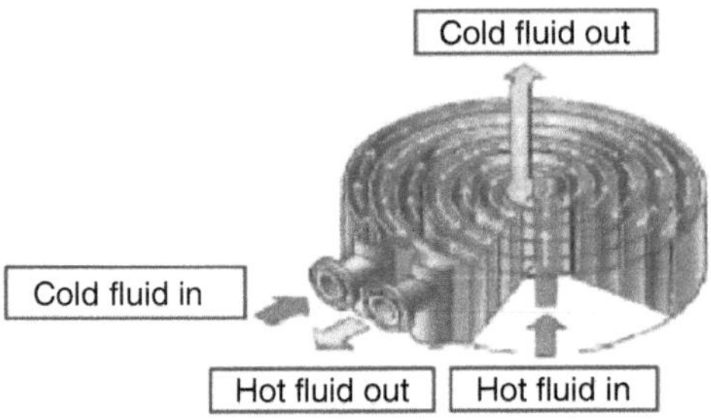

FIGURE 14.10 Internal view of a spiral heat exchanger from Alfa Laval.

14.2.7.12 Spiral Heat Exchanger

Fig. 14.10 shows the interior of a spiral heat exchanger. The hot fluid enters through the center of the vessel and moves toward the periphery. The cold stream enters from the shell side and moves toward the center and then exits. The heat transfer rate is high but as mentioned earlier, it is not as high as that of the plate and frame heater. The heat exchanger can retain a lot of liquid. Therefore, the installation must provide the capability to drain the heat exchanger for product changeover to minimize cross-contamination of two incompatible products. Proper drainage capability is also needed for minimizing oil loss at the time of clean out.

14.2.7.13 Proper Installation Guidelines for Heat Exchangers

The heat exchanger must always be installed with pressure and temperature indicators at the inlet and the outlet of both fluid streams. Fig. 14.11 shows the schematic diagram for the heat exchanger installation.

14.2.7.14 Fouling of Heat Exchangers

Heat exchanger efficiency is reduced with fouling of the heat exchange surface. Following are the common symptoms of heat exchanger fouling:

1. Increased pressure drop across the heat exchanger on the oil side indicates fouling, as long as the product flow rate did not increase.

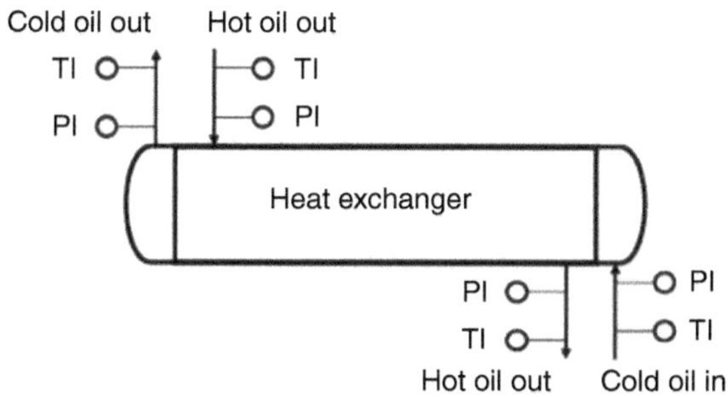

FIGURE 14.11 **Proper installation of a heat exchanger.**

2. The steam control valve is opening more to allow higher steam to flow through the heat exchanger while the oil flow has remained unchanged.
3. The oil inlet control valve will begin to open more to compensate for the reduced flow due to fouling.
4. The pressure drop across the heat exchanger increases with increased fouling.
5. Without the automatic oil flow control, the oil flow is required to be restricted to maintain the desired oil outlet temperature.

Note

1. There must be a no-flow switch on the oil side to shut off the steam valve. Without this, the steam can overheat the oil inside the heat exchanger and cause serious fouling. This can also cause an expansion of the oil trapped inside the heat exchanger. This can potentially damage the heat exchanger, especially if it is plate and frame type. Therefore, a relief valve should be installed on the oil line to relieve the excess pressure cause by the expansion of the oil.
2. Additionally, there must be a positive steam shut off valve before the modulating control valve.
3. This should be a plug type valve which is automatically actuated to close when the oil flow stops.

14.2.7.15 Frequency of Cleaning Heat Exchangers

The cleaning frequency has to be determined based on the points discussed earlier. The plant must observe the symptoms, and decide about the cleaning needs.

14.2.7.16 Troubleshooting Heat Exchangers

Table 14.2 summarizes the typical symptoms, the underlying cause or causes and suggested solution for troubleshooting heat exchanger fouling.

TABLE 14.2 Troubleshooting Heat Exchanger Fouling

Symptom	Probable cause	Recommended solution
For an oil to oil heat economizer: • Pressure differential between the inlet and the outlet has gone up on cold or the hot side.	• There is fouling or a blockage in the economizer. • The oil flow rate might have increased for some reason.	• Open the heat economizer to check for blockage or fouling. • Clean it with rotating metal brush cleaner. • Clean it with caustic or detergent and rinse it. • Check oil flow after restart. • Check and correct the oil flow rate.
The exit oil temperature is: 1. too low for the cold oil side, 2. too high for the hot oil side.	1. Possible blockage on the hot oil side restricting the flow. 2. Blockage on the cold oil side restricting the flow.	1. Check the oil flow on the flow monitor on the line. 2. Open the heat exchanger and clean it.
For a steam heater: • The oil flow has to be reduced to obtain the desired oil temperature.	• The oil side may be fouled causing poor heat transfer. • The pressure, flow, and/or temperature of the steam can be low.	• Clean the oil side. • Check the steam side for pressure, temperature and flow if the oil side is not fouled.
Water cooled heat exchanger: • Oil outlet temperature is too low. • Oil outlet temperature is too high.	• Cooling water temperature is low. • Cooling water flow rate is high. • Oil flow rate is low. • Incoming oil temperature is low. • Cooling water temperature is too high. • Cooling water flow is low. • Incoming oil temperature is too high. • Oil flow rate is high.	• Check and correct the cooling water temperature. • Check and reduce the cooling water flow rate. • Check and increase the oil flow rate. • Correct the low incoming oil temperature issue. • Check and correct the cooling water temperature. • Check and increase the cooling water flow rate. • Check and correct the high incoming oil temperature issue. • Check and correct the oil flow rate.

14.2.7.17 Piping

Following general guidelines are provided for the piping at the oil processing plant:

14.2.7.17.1 Oil Transfer Lines

- These are typically 40 schedule carbon steel pipes for transferring undeodorized and deodorized seed oils. Some plants use 304 stainless pipes (Schedule 10) for the deodorized oil.
- If the plant produces emulsifiers, all piping should be of either 304 or 316 stainless steel.
- For palm kernel, coconut, and palm oil the piping must be made of 304 or 316 stainless steel material.

14.2.7.17.2 Steam Supply Lines

All steam lines are made of Schedule 80 carbon steel.

14.2.7.17.3 Condensate Return Lines

These are Schedule 80 carbon steel pipes.

14.2.7.17.4 Compressed Air Supply Lines

These are Schedule 40 carbon steel pipes. It is better to use galvanized pipes for compressed air to prevent any rust plugging the air filter, located before the instrument.

14.2.7.17.5 Nitrogen Supply Line

These are made of Schedule 40 (or 80) carbon steel. In case of a cryogenic source with atmospheric vaporizer, one must be careful that no liquid nitrogen can spill over into the nitrogen distribution line. At the extreme low temperature of the liquid nitrogen, the carbon steel becomes brittle and the line may rupture, which can be a serious personnel safety issue.

14.2.7.17.6 Hydrogen Gas Supply Line

These are Schedule 40/80 carbon steel pipes with welded joints.

14.2.7.17.7 Steam Tracing

These are made of copper tubing. Sometimes stainless steel tubing is also used.

14.2.7.17.8 Caustic Lines

These are made of Schedule 30 or 40 carbon steel. However, 316 L stainless steel is preferred. Caustic may contain chloride ion which can damage 304 and the regular 316 stainless steel.

14.2.7.17.9 Citric Acid

This must be made of 316 stainless steel.

14.2.7.17.10 Water Lines (Hot or Cold)

These are Schedule 40 carbon steel pipes.

14.2.7.17.11 Concentrated Sulfuric Acid (Used in Acidulation of Soap Stock)

Carbon steel is suitable for storage and transferring concentrated sulfuric acid.

> **Caution**
>
> Sulfuric acid may produce hydrogen gas under certain conditions. This is especially observed after the tank has been washed to perform some maintenance work on the tank. Therefore, the sulfuric acid storage tank must be checked for explosive (or combustible) gas if any welding is to be performed inside or outside of the tank. This means the tank will require complete inspection with an explosimeter, a signed confined tank entry permit from the plant safety and operation before any maintenance work can be performed on the tank.

14.2.8 Process Accessories

14.2.8.1 Vacuum Ejectors

Vacuum ejectors are used to create vacuum at three different locations in an oil processing plant:

1. vacuum dryer in refining,
2. vacuum bleacher in bleaching, and
3. deodorizer.

Maintaining proper vacuum is a critical control point in vacuum drying, vacuum bleaching, and deodorizing. Therefore, some detail discussions are needed regarding this vital piece of process accessory.

In some plants, vacuum pumps or a combination of vacuum ejector and vacuum pump are used.

Vacuum ejectors are operated by the motive power of *dry saturated steam.*

Cold process water or greasy water is used to condense the steam from the ejector as well as the condensable vapor material removed from the oil by the stripping steam in the deodorizer. None of the dissolved air is condensed and is released by the last stage (atmospheric stage) of the ejector.

The vacuum ejector is designed to maintain a certain amount of vacuum (maximum operating pressure) in the system by removing the condensable and noncondensable vapor and the stripping steam. All free fatty acids, oil decomposition products and the stripping steam are condensable vapors. Any amount of dissolved air in the oil is considered noncondensable material. Thus, the vacuum ejector will not be able to maintain proper vacuum if for any reason the amount of the condensable and noncondensable vapor exceed the designed amount for a fixed flow of the dry saturated motive steam at the designed steam pressure.

The criteria used for designing vacuum ejectors are:

- Amount of condensable and noncondensable vapors coming from the oil per hour.
- Pressure of dry saturated steam.
- Amount of ejector motive steam per hour.
- Amount of stripping steam used per hour.
- Maximum temperature of the condenser water.
- Amount of condenser water needed per hour.

14.2.8.2 Noncondensing Versus Condensing Type of Vacuum Ejectors

There are two basic types of steam ejectors:

1. condensing type
2. noncondensing type

The condensing type of steam ejectors can be of two types, namely:

- direct-contact condensing type
- noncontact condensing type

Comparison between the condensing and noncondensing type of steam ejectors are shown in Table 14.3.

TABLE 14.3 Comparison Between Condensing and Noncondensing Type Steam Ejectors

Condensing type	Noncondensing type
Cold water is used to condense the steam and the distillates coming from the oil.	Contains no water condensers.
Can produce high vacuum. A four-stage steam ejector can produce a vacuum of 757 mm of mercury or an absolute pressure of <3 mm of mercury.	The minimum amount of vacuum produced is 710 mm of mercury or a maximum operating pressure of 50 mm of mercury (28 in. of mercury).
• Operating cost is higher because of using both steam and the condensing water. • A cooling tower with recirculating water system is required to cool the condenser water. This requires higher installation as well as operation cost.	• Lower operating cost. • No cost for the condenser water.
Suitable for producing high vacuum (lower absolute pressure), needed for deodorizers.	Not suitable for deodorizer because it does not produce sufficient vacuum. Can be used on vacuum bleacher.

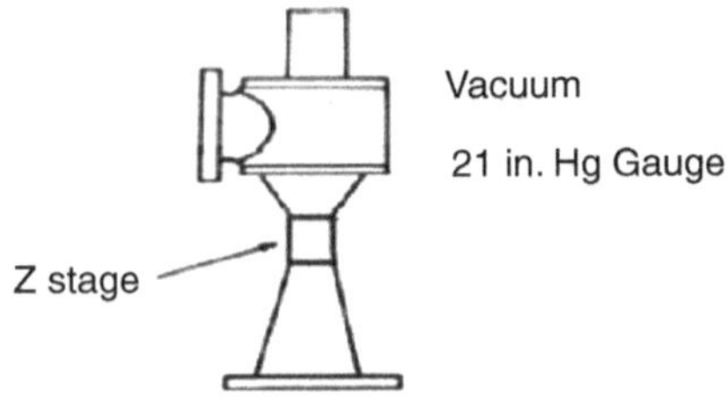

FIGURE 14.12 Single-stage steam ejector.

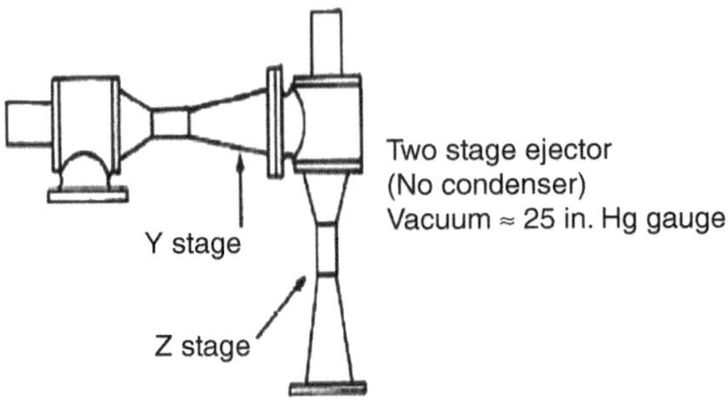

FIGURE 14.13 Two-stage noncondensing type vacuum ejector.

Fig. 14.12 shows the single-stage steam ejector which is capable of producing an vacuum of 21 in. (53.3 cm) of mercury. Fig. 14.13 shows a two-stage steam ejector without intercondenser. This ejector is capable of producing a vacuum of 25 in. (63.5 mm) of mercury.

14.2.8.3 Direct Contact Condensing Versus Nondirect Contact Condensing Type Steam Ejectors

14.2.8.3.1 Direct Contact Condensing Steam Ejector

Direct contact type condensing steam ejector is where there is a barometric condenser between every stage except at the end of the last stage. The water comes in direct contact with the stripping steam and the distillate from the deodorizer. The water from each condenser accumulates in a reservoir at the bottom of the barometric leg. This water is called greasy water because it contains the fatty components from the distillate. The water is pumped from this condenser water reservoir to a greasy water tower for cooling and recirculation back to the intercondensers of the ejector. The bottom of the barometric legs must always be submerged under the liquid surface while the ejector is in operation. Fig. 14.14 shows a two-stage steam ejector with the direct contact condenser. This type of steam ejector can produce a vacuum up to 28.5 in. and is suitable for the vacuum bleachers.

Fig. 14.15 shows the schematic diagram for a three-stage direct contact condensing steam ejector. This system can produce a vacuum of 754 mm of mercury or an absolute operating pressure of 6 mm of mercury.

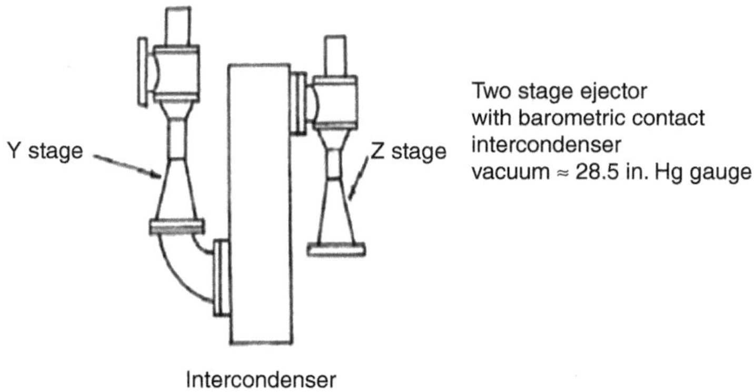

FIGURE 14.14 **Two-stage vacuum ejector with barometric contact condenser.**

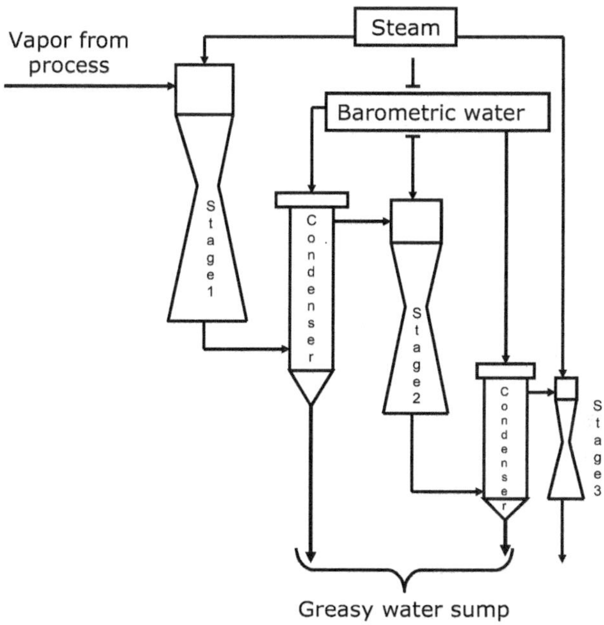

FIGURE 14.15 **Three-stage direct contact condensing steam ejector system.**

14.2.8.3.2 Nondirect Contact Condensing Type Steam Ejector

A nondirect contact condensing type of ejector uses cold process water to cool the steam and the vapor from the oil in a heat exchanger, very similar to a shell and tube heat exchanger. The cooling water does not come in direct contact with the steam and the vapor. Fig. 14.16 shows the schematic diagram of a four-stage steam ejector. This ejector system is capable of producing 1 mm of mercury vacuum.

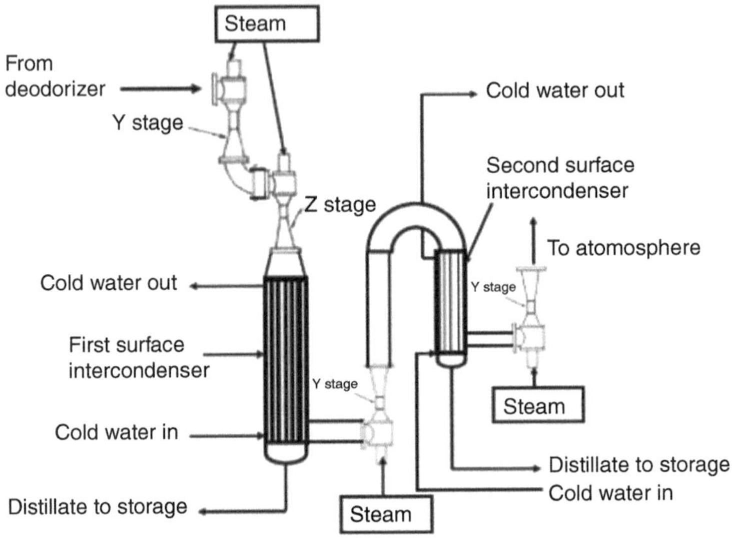

FIGURE 14.16 Schematic diagram for a four-stage vacuum ejector system—nondirect contact condensing system.

14.2.9 Troubleshooting Ejectors

It is important for the plant supervisor and the operators to understand the issues that can impact the performance of the steam ejectors and how to troubleshoot.

Table 14.4 lists the step by step troubleshooting procedure for the steam ejectors.

14.2.10 Freeze-Condensing Vacuum System

Graham Engineering offered a solution to obtain improved vacuum or reduce the absolute pressure (close to 1 mm of mercury) in the deodorizer by using the freeze-condensation technique shown in Fig. 14.17. Similar techniques have been applied by other equipment manufacturers.

The concept in this technique is to condense the stripping steam and the distillate from the oil before they even reach the steam ejector. This is done through a cold heat exchanger. This leaves only the air to be removed by the ejector. Therefore, the amount of motive steam requirement drops significantly.

14.2.10.1 Advantages

For the same deodorizer:

1. The motive steam consumption is reduced to approximately 10% of the conventional steam ejector for the same deodorizer.
2. Cooling water requirement is reduced to less than 10%.
3. The size of the ejector is reduced significantly.

TABLE 14.4 Step by Step Procedure for Troubleshooting Steam Ejectors

Symptom	Probable cause	Recommended solution
Poor vacuum (or high absolute pressure).	• Water vapor load is high because of higher volume of stripping steam. • Motive steam pressure is high. • Condenser water temperature is high. • Low flow of condenser water. • There may be a blockage in the condenser water line.	• Reduce stripping steam flow if it is high. If this does not improve the vacuum then reduce oil flow in a continuous deodorizer or reduce the oil volume per tray slightly in a semicontinuous deodorizer and reduce the stripping steam flow accordingly. • Reduce motive steam pressure if it is high. • Check and correct water distribution in the cooling tower. • Check and correct the flow of water to the condenser. • Check for any blockage and remove it.
The greasy water temperature is okay but the barometric leg is cold and the vacuum is poor.	• There is a blockage in the condenser or the barometric leg.	• The water flow may be high. Check the water flow and reduce it if it is found to be high. • If this is not the cause then shut off the greasy water and let the steam run through the barometric leg for 15 min. This will clean any grease build up. Restart the greasy water flow. • If this does not work, the blockage must be serious. Shut down the system and clean it.
Poor vacuum or high absolute pressure continues even after some of the above measures have been taken.	• There may be a leak on any of the joints, flanges, man heads, etc. • Leaky steam coil inside the deodorizer. • Leaky water coil in the cooling tray. • Leak in the ejector system. *Any of the above situations will require the deodorizer to be shut down for inspection and repair.*	• Shut off water into the cooling coil. The vacuum will improve if there is a leak in the cooling coil. • Each condenser stage is designed to show some specific pressure as indicated by the ejector manufacturer. • Blank off each stage at a time and check the pressure. An air leakage will produce a higher pressure at the ejector stage. • Identify the stage that produces higher than the designed pressure and repair the leak.
Poor vacuum or high absolute pressure continues even after some of the above measures have been taken.	• The ejector nozzle can be eroded, or steam leaking from the joint surrounding the nozzle of the first stage. • There can be blockage in the ejector nozzle.	• Repair any steam leak or replace the nozzle. • Open and clean the blockage.
Vacuum is unsteady and fluctuating.	• The atmospheric stage nozzle is eroded. This occurs if the motive steam is wet or if the nozzle is made of carbon steel and eroded by the wet steam.	• Check if the steam purifier is working properly. • Replace the nozzle with a 304 or 316 stainless steel nozzle.

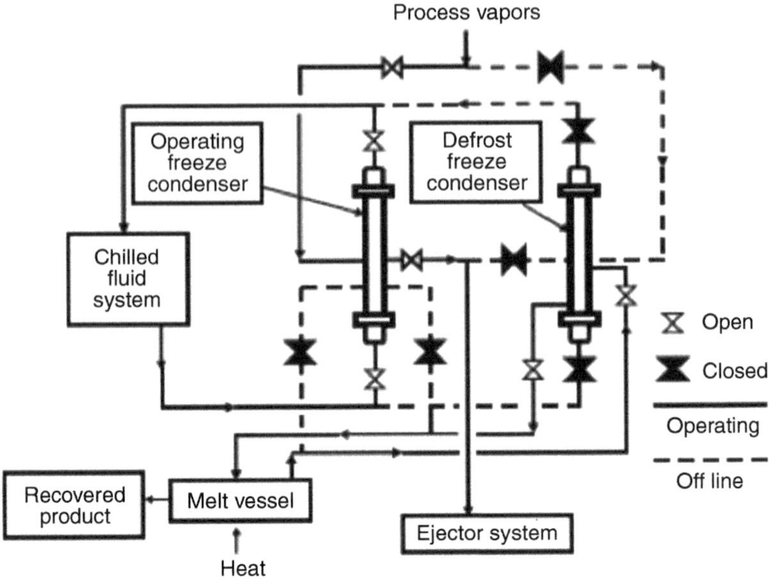

FIGURE 14.17 **Schematic diagram for a freeze-condensing vacuum system.**

4. Reduced pressure in the deodorizer can improve tocopherol recovery.
5. The system is more eco-friendly.

14.2.10.2 Disadvantage

1. The system requires a refrigeration system for chilling the deodorizer distillate and the stripping steam and also a heat (recovery) system for distillate recovery.
2. The initial capital cost can be 3–4 times higher than that of a conventional steam ejector.

However, the savings on steam, water, and avoidance of odor emission can compensate for the higher initial capital cost differential.

14.2.11 Agitators

Mechanical agitators are very common equipment used in the oil processing plant. Examples of most commonly used mixers are shown in Table 14.5. These are used for various purposes, such as:

1. mixing two or more miscible liquids,
2. dissolve a solid into a liquid,
3. dispersion of a nonmiscible liquid into other using high shear mixers, and
4. suspend and create an intimate contact between solids and liquid.

TABLE 14.5 Types of Mixers Used in an Oil Processing Plant

Impeller type		Application
Marine type	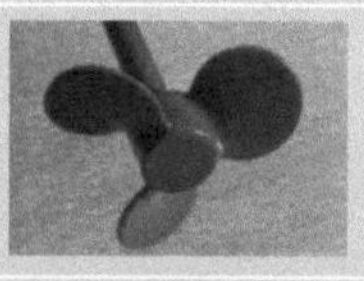	Used in tanks for blending oils for product formulation.
Axial type		Used in bleaching and hydrogenation to keep solids in suspension. This makes the oil and the suspended solids move vertically along the agitator shaft.
In-line static mixer	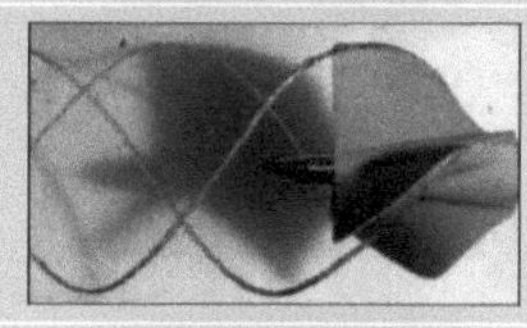	For dispersing caustic solution or citric acid into oil. This is not an efficient process.
Radial type		Used in bleaching to intimately mix bleaching clay and oil. Used in hydrogenation to disperse hydrogen gas in tiny bubble form into the oil for improved reaction
High shear mixer		Used to get better dispersion of caustic and citric acid into oil.
Ultra high shear mixer (a new entry)		Used to do the same as the high shear mixer but is far more efficient than the common high shear mixers.
Scrape wall mixer		Used in winterization in the chill tanks.

14.2.11.1 Examples

14.2.11.1.1 Category #1

This is most commonly used in blending two or more oils to formulate shortening or margarine product.

14.2.11.1.2 Category #2

Dissolving solid caustic pellets into water.

14.2.11.1.3 Category #3

The common examples are:

- Dispersion of NaOH solution into the crude oil in refining.
- Dispersing citric acid solution into the deodorized oil.
- Dispersion of oil soluble and water soluble components in making margarine emulsion.
- Dispersion of dimethylsiloxane in frying oil.
- Dispersion of antioxidants into the oil.

14.2.11.1.4 Category #4

The common examples are:

- Dispersion of bleaching clay into the oil in bleaching.
- Dispersion of nickel catalyst in the hydrogenation reactor.

14.2.12 Types of Mixers Used in an Oil Processing Plant

Table 14.4 lists the various types of agitators used in an oil processing plant.

14.2.13 Design Considerations for Selecting an Agitator

A mixer or an agitator is selected for a particular service. The mixer manufacture must be provided with the following information to specify an agitator for all applications:

14.2.13.1 Tank Information

This must include the following:

1. tank diameter,
2. height of the liquid,
3. height of the tank,
4. need for top or side entering agitator, and
5. desired turnover time.

14.2.13.2 Property of the Liquid

This must include the following information regarding the liquid:

1. nature of the liquid—corrosive or noncorrosive
2. temperature
3. density
4. viscosity

14.2.13.3 Service Application

The plant must provide the specific service information as listed:

1. Blending two or more miscible liquids.
2. Dispersing an immiscible liquid into another in fine dispersion.
3. Keeping solids in suspension and bringing the solid and the liquid into intimate contact.
4. Disperse gas into a liquid.
5. Produce gentle agitation, scrape the tank wall, and keep the crystals from depositing there.

An agitator must not be transferred from one specific service to another unless all of the aforementioned criteria are checked and found to be identical to its present application.

14.2.14 Pumps

Pumps are one of the vital process accessories at the oil processing plant. These are used to transfer oils between tanks, pump oil through the process, unload and load trucks and rail cars, transfer oil out of various process unit operations for processing.

There are two groups of pumps used in the oil processing plants, namely

1. nonpositive displacement pumps
2. positive displacement pumps

The nonpositive displacement pump generates the motive force through the centrifugal force created by an impeller, for example, centrifugal pump.

A positive displacement pump pumps liquids with the help of the motive force generated by the rotating action of the gears or by the reciprocating motion of pistons in the pump. Examples of positive displacement pumps are:

1. gear pumps
2. piston pumps
3. diaphragm pumps

Fig. 14.18 shows the cut-away view of a Durco centrifugal pump. The other brands will also have very similar internal construction.

Table 14.6 lists some of the important differences between the two categories of pumps. The comparison will indicate the specific contrasts between the two types of pumps.

14.2.14.1 Guidelines for Proper Pump Installation

There are certain guidelines that one must follow for all pump installations, for example,

1. There must be a strainer at the pump suction to protect the pump from any metal or hard material that could damage the pump.
2. There must be a pressure gauge at the pump discharge.

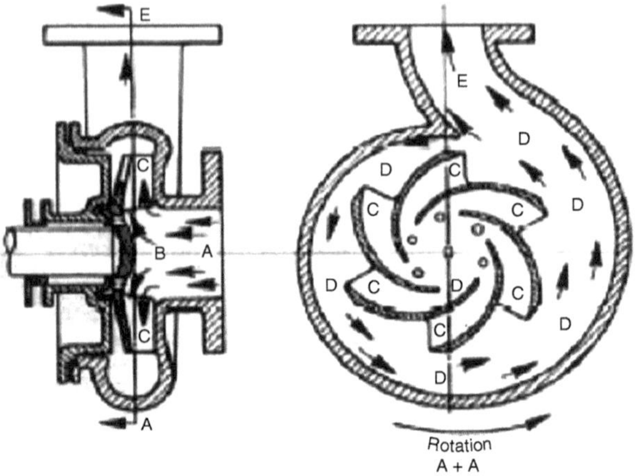

- The liquid enters at A
- It enters the impeller B
- Impeller vanes C picks up the fluid
- The vanes accelerate the liquid flow
- The liquid velocity reaches that of the impeller vane tips
- The liquid slides up the casing and discharge through E
- Some of the velocity head is converted to pressure head

FIGURE 14.18 Cutaway view of centrifugal pump.

3. The suction side opening is always larger than that at the discharge. This is to ensure proper flooding of the pump.
4. There must be a flow switch on the suction line of the pump. The flow switch turns off the motor of the pump if there is no liquid flow into pump suction.

14.2.14.2 Guidelines for Proper Pump Operation

Following are some recommendations regarding proper pump operation:

1. The pump suction must always be flooded. The centrifugal pump cavitates when the suction is not flooded. This damages the pump impeller and the seal.
2. Previous recommendation applies for positive displacement pump which can get overheated or even may siege. Waukesha pump can run for a while without any feed because there is no metal to metal contact in the gears.
3. The pressure drop in a modulating control valve must be taken into account in designing the pump size and the discharge head and motor horse power requirement. The pump must be started with the controller set at the desired flow rate on the flow controller otherwise the pump will discharge higher volume of liquid. This will draw more horse power for the motor. This may

TABLE 14.6 Comparison Between the Positive Displacement and Nonpositive Displacement Types of Pumps

Nonpositive displacement pump	Positive displacement pump
The pump moves fluid by converting the centrifugal velocity into discharge head.	Pumps fluids by physical displacement using: • Pistons—reciprocating, rotary, etc. • Gears (e.g., Viking, Waukesha, Triclover). • Sliding vane (e.g., Blackmer) • Progressive cavity (e.g., Moyno) • Diaphragm (e.g., air operated Weldon Pump)
• Good up to a discharge head of 100 psig. • Multiple stages are required to attain higher discharge head.	Fluid transfer is effective against very high resistance and discharge head.
• Supply tank cannot be emptied in every case. • It requires special design for the pump location, tank bottom, and the piping layout for pump suction.	The supply tank can be completely emptied by the pump.
It is necessary to select the pump with the correct NPSH (net positive suction head) to achieve the desired fluid flow rate to the desired discharge head.	NPSH is not critical.
• A pressure relief valve is not essential in most applications. • However, it could be installed with a recirculating line to protect the pump from overpressure from any sudden blockage on the discharge side. The recirculating line simply returns the oil from the discharge line to the pump suction.	• A pressure relief valve must be installed at the pump discharge to protect the pump from overpressure. • It is always recommended to install a pressure relief valve with the discharge run into an open bucket so the operator will notice oil is discharged. • In addition, the pump must have an internal relief valve to recirculate the fluid if any restriction develops on the pump discharge line.
• While the pump is running, the valve at the discharge end of the pump can be closed for a short time without causing any damage to the pump. • Prolonged closure of the valve will damage the pump.	• The valve at the discharge of the pump must NOT BE CLOSED while the pump is running. This will destroy the pump.
• The valve at the discharge side of the pump can be throttled (or partially closed), without damaging the pump. • In many applications fluid flow is regulated within a small range by using a control loop that modulates a flow control valve located at the discharge side of the pump.	• The valve on the discharge side must not be throttled. • Fluid flow with the positive displacement pump is controlled by variable frequency drive (VFD) that regulates the revolutions per minute (rpm) of the motor.
Although the pump could run dry for a minute or two, this practice can destroy the pump seal.	The pump must never be run dry because this will destroy the pump seal and eventually the pump.

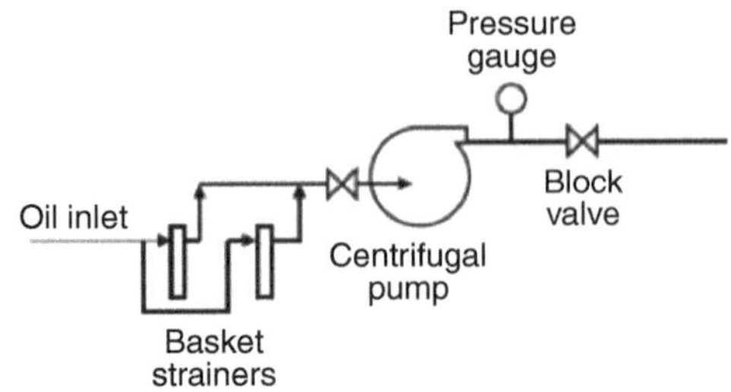

FIGURE 14.19 Schematic diagram for typical installation of a centrifugal pump.

trip the circuit breaker, burn the fuse, or may even burn the motor due to overloading.

The typical piping around the centrifugal pump installation is shown in Fig. 14.19 and that for a positive displacement pump is shown in Fig. 14.20.

Both types of pump need a pressure gauge at the discharge. This is very useful for troubleshooting any flow issues. The block valves at the pump inlet and discharge are for maintenance purpose. These valves must remain open all the time.

14.2.15 Valves

Numerous types of valves are used in an oil processing plant. Table 14.7 lists the types of valve and their normal applications.

14.2.16 Cooling Towers

14.2.16.1 Application

Cooling towers are used for cooling both clear process water and the greasy water. Clear water is used for process cooling, including the noncontact condenser of steam ejector system. Greasy water is used in the barometric condenser of a direct contact steam ejector.

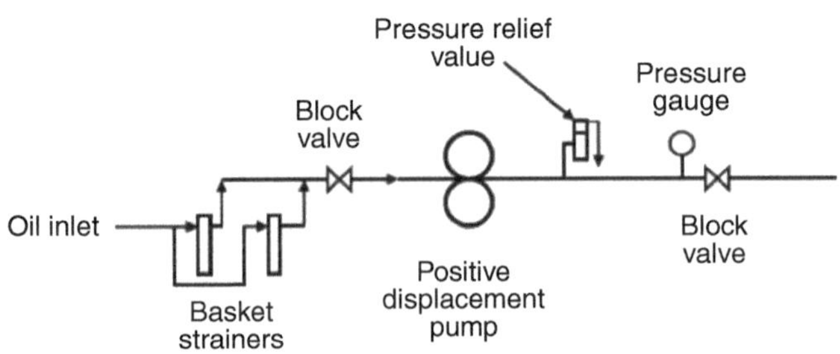

FIGURE 14.20 Schematic diagram for typical installation of a positive displacement pump.

TABLE 14.7 Valves Used in the Oil Processing Plant

Type of valve	External view	Brief description and application
Globe valve		Used in most liquid transfer lines. These valves can be used for throttling the flow of liquid.
Gate valve		• These valves are used for higher flow of fluids. The valves can be: 1. rising stem 2. nonrising stem • The gate valves cannot be throttled because that can seriously damage the valve gate.
Needle valve		These valves are used for fine control of fluid—gas or liquid.
Ball valve		This is an on–off valve that must be either fully open or closed. This type of valve is used on the steam supply line for steam traps, steam tracing, water line, etc.
Plug valve		Very similar to the ball valve but it got the shape of a plug. This valve is used to direct the flow of fluid manually or via automation.
Butterfly valve		Used on the gas or liquid flow line. This is an on–off valve. Should not be used for throttling.
Check valve		Used on both gas and liquid lines to maintain the flow of fluid in only one direction. This is essential whenever two liquid streams are metered into a common process unit, such as caustic and crud oil at the high shear mixer, citric acid and oil to the high shear mixer, air or nitrogen gas introduced into the oil line.

(*Continued*)

TABLE 14.7 Valves Used in the Oil Processing Plant (*cont.*)

Type of valve	External view	Brief description and application
Feeder valve	 Rotor	Used to feed solids. This type of valve is used in handling bleaching clay, activated carbon, diatomaceous earth, etc.
Three- or Four-way valve	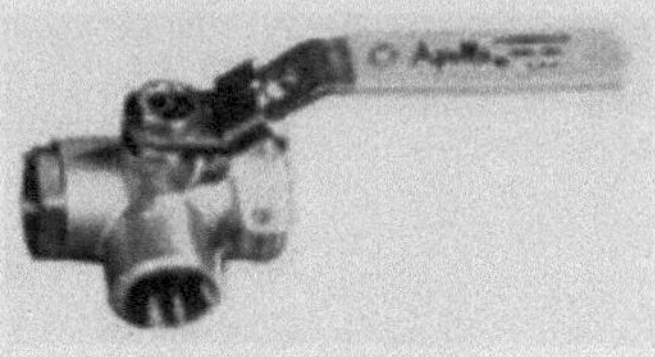	These are modified globe or plug valves used for changing the direction of fluid flow in more than one way. These valves can be manual or operated through automation.
Control valves		Modulating or on–off valves used to control flow, temperature, or pressure.
Solenoid valves		These are on–off valves, used either to turn on or shut off the flow of gas or liquid automatically.
Pressure relief valve		These valves are used to relieve the system from overpressure. The valves are used in several areas.

14.2.16.2 Mechanism of Cooling Water in a Cooling Tower

Cooling of water is accomplished by:

- Distributing the water returned from process through manifold and a number of spray nozzles down the slats (also referred to as louvers) located along the four sides of the cooling tower.
- The water trickling down the tower slats meets with ambient air draft created by the large suction fans located at the top of the tower.
- The air evaporates a portion of the water and, thereby, cools the water by losing the latent heat of vaporization.

14.2.16.3 Cooling Tower Design

- A cooling tower is designed for a certain amount of water flow gallons per minute.
- A designed volume of ambient air is pulled by the suction fan or fans at the top of the tower (cubic feet per minute or cubic meters per minute).
- The temperature drop for the water is governed by the following factors:
  - rate of water flow,
  - rate of air flow through the tower,
  - relative humidity of the ambient air,
  - distribution of the water at the top of the tower and down the slats,
  - degree of intimate contact between the water and ambient air,
  - cleanliness of the distribution nozzles, and
  - cleanliness of the slats.

14.2.16.4 Efficiency of Cooling the Water

It is always a good practice to design the cooling tower for the hottest temperature in the area and the prevailing relative humidity in the region.

This is because the air pulled through the cooling tower at the perfect air/water encounter can pick up moisture based on the moisture content in the ambient air and the relative humidity of the air. The amount of water that can be evaporated by the ambient air depends upon the following:

1. Ambient temperature (also can be called the dry bulb temperature).
2. The relative humidity of the ambient air or in other words how much of additional water vapor in the air would make it saturated with moisture at the ambient temperature or what is the maximum theoretical water evaporation that could be expected at the cooling tower.
3. In reality, the cooling tower cannot realistically evaporate the maximum potential amount of water as the air draft moves up the cooling tower.
4. Therefore, the cooling tower that works well for average weather condition may not provide the necessary cooling of the water when the weather is hot and humid. This means, evaporation of water is slower when the air is already full of water vapor.

14.2.16.5 Inadequate Cooling of the Water

Listed in the following are some of the factors that may interfere with cooling of the water:

- High ambient temperature.
- High humidity. This makes the wet bulb temperature higher and closer to the dry bulb temperature. Thus, lesser amount of water is removed by the same volume of air pulled through the cooling tower.
- Poor water distribution due to clogged nozzles.
- The slats are not clean or broken, making poor air and water distribution.
- The suction fan is not pulling enough air up the cooling tower.

14.2.16.6 Consequence of Inadequate Water Cooling at the Cooling Tower

The amount of cooling water used at the ejector is based on the calculated thermal load from the condensable and the noncondensable vapor from the system. The condenser size is determined by the volume of water flow, the water temperature, and the rate of heat extraction by the water from the vapors. A high water temperature may not condense all the vapors leaving the deodorizer. This will produce poor vacuum in the deodorizer.

14.2.16.7 Tower Cleaning Frequency

It is recommended that the cooling tower be cleaned once per year. This includes the following steps:

- Clean all nozzles.
- Replace any defective nozzle.
- Check and clean the water manifold at the top of the tower.
- Cleaning and repairing and replacing any broken or missing slats.
- Check the suction pump and make sure that it is pulling the designed volume of air through the tower. This will need the following:
    - Check the amperage on the motor to verify that it is carrying proper amount of load.
    - The rpm of the fan—it can be lower than the designed value due to wear on the drive mechanism.
    - One or more blades might not be at the right pitch after long use or might be damaged.
    - There may be some restriction to air flow through the tower.

14.2.17 Motors, Starters, Switches, Fans, and Blowers

Electrical motors are used in every rotating unit in an oil processing plant, such as:

1. pumps
2. air and hydrogen compressors

3. agitators
4. centrifuge motors
5. fans and blowers
6. air conditioner compressor
7. fat crystallizers on the shortening and margarine lines

Most motors at the plant are operated with 110 V/60 H power (Except the United Kingdom and in some of the former British colonies 50 H is still used).

Motors with high horse power are operated at 440 V 3 Phase AC power supply.

The control circuit is operated at 24 V, reduced from 110 V supply line.

The process supervisor should be familiar with the following:

1. Horse power and the operating voltage for every motor in the area.
2. Location of the control panel (motor control center, MCC).
3. Location of the electrical switch and cut-off for every motor.
4. How to check whether a motor has tripped for some reason and how to restart it at the motor control panel.
5. When replacing a motor in an explosive environment, the new motor must be of explosion-proof type.

14.2.18 Compressors

Compressors are used in an oil processing plant for various services, such as:

1. Supply compresses air to the process.
2. Supply compressed air to process instruments.
3. Compress hydrogen gas for the hydrogenation process.
4. Compress ammonia for shortening and margarine production and other refrigeration services required at the plant.

It is essential for the process supervisor to be familiar with the following:

1. type, make, and capacity of each compressor;
2. the type of service for the compressor;
3. operating and maintenance procedure;
4. material of construction;
5. proper procedure for lubrication;
6. proper function of the aftercoolers; and
7. maintenance frequency recommended by the manufacturer.

14.2.18.1 Special Notes on Compressors

1. Air compressor must be nonlubricated for instrument air.
2. Hydrogen compressor must be nonlubricated.

When any gas is compressed, heat is released. This heat of compression must be removed from the gas in an aftercooler. Moisture is also released from the gas in the aftercooler. This moisture must be drained. The remaining moisture from the compressed gas must be removed from the air or hydrogen gas through a dryer.

14.2.19 Air Dryers

Air dryer is essential in an oil processing plant. As mentioned earlier, compressed air is used for:

1. clearing lines in the process,
2. pneumatic instruments, and
3. clearing oil from the transfer lines.

Instrument air or the process air if used for clearing the oil lines must come from a compressor separate from that supplies compressed air for general process use due to the reasons listed as follows:

1. Compressed air is generally contaminated with the lubricating oil. Although this lubricating oil is approved for occasional contact with food, it is harmful to the process instruments and must not be allowed to come in contact with the oil.
2. The entrained oil can plug the air filter of the instrument.
3. The entrained moisture in the compressed air can also damage the instruments as well as the oil.

Instrument air must have the following features:

1. a nonlubricated type of compressor and
2. an air drying system, either a desiccant or a refrigerant type.

The process supervisor must set up a program for the routine inspection of the following items:

1. The lubrication system—make sure that the lubrication pot on the compressor is full.
2. Make sure the aftercooler is operating properly.
3. The desiccant dryer is properly functioning and the desiccant is being replaced as per manufacturer's recommendation.
4. The refrigerant drying system is operating properly.
5. All safety relief valves are properly functioning.

Note

One must never crossover between the general process air and the instrument air at any time.

14.2.20 Steam Tracing

14.2.20.1 Purpose of Steam Tracing

Steam tracing is used in the oil processing transfer lines to prevent any of the higher melting-fraction to separate from the blend and get deposited in the pipes and tanks. All steam traced pipes, tanks, pumps, filters, etc. must also be insulated to prevent heat loss. Some operations use electrical tapes and insulation in place of steam tracing.

14.2.20.2 Basics of Steam Tracing

1. Low pressure steam, steam 10–15 psig (1.79–2.16 kg/cm^2) is recommended for using in steam tracing. The saturation temperature of steam at this pressure is relatively low and yet it is much higher than the melting point of the highest melting fat fraction in the product.
2. The steam passes through copper tubing either wrapped around the pipes and vessels or the tube is run straight along the length of the pipe with insulation.
3. A thermodynamic steam trap is used at the end of the tracing line to let the condensate be discharged at a preset temperature. The condensate discharge temperature can be adjusted on the thermodynamic steam trap.
4. The total tracing for the process area should be divided into several segments.
5. The steam inlet must be at the highest point in any particular segment (also called loop) and the condensate discharge should be at the lowest point of the same loop.
6. It is a good practice to follow the design procedure outlined as follows:
   a. Number all the tracing loops.
   b. Install a common header for a certain number of loops for steam supply to the individual loops.
   c. The condensate returns to a bank of thermodynamic steam traps connected to each loop and properly identified.
   d. The condensate drains into a common condensate return line.
   e. There should be ball valves for steam supply line to the trap.
   f. There are also for ball valves to isolate any particular steam trap for inspection or replacement.

Note

Steam tracing can fail. Therefore, on a long transfer line, one must install steam hook-ups and wrap around tubes, in addition to the steam tracers. These additional tubes and the wrap around are also located inside the insulation. The temporary hook ups are outside the insulation. This arrangement enables one to connect steam to these hook-ups using a flexible hose to heat the pipe and melt the solidified material inside the pipe.

14.2.21 Steam Traps

Steam traps are essential for energy utilization at the plant. Without steam traps the plant's energy consumption will be extremely high and the oil in the system can be overheated and damaged.

Steam has two components of energy:

1. sensible heat
2. latent heat of condensation

The sensible heat is the amount of thermal energy that the steam possesses at the saturation temperature at a given pressure for the steam. The standard temperature of 70°F or 20°C is used as the point of reference for the calculation of the sensible heat.

The latent heat of condensation is the amount of thermal energy the steam releases when it is transformed from the vapor form to the liquid form at the saturation temperature. The latent heat of steam is several times higher than the sensible heat. Following explanation will make this point clear.

Steam pressure (psig; 1.79 kg/cm^2)	Saturation temperature (°F)	Sensible heat (BTU/lb)	Latent heat (BTU/lb)	Total energy (BTU/lb)	Latent heat as percent of total energy
10	239	207	953	1160	82.15
150	366	339	857	1196	71.65

As one can see the latent heat of condensation constitutes the major part of the total thermal energy of the steam. Therefore, for heat transfer, it is always desirable to condense the steam instead of letting live steam pass through any heater. When the live steam is allowed to pass through the heater, the only energy transferred is part of the sensible heat of the steam and this temperature can be too high.

Additionally, the saturated steam at 10 psig (1.79 kg/cm^2), as compared to 150 psig (12.02 kg/cm^2), has:

1. 127°F (52.8°C) of lower saturation temperature,
2. 123 BTU of lower sensible heat, and
3. 96 BTU of *higher* latent heat.

The previous data demonstrate several important properties of steam:

1. At 150 psig, the steam contains 3% more total energy than at 10 psig.
2. At 10 psig, the steam has latent heat that is 82.15% of the total energy, whereas, at 150 psig, it is only 71.65%. This justifies full condensation of the steam as opposed to blowing steam through the heating coil for heating the oil.
3. Steam at 10 psig is more desirable for steam tracing because of its lower saturation temperature (239°F), which will not scorch the oil in contact with the heating coil and higher latent heat (953 BTU/lb) for heating.

Thus, one can see the benefit of using lower pressure steam and fully condensing it for the purpose of steam tracing and heating. The basic concept applies for higher temperature requirements, such preheating the oil for heat bleaching and deodorization. In this case a higher pressure steam is required because the oil has to attain a higher temperature.

A steam trap is the only equipment that allows the maximum utilization of the thermal energy of the heating steam by allowing it to condense and release the maximum amount of its full thermal energy to the oil.

14.2.21.1 Types of Steam Traps

There are different types of steam traps and each has very specific application as described in Table 14.8.

14.2.21.2 Managing Steam Traps

It is very important to maintain the steam traps in good operating order. In the absence of such an effort, there can be a tremendous amount of steam loss at the plant that can go unnoticed. Therefore, it is necessary to implement an active program for managing the steam traps at the plant. This involves proper installation and monitoring of the traps.

A steam trap discharges a certain amount of condensate at some frequency. A small amount of vapor may be visible during condensate discharge. The trap must shut off immediately after the condensate has been discharged. If the trap does not close after the condensate release, the trap begins to discharge live steam. This is what contributes to a great loss of energy and capital.

It is not possible to visually determine any steam loss due to a malfunctioning trap if the condensate discharges into an enclosed condensate return header. In this case, there are some techniques that can be applied to detect any malfunctioning steam trap. Table 14.9 lists the various detection methods and the suggested precautions and action steps in testing the traps.

14.2.21.3 Proper Steam Trap Installation

A schematic diagram for steam trap installation is shown in Fig. 14.21. Proper installation of a trap is important. The important guidelines are listed as follows:

1. All piping must be of Schedule 80 carbon steel.
2. The ball valves in the diagram must have reinforced polyfilled seats.
3. Condensate discharge line must be at a lower level than the steam inlet line.
4. Any insulation around the trap must be in accordance with the trap manufacturer's recommendation.
5. The condensate discharge from a trap should be at the eye level or lower so an operator can actually observe the condensate discharge from the trap.

TABLE 14.8 Most Commonly Used Steam Traps in Oil Processing Plants

Type of steam trap		Typical application	Example of specific application in oil processing
Inverted bucket		• Used in a continuous process. • Not suitable for intermittent start up and shut down.	Continuous or semicontinuous deodorizer
Float and thermostatic	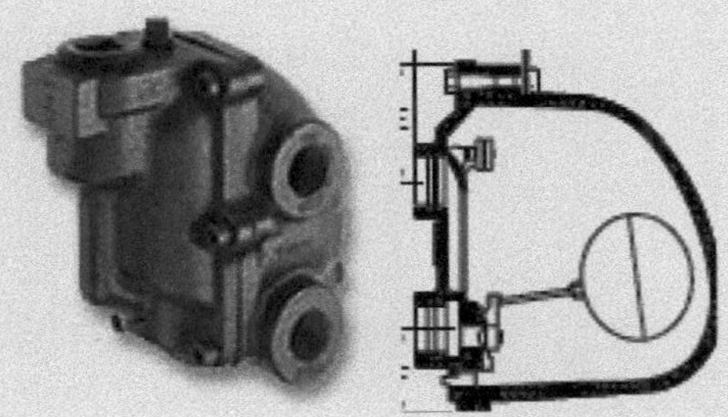	Large initial discharge of condensate followed by controlled discharge of condensate at start of the process.	Heat exchangers. At start the residual condensate in the system must be discharged rapidly to have a quick response from the heat exchanger.

Thermodynamic	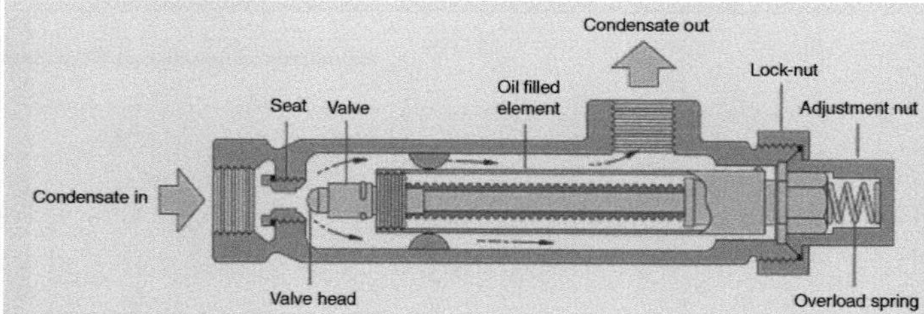	Excellent for temperature control and for controlling the condensate discharge temperature.	• Inside or outside oil storage tanks. • Steam tracers. • Melting of solidified fat in a truck or a rail car.
Thermostatic steam traps	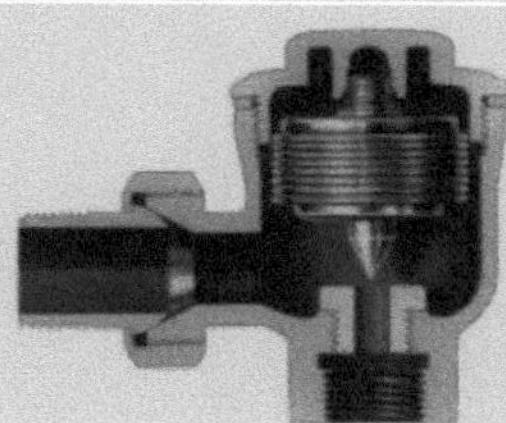	Temperature control.	Inside oil storage tanks where the room temperature is fairly uniform. The trap control can chatter if it is used outdoors and the temperature gets cold.

TABLE 14.9 Detection Methods for Steam Leaks From Steam Traps

Detection method	Description	Precautions/action steps
Physical inspection	• Steam leak from a steam trap is observed visually. • Steam discharge does not stop after a few seconds.	• One must be careful that there is always a small amount of steam released along with the condensate discharge but normally this is for a few seconds. • Close the steam supply, open the trap and check for dirt or rust on the seat. Clean the trap and recheck. *If not found dirty, the trap is bad, replace it.*
Use an ultrasound detection device	Detects continuous steam flow through a trap. This is especially good for an enclosed condensate discharge line.	Clean or replace the trap if the test detects a leak.
Infrared surface temperature detector	Checks the surface temperature of the trap. A high temperature will be indicated by the detector if there is a steam leak.	This method can be tricky and can be performed only by an experienced person.

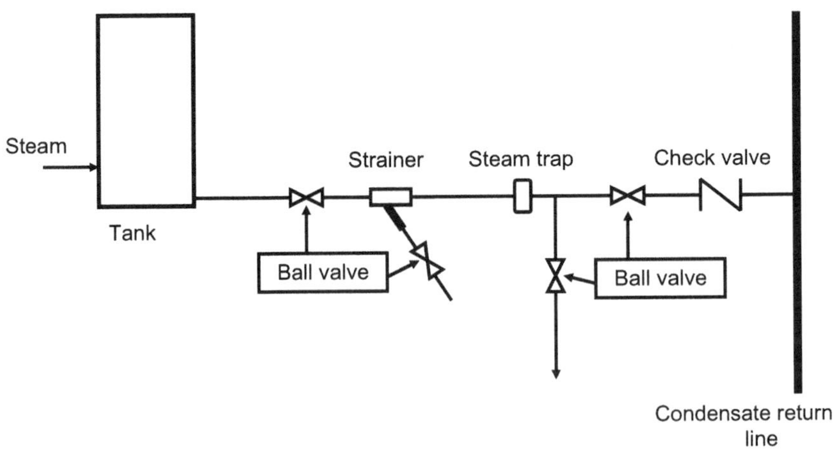

FIGURE 14.21 Schematic diagram for steam trap installation.

14.2.22 Steam Purifier

A steam purifier is used to separate the condensate from the steam. Process application, such as steam ejector requires dry saturated steam. Any amount of condensate in the steam can reduce the performance of the ejector. It can also cause erosion on the nozzles, especially the atmospheric stage.

Steam can generate some amount of condensate in the distribution line due to the following reasons:

1. The steam may be slightly supersaturated as it leaves the boiler house so the excess moisture is deposited in the line as the steam loses some pressure.
2. The steam pressure may drop in the distribution line releasing some condensate.
3. The condensate is carried along by the steam to the process equipment.

The steam purifier, which is a specially designed wide spot on the steam line, separates the condensate at the bottom of the vessel. The condensate is periodically discharged into the condensate return line with the help of a level controller and an on–off control valve. Fig. 14.22 shows the schematic diagram for a steam purifier.

14.2.23 Seals

Seals are used to maintain proper alignment of shafts and also to prevent leakage of fluids around a rotating shaft. The fluid can be gas, water or oil.

Seals are found in the following equipments:

1. pump shafts
2. piston rods on the reciprocating pumps

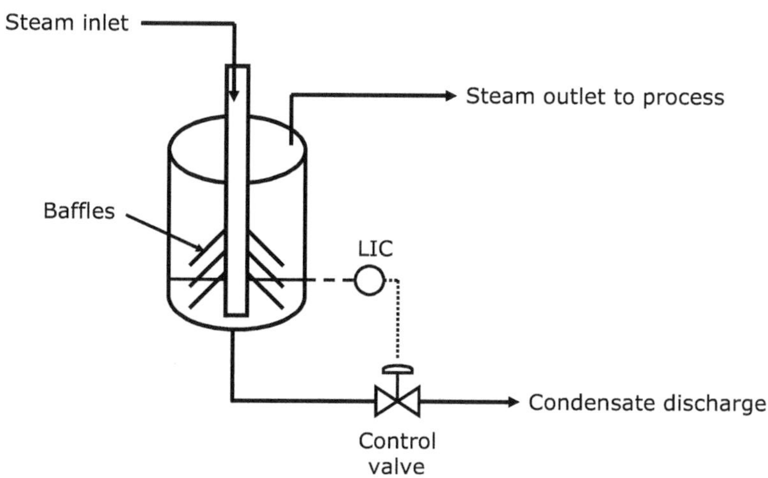

FIGURE 14.22 Schematic diagram for steam purifier.

FIGURE 14.23 **Packing for rotating or reciprocating shafts.**

3. motor shafts
4. agitators
5. centrifuges
6. heat exchangers (gasket)
7. valves

Seals around the rotating shafts are of two types:

1. packed glands
2. mechanical seals

Packing glands in old days used to contain asbestos packing. The modern packing material is made of nontoxic, nonstaining fiber blended with lubricant approved by the Unites States Food and Drug Administration (USFDA). The packing is suitable for both rotating as well as reciprocating shafts. Fig. 14.23 shows a picture of this packing material distributed by the United Products & Service Inc.

Mechanical seals have replaced the packing glands in many cases. These are made from special alloy capable of withstanding constant rubbing action from the shaft, whether it is a rotating or a reciprocating shaft. There are various designs for mechanicals depending on the pressure, temperature, and the severity of the service. Fig. 14.24 is showing the assembled view of the simple mechanical seal.

These mechanical seals always show a small amount of oil leak because the oil is used as lubricant to prevent wearing or seizing of the shaft. Some mechanical seals on centrifugal pumps on high oil temperature are designed to recirculate a small amount of the oil through the seal to keep them lubricated.

Mechanical seals are damaged when:

1. the side mounted agitator is operated with no oil in the tank, and
2. the pump is allowed to run with no oil feed.

14.2.24 Process Instruments

There are four different process parameters that need to be controlled in an oil processing plant. They are:

FIGURE 14.24 **Packing for rotating or reciprocating shafts.**

1. temperature
2. pressure
3. flow
4. oil inventory in a tank

Process instruments are used to measure and control the aforementioned process parameters. The most common process instruments used in an oil processing plant are listed as follows:

1. temperature indicators/controllers,
2. pressure indicators/controllers,
3. flow indicators/controllers,
4. level indicators/controllers,
5. automatic control valves,
6. gas purity indicators,
7. oxygen monitors in the deodorized oil storage tanks, and
8. alarms.

In the older plants, the process control instruments are operated with instrument air. In the 1960s, the relay system was used for process control in the oil processing plants. The modern plants use the programmable logic control technique to operate all control points in the process. The positive displacement pumps are controlled by VFD.

Chapter 15

Loss Management

15.1 INTRODUCTION

Managing product and material losses is an important aspect of controlling production cost at any plant. Vegetable oil refineries are no exception. Losses at the plant can become like a cancer that goes unnoticed but chips away the company's profit, without providing any clue to the management.

Managing losses at the plant is every manager's and supervisor's responsibility. The concept of loss control should become part of the culture of the company's management team because their paychecks and bonuses (if any) are also affected by the profitability of the operation.

Losses can occur in a plant in many areas, such as raw material, finished products, packaging material, processing, off-quality products that require reprocessing or discarding, etc. Most of these go unnoticed because only a clerical staff collect all the data and publish it at the end of every month for product costing. The supervisors usually look at these numbers but they do not have any appreciation for them because they do not have the time to get involved and try to understand the cause or causes behind any losses. The author is speaking about this from his personal experience. As a production manager I did not take active interest in losses reported by the clerk. However, I discovered a gold mine for the plant once I was assigned as loss control manager. I found out that many of the areas reported by the clerk earlier provided unique opportunities for savings.

In today's environment of high-tech and the new management style, the concept of loss management may sound very mundane—but the author is aware of the fact that, after identifying the sources of losses, one can implement high-tech solutions for controlling the system designed for loss prevention. It goes without saying that in some operations, the profitability can be significantly improved if the losses are brought under control. It is certain that the CFO and the COO will be very interested in the savings that can result from the loss prevention program.

The object of the discussions in this chapter is to highlight the issues related to plant losses, their implications, and the measures that one can take to control them. This chapter will also discuss the proven methods showing how to implement a loss control program and manage it.

Practical Guide to Vegetable Oil Processing. http://dx.doi.org/10.1016/B978-1-63067-050-4.00015-5
Copyright © 2017 AOCS Press. Published by Elsevier Inc. All rights reserved.

15.2 DEFINITION OF LOSSES

There are different ways to designate losses:

1. Some prefer to call this *L&D*, or *losses and degrading*
2. Others prefer to call it *D&V*, or *degrading and variations*
3. Some prefer to refer to this simply as *losses*.

One could use any of the aforementioned titles to define plant losses. The idea is to understand the concept and be able to apply it for obtaining results.

15.2.1 Degrading and Variations

In this chapter the Degrading and Variations (D&V) will be used.

15.2.1.1 Degrading

The value of the oil increases as it passes from the crude stage to the finished product. There are elements of cost involved in processing the oil at each step of the process. Packaging the product is the most expensive of all because it includes the cost of packaging in addition to that for processing.

Conversely, the value of a product is lowered as it is returned to its previous processing step for reprocessing. This is referred to as degrading the product. A product is considered degraded when it cannot be used in its present condition or quality for sale, packing, or taking it to the next step in the process. For example:

1. The product from the warehouse is returned to the packaging department because of unacceptable quality, damages in handling, etc.
2. Deodorized oil is returned by packing department to the deodorizing department for reprocessing.
3. Finished product is rejected by packing for darker color and returned to bleaching for color removal.
4. Refined and bleached oil (RB) is returned to the refinery for rerefining.

One must appreciate the fact that there is financial loss in each of the previous listed cases.

15.2.1.2 Variations

Variations in product accounting occur when the products are not properly accounted for or they are physically lost. The accounting errors or deviations can result in a gain or in a loss of material that cannot be explained. The variations can also result from true losses, such as:

1. overpacking of products
2. loss of oil through the sewer because of a spill
3. pilferage of products

This can also occur for some reason that is not quite obvious, such as:

- Incorrect record keeping on high-price ingredients added to the product, such as emulsifiers, antioxidants, coloring, defoaming agents, etc.

It is quite clear that all of the previous factors can be costly and reduce company's profit.

Variation in any raw material, finished product, packaging material, or in-process oil can be determined as follows:

Variation (V) = Beginning inventory (BI) + Receipt (R) + Production (P) − Use (U) − Ending inventory (EI)

$$V = BI + R + P - U - EI$$

A large gain or loss in the monthly variation number indicates poor accounting procedure.

Ups and downs or a steady high monthly variation cost in process degrading indicates a serious issue in the production department. This affects the unit cost for the product, whether it is real or it is a paper variation.

15.3 FACTORS CONTRIBUTING TO HIGH PLANT LOSSES IN DEGRADING AND VARIATIONS

High degrading and variations at the plant can be attributed to the lack of management attention and lack of desire to be involved in the accounting procedures. The areas of major degrading and variations should be identified and procedures implemented to curb and control this. The various contributing factors leading to high degrading and variations are listed as follows:

1. accounting deficiency
2. improper receiving and storage procedure for the packaging or raw material
3. improper operating practices resulting in high rate of reprocessing or losses
4. equipment limitation
5. insufficient knowledge of cost elements and their impact on the cost degrading and variations
6. other factors

1. Accounting deficiency

   This does not mean that the deficiency lies with the accounting department only. This also may refer to the process department's inability or lack of attention in accounting for all ingredients, packaging material, process material, packaged weights of the finished products, etc. The following are some examples in this area:

a. Poor inventory of stocks in the tanks due to faulty gauges or improper taping procedure in the cases of manual inventory checks.
b. Improper temperature measurement for the oil. This can create variations because the density of the oil depends on temperature. This can result in incorrect inventory measurement, whether it is done manually or with the help of instruments.
c. Receiving reports of materials are not properly filled out for shortage or damages (OS&D report, which stands for overage, shortage and damages).
d. Claims are not filed against the vendors for shortage or damages on time.
e. Receiving reports are submitted late.
f. Transfers of in-process oils are not properly recorded on the transfer forms.
g. Mix formulation sheets are not properly filled out.
h. Spills and disposition reports are not filled out or incompletely filled out.
i. Misidentification of stocks.
j. The plant supervisors, the operators, and the cost accounting personnel need to understand proper method for accounting for the materials.
k. The supervisors are not taking any initiative in the loss management program.

2. Improper receiving and storage procedure
   The variations can be very high at the receiving department. In order to bring this under control the plant needs to do the following:
   a. Inspect all receipts for overage, shortage, and damages (OS&D).
   b. The plant must have a form for OS&D.
   c. The operator must fill out the OS&D report if there is any incident of shortage, damage, or overage and report it to the supervisor.
   d. The report must be sent promptly to the purchasing department for receiving proper credit from the vendor.
   e. The same applies to crude or degummed oil received at the plant.
   f. Oils might be left in the truck after the unloading pump stops pumping and unloading process is presumed to be complete. The interior of the truck must be inspected at the end of the unloading process. There must be no oil left in the truck.
   g. After receiving, the process materials must be stored with proper protection from rain or other damages from vermin. These materials can become unusable if they get damaged by rain or otherwise.
   h. The process additives, ingredients, and catalysts must be stored in a cool dry place. These materials can become unusable if they are not handled properly.
3. Improper operating practices
   Improper processing can result in serious losses that may go unnoticed. The following are some selected examples of variations due to improper process conditions.

a. Using higher caustic dosage or more concentrated caustic solution can significantly increase refining loss.
b. Higher than required back pressure on the primary separator can result in high neutral oil loss in the heavy soap phase.
c. Incomplete bleaching causing process difficulties in hydrogenation, winterization, and deodorization may force the oil to be rebleached.
d. This can slow down the hydrogenation reaction, plug the filter, and shorten filter cycle time in both hydrogenation and winterization.
e. Storing deodorized oil at high temperature or without nitrogen protection may cause the flavor or the color of the oil to become unacceptable, requiring reprocessing.
f. Deodorizing under improper conditions, such as high temperature and/or poor vacuum may produce unacceptable product that requires reprocessing.
g. Lack of weight control checks at the packaging line may lead to large losses due to overfilling of the products or legal issues with the regulatory agencies for underfilling or overfilling.

4. Equipment limitation

Inadequate equipment design or plant facility can cause high variations. For example:

a. An inadequately designed unloading facility can leave large amounts of oil in the truck or rail car, resulting in high variation or an outright loss.
b. Unreliable tank gauges can produce inaccurate inventory.
c. Poorly calibrated or unreliable flow recorders may produce high variations by reporting overfilling or underfilling of tanks, trucks, or rail cars.
d. The heat exchangers on the deodorizer may hold a large volume of oil at shutdown. This oil can cause high variations if the units are not properly drained before their cleaning and the oil is recovered.

Examples of inadequate equipment design

Example 1: In many instances, engineers design a piece of equipment using the average flow rate to size the unit. At this particular case the plant had a batch reactor that was used to make propylene glycol monoester by reacting fully hydrogenated fat and propylene glycol. The reaction was carried out under pressure in order to prevent any loss of propylene glycol at the reaction temperature of 375–425°F (190.6–218.3°C). The reaction mixture was partially cooled to <300°F (148.9°C) to neutralize the catalyst. The reaction mixture was then further cooled down to 260–280°F (126.7–137.8°C), and the pressure was released and the vacuum was applied with an absolute operating pressure of 4 mm of mercury. There was a sudden surge of propylene glycol vapor rushing into the distillate condenser. The cooling capacity was designed assuming that the propylene glycol vapor would be released at a steady rate matching the total distillation time, which was normally 90 min. Flash evaporation for the propylene glycol was overlooked. This overloaded the vapor condenser by a factor of 10. Thus, a large portion of the uncondensed vapor of propylene glycol went out through the vacuum ejector.

In addition, the collection tank for receiving condensed propylene glycol had insufficient capacity. Therefore, some of the recovered propylene glycol had to be drained for lack of storage capacity. This design deficiency cost the company nearly $20,000 per month due to the loss of propylene glycol.

Example 2: At this plant, coconut oil was brought in. Crude coconut oil can contain 5% or higher FFA. The distillate condenser was originally designed for seed oils, which contained much lower FFA 0.6–2%. As a result, most of the coconut oil distillate went to the greasy water tower. The temperature in this region could reach −40°F (−40°C) with the wind chill factor. This caused instant solidification of the coconut oil distillate. The plant had to be shut down to clean the cooling tower. This essentially shut down the operation because the greasy water became overloaded with the coconut oil distillate.

Example 3: A high capacity self-cleaning centrifuge was installed where the refinery required product that the centrifuge could deliver in one shift per day. Instead of running the centrifuge for five consecutive shifts the plant ran it only one shift of per day. Every day the centrifuge was started, operated for 8 h and shut down. The refining loss was astronomical.

The cause for this catastrophic failure was attributable to the fact that top management selected this particular centrifuge without consulting the project engineer who would not have chosen this machine for the required service.

When I was called in for a consultation for high-refining loss, I was told by the top management that this was one of the best centrifuges that money could buy. I had to tell the top-level manager that yes, this machine was one of the best that money could buy. The statement was correct on absolute basis but the centrifuge was not the right one for this production volume. Frequent start up and shut down of such a unit caused high refining loss.

5. Insufficient knowledge of cost elements

   It is necessary that every process manager and supervisor understand how the product manufacturing cost is determined. They must know all the components behind product costing and the impact of each individual component on the total value of the product (VOP). It is also essential that every manager be familiar with the process and formulation detail so he/she can make the right decision and choose the best alternative when certain finished product or in-process oil has to be degraded.

   The basic elements of cost are:

   a. raw material—cost at receipt (purchase price + freight)
   b. cost of unloading and storage (labor, electricity, plant air, etc.)
   c. processing cost (direct labor, direct supervision, raw material, utilities, such as steam, electricity, water, process air, etc.)

d. packaging cost (packaging material, labor, utilities, etc.)
e. warehousing and shipping
f. general plant overhead (GPO) and other overhead costs, such as sales and distribution costs
g. degrading and variations
h. employee benefits—bonus, profit sharing, medical insurance, etc.

The aforementioned cost elements and their impacts on the unit cost for the product should be clearly understood by the plant supervision. A thorough understanding of the subject can help the plant supervision exercise better judgment to control the overall cost of products.

The following are a few examples to illustrate how the manufacturing cost can be affected.

Example 1: Unfavorable purchasing contract

a. In this particular example crude oils were received daily by the company in trucks from the oil supplier.
b. The weight of the shipping load was determined by the shipper's scale.
c. Receiving was done using the company's own scale.
d. Any difference between the two was split between the supplier and the company.

The previously mentioned system was flawed because the company was paying for half of the oil that it did not receive due to underweight trucks. This, in reality, increased the price of the oil. The following calculation will make it clear:

Invoiced weight from the oil supplier	48,000 lb
Receiving weight at the company's plant	47,500 lb

Therefore, the oil was either underloaded or not completely unloaded by the supplier by an amount of 500 lbs or the receiving plant did not fully empty the truck at unloading.

The receiving report by the receiving plant showed and paid the supplier for:

48,000–½(500) =	47,750 lb
The plant actually received	47,500 lb
At $0.30/pound of oil, the loss per truck was	$0.30 × 250 = $75.00
The plant received 5 trucks/day for 250 days/year	
At this average loss of $75.00/truck, the average estimated loss was:	
5 × 250 × $75.00 = $93,750.00	

Example 2: Improper truck unloading

A truck arrived with 50° Be caustic solution. The invoice showed the following information:

Gross weight	74,000 lb
Tare (Empty) weight	28,000 lb
Difference	46,000 lb

- During the course of unloading the hose developed a leak. Pumping was stopped.

- The material was contained inside the spill protection area and was properly disposed.
- The hose was fixed and the unloading process was resumed.
- The final receipt was recorded as 45,000 lb.
- The loss was 1000 lb

This increased the cost of the 50% caustic receipt by:

$$\left(\frac{46,000-45,000}{46,000}\right)=2.17\%$$

Example 3: Incomplete unloading of oil from trucks

A centrifugal pump was being used to unload the incoming trucks. Owing to the inherent nature of centrifugal pumps or improper selection of right NPSH, the pump would leave oil in the truck. The amount would range from 300 to 800 lb of oil left in the truck per load.

The average loss at the plant	500 lb × \$0.30 per pound = \$150 per truck
Average truck load	48,000 lb
Increased cost of oil	(500/48000) × 100 = 1.04% (or \$0.312 per pound)

Replacing the centrifugal pump with a Viking pump solved the issue.

6. Other factors

   Examples in this category either occur infrequently or they go unnoticed by the plant personnel. The common examples are:

   a. Leaky agitator or pump seal can cause oil loss.
   b. Overfilling a tank can run the oil to the ground.
   c. Incompletely or improperly completed spill reports do not allow proper accounting for the loss.
   d. Mis pumping of oil can cause degrading of the contaminated stock or the contaminant, depending on the value of the components involved in MIS pumping.

15.4 ELEMENTS OF GOOD LOSS MANAGEMENT

The production management needs certain training, some resources within the company, and most of all the desire to control losses. The various essential elements that the plant supervision must know about are listed as follows:

1. the operation
2. the design and proper operating conditions for all of the equipment, their capacities, and process standards
3. the accounting points for all products, starting from the crude oil to finished product, as shipped
4. the value of each product as it flows through the various steps of processing in the plant

5. the key loss points, the cause/causes for losses, and the methods for their prevention

It is imperative that the plant supervision be knowledgeable in the earlier listed areas. The manager must be aware of the process standards, product quality, equipment capability, and the economic consequences of producing off-quality (nonshippable) products.

The accounting points can be in meters when oil is transferred from one department to another. However, the accounting points do not always involve meters. In most cases they indicate the different stages of process for the products.

The manager needs to be aware of the key loss points in the department, the cause/causes for the losses and the corrective action needed for the prevention or control of losses.

Additionally, the plant supervision needs the following:

1. An internal resource on D&V.
2. A loss prevention program and total employee involvement in the program.
3. A sound inventory procedure.
4. Accurate and timely reporting of inventory by the person/persons assigned to the task.

The production supervisor must work closely with the D&V resource to understand and reduce losses. The close working relationship leads to a successful loss prevention program.

Additionally, the area of D&V requires total involvement of the managers and the employees. The results of the loss prevention effort can be seen and measured readily. This is a nonthreatening task for everyone at the plant. The savings achieved can be observed and enjoyed by all as they reduce the D&V cost at the plant. In many cases, this can be 4–6% of the operating budget of a plant, which can be quite significant in terms of company profitability.

15.5 GUIDELINES FOR MANAGING D&V

So far the discussions have been around the following topics:

1. definition of D&V
2. cause/causes for high D&V
3. the tools required to control D&V

This section will be devoted to the discussion about certain guidelines that are useful for the proper management of D&V. The principle behind D&V management is no different from any other cost control procedure used at the plant. It requires the manager to proactively pursue the guidelines that are presented in this section. It is possible for any plant to formulate its own plan for D&V control. However, the seven-step approach illustrated in this section can be used as a template for managing D&V at any plant. These are:

Step 1: Identify all material flows at the plant.
Step 2: Identify key loss points.
Step 3: Determine the causes for the losses at each location.
Step 4: Define solutions to prevent losses.
Step 5: Define goals.
Step 6: Set priorities for the improvement activity.
Step 7: Define action steps, target dates, milestones, the success criteria, and the method used for measuring progress.

15.5.1 Step 1: Identify all Material Flows at the Plant

Determine the volumes and the values of all materials that flow through each step of the process. This allows the manager to be aware of the plant's production activity, material usage, and the amount of finished products made.

15.5.2 Step 2: Identify Key Loss Points

These are the areas in the operation where most significant losses can or do occur. Fig. 15.1 shows the loss points in a typical oil processing plant. There are dotted lines showing the demarcation for the accounting points and they are designated with different letters, for example,

A. Oil refinery
B. Hydrogenation
C. Formulation

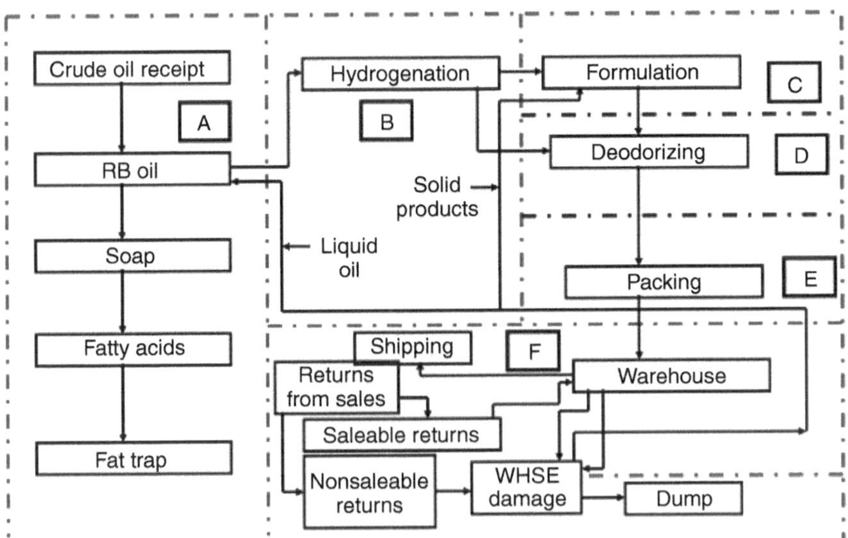

FIGURE 15.1 Accounting points at an oil processing plant.

D. Deodorizing
E. Packaging
F. Warehousing and shipping

A. Oil refinery

1. *Activity*
   - Crude oil is received, unloaded, and stored.
   - Crude oil is refined and bleached to produce RB oil using bleaching clay and possibly activated carbon.
   - By-products, such as soap and fatty acid from acidulation, are produced.
   - Solid waste (spent bleaching clay) containing approximately 33% oil is disposed.
2. *Accounting points*

   Crude oil receipt

   The various cost elements in this area are:
   - price of crude oil
   - freight
   - unloading cost
   - cost of underreceipt due to shortage or incomplete unloading of the truck or rail car
   - analytical (quality control, QC)

   RB oil made

   The various cost elements are:
   - crude oil received
   - refining cost (includes caustic, citric or phosphoric acid, labor, water, plant air steam, and power)
   - refining loss (includes moisture and impurities, FFA, phospholipids, neutral oil loss in the soap, and water in the water wash)
   - bleaching cost (includes bleaching clay, filter aid, activated carbon, citric or phosphoric acid, labor, steam, power, and nitrogen)
   - oil loss in the bleaching clay
   - cost of acidulation (sulfuric acid, caustic, steam, water, and power)
   - credit for the fatty acids
   - loss of fatty acid in acidulation (going into the process sewer)
   - cost of disposal of the spent clay
   - variations in the refining loss
   - analytical (QC)

   Credits (if any)

   The potential credit items are:
   - fatty acids, if there is an outlet
   - soap as a feed-stock for animal feed, if there is an outlet

B. Hydrogenation

The cost elements in hydrogenation are:

- RB oil

- catalyst
- filter aid
- hydrogen gas
- utilities (power, water, steam, plant air, instrument air, nitrogen, etc.)
- analytical (QC)
- gain due to the reacted hydrogen in the oils
- variations

C. Formulation

Hydrogenated oils are blended together to formulate various product bases. The value of the mix is determined by the following:

- cost of hydrogenated oil
- labor for mixing
- utilities (power, steam, plant air, nitrogen)
- analytical (QC)
- variations

D. Deodorizing

The product blends are deodorized under vacuum and at elevated temperature. The cost elements in the deodorizing process are listed as follows:

- mixed oils
- labor
- utility (steam, power, water, process air, nitrogen)
- loss of FFA and some neutral oil along with other volatile matter
- analytical (QC)
- variations

E. Packing

Deodorized oil is delivered to packaging, where the product is packed. Production records are kept in the form of number of cases packed for each brand in a day's production.

The total production is computed in the form of case count and is reconciled with the physical count of cases received at the warehouse. Any discrepancy between the production and the warehouse count creates variations in the deodorized oil and the packaging material. The cost elements related to the packed product are:

- deodorized oil
- labor for packing
- packing material (includes any variation for the packing material)
- utilities (nitrogen, Freon or brine, water, plant air, and power)
- labor and utilities for handling off-quality product
- analytical (QC)
- variations

F. Warehousing and shipping

Packed products are stored in the warehouse until they are shipped. Warehousing and shipping adds some additional cost to the product. The value of the product as shipped includes the following cost elements:

- packaged product cost
- warehousing and shipping expense (labor, power, steam, water, direct overheads, etc.)
- GPO
- transportation cost
- sales and distribution cost
- analytical (QC)
- variations

The GPO and the sales and distribution costs are generally fixed and are distributed over the units of production in a month or a year. Therefore the plant must aim to produce and ship more products to reduce the overall cost of the overheads.

The total D&V cost is added to the product at this stage to determine the profitability for the plant. Not all D&V expenses are incurred at the warehouse. However, the D&V expenses incurred at the warehouse can result from all of the following sources:

- finished product variations
- variations in the reclaiming accounts
- rejected finished product from degrading factors that include:
    - warehouse damage
    - degradation due to quality reasons
    - returns from sales
    - products dumped for quality or sanitation reasons

15.5.2.1 Bulk Receipts of Refined Oils

Some plants receive refined oils in bulk and package them. The loss points in this case are:

- shortage in the receipt
- incomplete unloading of the trucks or rail cars
- losses in handling (including spills)
- overfilling in the package

15.5.2.2 Finished Product Variations

This includes all finished products, both regular and interplant receipts. The variation is computed as follows:

Finished product = (Beginning inventory + Production + Return from sales + Interplant receipts + Reclaiming) − (Month-end inventory +Shipment)

Any variation in the aforementioned accounting can be created by one or all of the following factors:

- Inaccurate production report by packaging.
- Inaccurate finished product count at the warehouse.
- Interplant receipt not accurately recorded or not recorded at all (receiving report not filled out) by the warehouse.

- Return from sales not accurately recorded by the warehouse.
- Shipment papers not completed by the warehouse.
- Warehouse damages not accounted for.

15.5.3 Return from Sales

Products are returned from sales for various reasons, such as:

- The product did not meet customer specification and was refused.
- The product got damaged in transit.
- The amount of product shipped was more than the customer had ordered.
- The product returned are separated into the following categories:
  - saleable returns
  - nonsaleable returns

The saleable product is returned to the warehouse and included in the warehouse inventory.

The nonsaleable returns are divided into two categories, namely:

1. The product that can be reprocessed, which is sent back to refining, formulation, or deodorizing.
2. The product that cannot be reprocessed because of excessive damage and contamination with dirt and other nonedible materials.

15.5.4 Dump

The product from category #2 aforementioned and similar ones from the warehouse are physically destroyed.

Fig. 15.2 shows the various accounting points that can be set up at the warehouse for proper control and accounting of the product movement in the area.

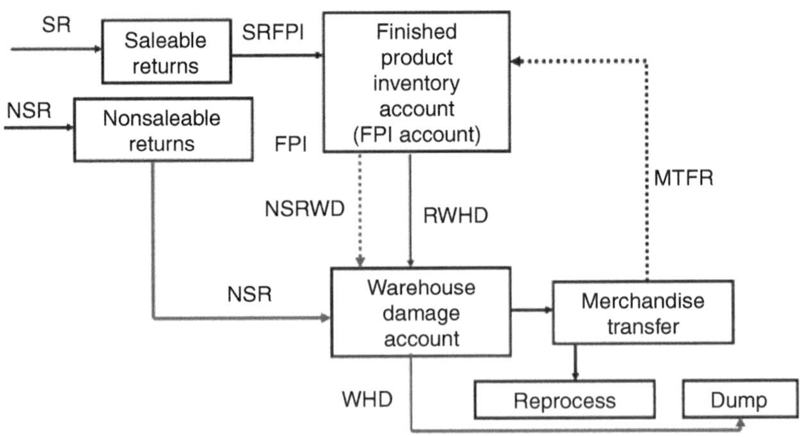

FIGURE 15.2 Warehouse accounts.

- The main accounting point here is the finished product inventory account (FPI Account).
- Saleable returns are accepted under the account SR.
- The products go to the warehouse inventory (FPI) using the transfer account SRFPI.
- If any of these are found to be damaged or unfit to go to the regular inventory, they are transferred to the Warehouse Damage Account, using the transfer form NSRWD.
- The nonsaleable products go directly to the damaged product storage using the transfer account NSRFPI. This is shown by the solid line.
- The nonsaleable return goes to the warehouse damage account using the transfer account NSRWD.
- Damages created at the warehouse are transferred to the warehouse damage account using the transfer account RWHD.
- Reclaimable products are sent to process via the merchandise transfer account (MTFR).
- The MTFR is shown as going to the FPI account via a dotted line because that is where all warehouse accounts are maintained.
- Nonusable products are destroyed or sent for nonfood applications or clients using transfer account WHD.

15.5.5 Step 3: Determine the Causes for the Losses at Each Location

Through careful observations, determine the underlying cause(s) for these losses. This will allow the manager to solve the loss issue. .

15.5.6 Step 4: Define Solutions to Prevent Losses

Define or determine the action steps that are required to eliminate the causes for the earlier losses. This plan must include the following:

- What is the nature of the loss?
- What are the action steps?
- Who is assigned or accountable for taking the steps?
- When will the action steps be taken?
- How involved is the job?

15.5.7 Step 5: Define Goals

Define D&V goals based on the estimated savings to be achieved and financial and human resources required to accomplish the desired goals.

15.5.8 Step 6: Set Priorities for the Improvement Activity

From the key loss point analysis, establish the priority based on:

- Where is the largest source of loss?

- What is its economic value?
- Which of the losses can be corrected with a high probability of success?
- Does the action step require any outside resources?

15.5.9 Step 7: Define Action Steps, Target Dates, Milestones, the Success Criteria, and the Method Used for Measuring Progress

- Form a D&V team in each area of the operation. This must include operators, clerical staff, quality assurance personnel, and a member from the management.
- There must be one person assigned as the leader for each group.
- Define the roles of each individual on the team and assign accountability.
- Prepare the list and the state of each identified area to be tackled.
- Prepare a checklist, goals, and follow-up plans for each project.
- Meet at a scheduled frequency to review each project to discuss the following:
  - Progress made during the period.
  - List of the follow-up items.
  - Identify any new finding that might require a change or modification of plans.
  - Develop the revised plans.
  - Proceed and follow up on progress.
- At the end of the project, establish the following:
  - The degree of overall success.
  - Any need for equipment repair or modification.
  - Any need for personnel training.
  - When and how to complete the equipment and personnel training.
  - Establish the new target and action steps on every project.

15.6 MANAGING PLANT LOSSES

There are two types of losses encountered at any plant. They are:

1. known losses
2. unknown losses

15.6.1 Known Losses

These losses are well identified in the plant operation and accounting. These are or can be:

1. completely defined
2. reasonably quantified
3. identified sources
4. reduced, controlled, or eliminated

Examples of some of the known losses are listed as follows:

- Standard losses of packaging materials for normal start-up and shutdown.
- Product sent to reprocessing due to:
  - start-up,
  - shutdown,
  - packaging line issues,
  - off-quality product,
  - oil quality going bad because of long storage time, and
  - product flavor is unacceptable in the fresh oil.

15.6.2 Unknown Losses

This form of losses is most damaging to the plant because:

1. The cause/causes for the losses are not clearly known.
2. The amount of loss can be quantified through the accounting process but may not be accurate.

Examples of some of the unknown losses are:

1. Daily inventory does not match up with the usage, whether it is the oil, process material, or packaging material.
2. Oil receipt does not match up with the tank inventory.
3. Warehouse finished product inventory does not match up with the production, return from sales, and warehouse damages.
4. Overpackaging of the products is not properly recorded.
5. Spills and other product losses are not recorded.

15.6.3 Key for Successful Loss Management

This is a very simple approach. Turn unknown losses to known losses. In order to accomplish this, one needs to conduct a loss point analysis of the process as discussed earlier. The step-by-step approach for loss point analysis is outlined as follows:

1. Prepare a complete mapping (drawing) of the process showing every step of the process and the equipment involved.
2. Identify every process location and piece of equipment by number (1, 2, 3, etc.), where the raw material or the product passes through.
3. List the probable cause/causes for disappearance or loss of material or product at each location.
4. Physically investigate and gather information on the actual loss of material or product at any of these locations.
5. Use Loss Point Analysis Worksheet as shown in Table 15.1 to record the findings.

TABLE 15.1 Loss Point Analysis Worksheet

Location No.	Source/ Equipment	Probable Cause/ Causes	Potential Solution/ Solutions
1			
2			
3			
4			
5			

6. Once the loss points are identified, it becomes easier to develop a potential solution to reduce the losses.

Once the loss points are identified and the losses (if any) are quantified, the plant supervision should use the seven-step approach discussed in Section 15.5, set goals priorities, define action steps, and diligently pursue measures to reduce losses.

15.7 FINAL COMMENTS ON LOSS MANAGEMENT

Loss management at an oil processing plant can be involved, boring, lacking high-tech challenge, glamour, etc. However, the savings achieved through a successful loss management program at the plant are real and can be financially rewarding. In addition, the solutions for the prevention of some of the losses can use highly technical tools, involving engineering design and high-tech automation.

The loss management manager should be carefully selected. The following are some of the suggested criteria for the selection of an individual for the assignment:

- The person must be intimately familiar with the process. This person cannot be hired from outside with no prior experience with oil processing.
- The person must have good command over product costing and accounting.
- The person must have good processing and engineering skills to initiate and work with the plant management and engineering to develop and implement solutions for loss prevention.

15.8 SAMPLES OF FORMS HELPFUL FOR TRACKING VARIATIONS

A few sample forms are listed later that might be used by the plant to track plant variations:

1. Table 15.2, overage, shortage, and damaged OS&D report
2. Table 15.3, spill report

TABLE 15.2 Overage, Shortage and Damage (OS&D) Report

Purchase Order No. Date: (Quantity Ordered)	P.O. Issue Date	Receiving Date	Shipping Document No. (Supplier)	Quantity Requested in the P.O.	Quantity Received in Good Condition	Quantity Received in Damaged Condition	Shortage	Remarks/ Signature

.... Signed, Supervisor

.... Dept. Manager

.... Loss Control Manager

TABLE 15.3 Spill Report

Date	Location	Type of Product/ Material	Quantity	Disposition Re process (Dump)	Estimated Cost	Recommended Corrective Action

.... Signed, Supervisor

.... Dept. Manager

.... Loss Control Manager

TABLE 15.4 Degrading Report

Date	Incident	Product (Amount)	Nature of the Incident	Product Degraded to	Estimated Cost	Recommended Corrective Action

.... Signed, Supervisor

.... Dept. Manager

.... Loss Control Manager

3. Table 15.4, degrading report
4. Table 15.5, mispumping report

The OS&D report has been discussed earlier.

The spill report should be filled out whenever there is oil loss due to any of the following reasons:

- tank leak
- broken pump seal
- tank runover
- dropping and destroying process material during handling

TABLE 15.5 MIS Pumping

Date	Incident	Product #1	Product #2	Disposition	Estimated Cost	Recommended Corrective Action

.... Signed, Supervisor

.... Dept. Manager

.... Loss Control Manager

The degrading report should be prepared when:

- The oil from any process area has to be reprocessed within or outside the same process area.
- The oil has to be returned to one step or more prior to the existing stage where the degrading has occurred. The examples are:
  - Packaged product has to be removed from its packaging and sent back to process.
  - Finished product is returned to process.

The mispumping report is used when a freshly formulated product has been made with incorrect oil that cannot be used in the original brand, and has to be blended or reformulated. Also, two incompatible products may be commingled such that the mixture cannot be used as either product without further reformulation.

All of the aforementioned forms are signed by the supervisor on duty, the department manager, and the loss control manager.

Using the same principle as outlined in the tables listed earlier, the readers can develop their own D&V forms to fit their own organizational need. In addition, with the help of automatic data collection, it can be quite convenient to monitor daily inventory movement. The only complication is that the measurement techniques, such as flow meters, tank gauges, gas meters, etc., need to be accurately calibrated and routinely maintained. The packaging material count and process material count may still require manual inventory checks.

Chapter 16

Plant Safety Procedures

16.1 INTRODUCTION

Oil refineries, like all other manufacturing operations, must make personnel and plant safety the primary focus of the organization. The cost of production is reduced when the plant operates safely without any incident or injury.

There are five golden rules that any manufacturing operation should follow:

1. safety
2. sanitation
3. quality
4. production
5. cost

A plant is built to manufacture products at the highest volume based on the design of the equipment, and the product must meet the high-quality standards specified by the company. However, at every step of production one must adhere to the safety rules.

The basic concepts are as follows:

1. Personnel and process safety should be of the highest priority in an oil refinery, as in any other manufacturing operation. A safe operating plant induces the workers to follow the safety procedures on the job as a normal habit. Cutting corners on safety practices should not be allowed because this can result in on-the-job accidents. Personal injury or loss of life is extremely hard on the morale of the personnel, in addition to the financial loss to the individuals and their families and also to the corporation.
2. The plant must be sanitary, and the floors and the equipment should be clean. Clean all oil spills immediately to keep the floor dry and this will help prevent injury to the plant personnel who may fall on the slippery floor.
3. High-quality standards set by the company are instrumental in keeping the plant safe and sanitary.
4. Productivity at the plant improves when the aforementioned three items are strictly enforced.
5. Cost becomes easy to control when the plant is accident free, sanitary, and maintains high productivity of good quality products.

Practical Guide to Vegetable Oil Processing. http://dx.doi.org/10.1016/B978-1-63067-050-4.00016-7
Copyright © 2017 AOCS Press. Published by Elsevier Inc. All rights reserved.

16.2 PLANT SAFETY

16.2.1 General

Plant safety means protecting plant personnel from injury and protecting the plant equipment from damages. This involves the following actions:

1. Injuries from falls or injury on the job from improperly designed or improperly maintained equipment, as well as from not following proper procedure.
2. Protection of the equipment to prevent excessive pressure or temperature, fire, and explosion.
3. Institution of an ongoing safety procedure to educate new employees, and offer refresher training to existing employees; provision of new information on personnel and equipment safety. Attendance to these meetings should be mandatory.
4. A weekly review of area safety involving personnel, sanitation, and equipment should be carried out. Meetings should include the following:
   a. A list of safety-related issues for the area.
   b. Assignment of accountability to both managers and employees to follow up on the safety-related items identified at this meeting.
   c. At the following meeting, the person assigned to work on an item must present a quick summary of his or her progress.
   d. If an item has not been completed, the person should explain the reason and offer the necessary action steps required to complete the job.

This type of meeting works best when the area manager and an employee are assigned to be the joint leaders for the safety meeting and follow-up. The meeting should be short (15 min), and there should be no finger pointing at an individual for failing to complete the task.

16.3 SAFETY AGENCIES

Several government and private agencies offer information on employee safety (e.g., NIOSH, OSHA, etc.). These agencies address all occupational hazards, including those due to prolonged exposure to any chemical and its potential long-term impact on employees' health.

16.3.1 Occupational Safety and Health Administration

The US Occupational Safety and Health Administration (OSHA) is a federal government agency in the US Department of Labor. The primary goals of OSHA are to save lives, prevent injuries, and protect the health of America's workers. OSHA employs over 2000 inspectors to ensure job site safety. OSHA's website is www.osha.gov.

OSHA was created by the Occupational Safety and Health Act (OHSA) of 1970. Many see OSHA as an intrusive government agency intent on enforcing arcane rules; the fact is that OSHA helps the industry to save lives.

OSHA's Voluntary Protection Plan (VPP) saves the industry money in the long run by reducing the cost of injuries, accidents, downtime, and litigations. OSHA's Hazard Communication Standard 1910.1200 provides guidelines to employers on how to establish hazard communication programs for their employees by means of the following materials and procedures:

- Labels on containers.
- Material safety data sheets (MSDS)
- Training programs.
- Implementation of these hazard communication programs will ensure that all employees have the "right to know" the hazards and identities of the chemicals they work with, and will reduce the incidence of chemical-related occupational illnesses and injuries.

16.3.2 American National Standards Institute

The American National Standards Institute (ANSI) is a private, nonprofit membership organization representing over 1000 public and private institutions, businesses, and government agencies. ANSI is engaged in seeking and developing technical, political, and policy consensus among various groups. ANSI's website is http://www.ansi.org/.

ANSI-approved standards are voluntary. However, it is possible that some of the content of the ANSI standards could be made into law by a governmental body.

ANSI is the official US representative to the International Organization for Standardization (ISO).

16.3.3 National Institute for Occupational Safety and Health

The National Institute for Occupational Safety and Health (NIOSH) is part of the US government's Centers for Disease Control and Prevention (CDC). NIOSH's website is http://www.cdc.gov/niosh/homepage.html.

NIOSH is the only federal institute responsible for conducting research and making recommendations for the prevention of work-related illnesses and injuries.

16.3.4 The National Fire Protection Association

The National Fire Protection Association (NFPA) is a private nonprofit organization. It is the leading authoritative source of technical background, data, and consumer advice on fire protection, problems, and prevention. NFPA's website is http://www.nfpa.org/.

The primary goal of NFPA is to reduce the burden of fire and other hazards on the quality of life worldwide by providing and advocating science-based consensus codes and standards, research, training, and education.

16.3.5 Workplace Hazardous Materials Information System

The Workplace Hazardous Materials Information System (WHMIS) is the product of Canadian legislation covering the use of hazardous materials in the workplace. Its responsibilities include assessment, signage, labeling, MSDS, and worker training. WHMIS closely parallels the US OSHA Hazard Communication Standard.

WHMIS is different from the HMIS, which is a hazard labeling system of the National Paint and Coatings Association.

WHMIS is the responsibility of plant management or the corporation management to maintain a safety organization in the corporate office. This group must be responsible for keeping abreast of all safety regulations and their implementation. In addition, there should be a safety manager at the plant, who is responsible for training the employees at the plant on the safety procedures.

16.4 AREAS OF SAFETY TRAINING REQUIRED AT THE PLANT

The primary areas of safety training for plant personnel are:

1. fire and explosion safety
2. compressed-gas safety
3. chemical safety
4. electrical safety
5. confined-space entry procedures

16.4.1 Fire and Explosion Safety

Fire can start with ignitable material under suitable conditions. The material can be wood, paper, combustible chemicals, or gas. It will not be possible to discuss every detail on this subject in a chapter. Therefore, the important points most pertinent to an oil-processing plant will be highlighted.

16.4.1.1 Types of Fires Encountered

The NFPA has classified fire into several types.

16.4.1.1.1 Class A Fire

This originates from ordinary materials, such as burning paper, lumber, cardboard, or plastics.

16.4.1.1.2 Class B Fire

This involves flammable or combustible liquids, such as gasoline, kerosene, and common organic solvents used in the laboratory.

16.4.1.1.3 Class C Fire

This involves energized electrical equipment, such as electrical switches, panel boxes, power outlets, power tools, hot plates, etc.

16.4.1.1.4 Class D Fire

This type of fire involves combustible metals, such as magnesium, titanium, potassium, and sodium, and pyrophoric organometallic reagents; such as alkyl lithium, diethyl zinc, etc. These materials, when ignited, produce intense heat and react violently with water, air, or other chemicals.

16.4.1.1.5 Class K Fire

This is a kitchen fire. This class was added to the NFPA Portable Extinguishers Standard 10 in 1998. Kitchen extinguishers installed before June 30, 1998, are "grandfathered" into the standard.

16.4.2 Selection of Fire Extinguishers

Some small fires can be put out at the plant using handheld fire extinguishers. There are different types of fire extinguishers designed for putting out fires that originate from different combustible materials.

The NFPA offers instructions for making the proper choice of fire extinguisher. The NFPA classifications for types of fire extinguishers are briefly discussed in the following paragraphs.

Water extinguishers or air-pressurized water (APW) extinguishers are suitable for *Class A fires only*. They should never be used on Class B, C, or D fires for the following reasons:

1. In a grease fire, the water can settle below the grease and not be effective. Also, the flame may instantly vaporize the water and cause an enormous spread of the fire. This is a common mistake people make when they encounter a grease fire in the skillet while cooking at home.
2. APW extinguishers can be dangerous with electrical fires because of the risk of electrical shock unless specialized water mist units are used, where there is no continuous water stream between the source of the electrical fire and the person with the extinguisher.
3. Water extinguishers are filled with water and pressurized with oxygen. Therefore, their use can be devastating with a Class D fire.

Carbon dioxide (CO_2) extinguishers are used for *Class B and C fires*. CO_2 extinguishers contain carbon dioxide, a nonflammable gas, and are highly pressurized. The pressure is so great that it is not uncommon for bits of dry ice to shoot out of the nozzle. They do not work very well on Class A fires because they may not be able to displace enough oxygen to put out the fire, causing it to reignite.

CO_2 extinguishers have an advantage over dry chemical extinguishers, as they do not leave a harmful residue, making them a good choice for an electrical fire in a computer or other electronic device, such as a stereo or TVs.

Dry chemical extinguishers come in a variety of types and are suitable for a combination of *Class A, B, and C fires*. These are filled with foam or powder and pressurized with nitrogen.

- *Type BC*: This is the regular type of dry chemical extinguisher. It is filled with sodium bicarbonate or potassium bicarbonate. The BC variety leaves a mildly corrosive residue that must be cleaned immediately to prevent damage to materials.
- *Type ABC*: This is the multipurpose dry chemical extinguisher. The ABC type is filled with monoammonium phosphate, a yellow powder that leaves a sticky residue that may be damaging to electrical equipment, such as a computer.

16.4.3 Hazards of Dry Chemical Extinguishers

Dry chemical extinguishers can be hazardous under certain circumstances:

1. They are corrosive to metals, such as aluminum, and are very abrasive.
2. Dry chemical ABC extinguishers are more corrosive than the type BC extinguishers because the ammonium phosphate can hydrolyze, forming phosphoric acid. This is why the type ABC extinguisher is not used on aircraft or electronic equipment, such as computers.
3. Dry chemical fire extinguishers are used on aircraft or around aluminum only if no other extinguishers are available.

16.4.4 Compressed Gas Safety

Oil-processing plants use compressed air, hydrogen gas, and nitrogen gas. The typical pressure of compressed air is 40–105 psig (3.99–7.63 kg/cm^2). Hydrogen gas pressure can be 100–200 psi (8.37–15.67 kg/cm^2) in the high-pressure storage tanks. The nitrogen line (after the vaporizer) is maintained mostly at 40–80 psig (3.99–6.91 kg/cm^2).

All compressors and high-pressure storage tanks are equipped with pressure relief valves. These must be maintained in good operating condition.

Hydrogen is a combustible gas and can explode in the presence of any source of ignition. Therefore, extreme care must be exercised before any welding is performed in or near the hydrogenation area to prevent an explosion.

16.4.5 Recommended Procedure for the Preparation for Welding or Hot Work (Using Gas Torch for Metal Cutting)

1. Turn off and chain the valve on the hydrogen gas supply line from the storage.
2. The hydrogenation plant, including the compressor and all lines originating from the chained block valve, must be vented and purged with nitrogen or steam for at least 24 h.
3. The hydrogenation room, all equipment, and all piping in the area must be checked for the presence of explosive gas with the help of an explosion meter (from the Mining and Safety Institute or other reliable supplier). The entire area must be free of explosive gas.

4. If the meter shows any presence of explosive gas, continue purging until the entire area and all the equipment are free of any explosive gas.
5. Shut off the steam (or nitrogen).
6. If welding is required in the interior of the hydrogenation reactor, purge the steam or nitrogen from the equipment. The workers inside the vessel cannot survive in a nitrogen environment. Check for explosive gas once more. Once the reactor is found to be free of explosive gas, it must be purged with air until the atmosphere in the reactor shows >19.5% oxygen.
7. The equipment is ready if the explosion meter shows no presence of explosive gas and the oxygen meter shows that the oxygen level is above 19.5%.
8. After the welding inside the reactor, steam out the equipment before a restart. This will prevent the formation of an explosive mixture between hydrogen and air.
9. Nitrogen purging is sufficient only if the work is to be done on the exterior of the reactor.

16.4.6 Chemical Safety

Chemicals used at the oil-processing plant can be:

1. corrosive
2. flammable
3. explosive

16.4.6.1 Corrosive

The most common corrosive chemicals used in oil-processing plants are caustic solution and phosphoric acid. Citric acid is not corrosive but it is irritating to the eyes and skin. The corrosivity of a chemical can vary with different metal or material of construction of plant equipment. A certain degree of familiarity with the basic principles of metal corrosion and resistivity to chemicals involved is necessary for plant supervision. For example:

1. Hydrochloric acid is corrosive to stainless steel except type 316L.
2. Sodium hydroxide is corrosive to aluminum.
3. Sodium hydroxide may contain some sodium chloride as impurity. This can cause pitting on 304 or 316 stainless steel, making it brittle.
4. High free fatty acid content in the oil is corrosive to carbon steel.
5. Dilute sulfuric acid is highly corrosive to stainless steel of any form. This is why Carpenter 20 alloy is used to build the acidulation reactor in soap acidulation process (refer to Chapter 5).
6. Phosphoric acid is an irritant to skin and eyes. It is also corrosive to carbon steel and 304 stainless steel. It is better to use 316L stainless steel.

This list is extensive. It is important that the plant supervisor consult with the appropriate person on the compatibility of various chemicals and the material of construction of the process equipment.

FIGURE 16.1 Chemical classification and hazard ratings by the National Fire Protection Association (NFPA).

Most of the chemicals are used in the QC laboratory. A great majority of them are solvents with low-boiling points.

OSHA defines a flammable liquid as a one with a flash point below 100°F (37.8°C). Flammable liquids are known as Class I liquids.

NFPA suggests the use of color-coded signs to indicate the hazardous property of a chemical and the magnitude of it is harmful effect. Fig. 16.1 shows the NFPA's diamond-shaped chemical classification scheme for a chemical.

16.4.7 Significance of the Color Code and the Numbers for the Chemicals and the Degree of Hazard

All chemicals received at the plant must have this information on their labels. The MSDS should also specify the type and the level of hazard. The storage tanks for the chemicals must also carry these signs. Each color and number has a special significance:

Color	Type of hazard
Blue or "no. 3"	Health hazard
Red or "no. 4"	Flammability hazard
Yellow or "no. 2"	Instability
White or "₩"	Special hazard

Levels of no. 3 or blue: health hazard

4	Very brief exposure could cause death or serious long-term injury even if prompt medical attention is given
3	Brief exposure could cause serious temporary or residual injury even if prompt medical attention is given
2	Intense or continued exposure could cause temporary incapacitation or possible residual injury unless prompt medical attention is given
1	Exposure could cause irritation but only minor residual injury even if no treatment is given

0 Exposure under fire conditions presents no hazard beyond that of ordinary combustible materials

Levels of no. 4 or red: flammability hazard

4 Will rapidly or completely vaporize at normal pressure and temperature, or is readily dispersed in air and will burn readily
3 Liquids and solids that can be ignited under almost all ambient conditions
2 Must be moderately heated or exposed to relatively high temperature before ignition can occur
1 Must be preheated before ignition can occur
0 Materials that will not burn

Levels of no. 2 or yellow: instability

4 Readily capable of detonation or explosive decomposition or reaction at normal temperatures and pressures
3 Capable of detonation or explosive reaction, but requires a strong initiating source or must be heated under confinement before initiation, or reacts explosively with water
2 Normally unstable and will readily undergo violent decomposition but will not detonate; also, may react violently with water or may form potentially explosive mixture with water
1 Normally stable, but can become unstable at elevated temperatures and pressures or may react with water with some release of energy, but not violently
0 Normally stable, even under fire exposure conditions, and not reactive with water

Levels of ~~W~~ or white: special hazard

This section is used to denote special hazards; there are only two NFPA 704–approved symbols:

OX: This denotes an oxidizer, a chemical that can greatly increase the rate of combustion/fire

~~W~~: Unusual reactivity with water; indicates a potential hazard if water is used to fight a fire involving this material

The quality control laboratory has to carry various solvents. The solvents must be stored in special fireproof metal cabinets as shown in Fig. 16.2. The maximum amount of solvent stored in each cabinet must be 60 gallons.

Many oil processors prefer to use the HMIS system. They find the information in this system to be more explicit.

FIGURE 16.2 Safety cabinets for solvent storage.

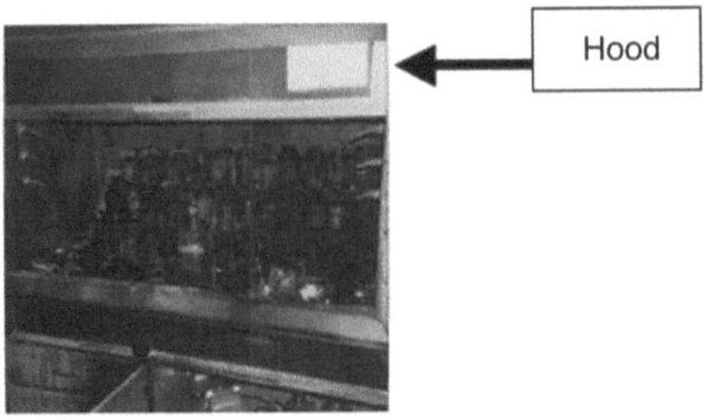

FIGURE 16.3 **Solvents stored under the hood.**

16.4.8 Improper Storage of Solvents

Many QC laboratories store solvents improperly. Fig. 16.3 shows how one laboratory stored solvents under the hood. This is a dangerous practice, and there are many reports of fires starting under the hood and getting out of control.

16.4.9 Electrical Safety

16.4.9.1 Electrical Shock

Most fatal electrical shocks occur to people who should know better. Here are some medical facts regarding electrical shocks.

It is the current and not the voltage that kills. The real measure of a shock's intensity lies in the amount of current (in milliamperes) passing through the body. Currents between 70 and 100 mA (0.07 and 0.1 A) are fatal. Any current in the neighborhood of 10 mA (0.01 A) is capable of producing painful to severe shock.

A list of the electrical safety laws and regulations can be obtained from OSHA. In addition, one must check with state and local agencies for the local codes and also check the national electrical codes. By following these laws and codes one can work safely with the electricity. The plant safety coordinator must have all updated information on file to assist an electrician who might have any question. Employees must report any unsafe conditions, equipment, or work practices as soon as possible.

Electrical safety at the plant requires one to follow a few simple instructions and procedures.

16.4.9.1.1 Fuses

Before removing any fuse from a circuit, one must turn off the power supply switch. An approved fuse puller must be used to break contact on the hot side of the circuit first. To replace the fuse, the fuse must be installed with the load side of the fuse clip first and then the into the line side.

16.4.9.1.2 Ground Fault Circuit Interrupter

Ground fault is any amount of current above the level that may deliver a dangerous shock. Any current over 8 mA is considered potentially dangerous depending on the path the current takes, the amount of time one is exposed to the shock, and the physical condition of the person receiving the shock.

A *ground fault circuit interrupter* (GFCI) is an electrical device that protects personnel by detecting potentially hazardous ground faults and quickly disconnecting power from the circuit.

GFCI protection may be installed at different locations within a circuit. Direct-wired GFCI receptacles provide ground fault protection at the point of installation. GFCI receptacles may also be connected to provide GFCI protection at all other receptacles installed downstream on the circuit.

Plug-in GFCl units provide ground fault protection for devices plugged into them. These must be used by personnel working with power tools in an area that do not include GFCI receptacles.

16.4.9.1.3 Lockout/Tagout

To ensure the safety of personnel working any electrical equipment, or in a tank or a reactor that has mechanical mixers or agitators, all sources of power must be removed and the equipment must be locked out and tagged out. In most cases, it is appropriate to turn off the power supply and remove the fuses from the main in addition to lockout.

OSHA regulations require the electrical power supply to be totally disconnected and fuses removed; the electrical main must be locked out and tagged out before any maintenance work is performed.

Lockout is the process of placing a lock on each electrical box that supplies electrical power to moving components, such as motors for mixers, and blowers located on the equipment to be repaired. This prevents anyone from turning on the power to the equipment.

Tagout is the process of placing a danger tag on the source of electrical power, which indicates that the equipment is not to be operated until the danger tag is removed (see Fig. 16.4 for lockout/tagout samples). This is also done to the other sources of utilities, such as steam, water (hot or cold), compressed air, and nitrogen.

A danger tag has the same importance and purpose as a lock and is used alone only when a lock does not fit the disconnected device. The danger tag is attached at the disconnected device with a tag tie or equivalent and has space for the worker's name, craft, and other required information.

16.4.10 Confined Space Entry Procedure

This is a very important part of the plant maintenance procedure. Workers need to enter tanks or other vessels that are not open to the atmosphere.

16.4.10.1 Definition of a Confined Space

A *confined space* is defined as follows:

1. It is a limited space or a vessel where an employee can enter and perform assigned work. It is not normally open to human traffic.

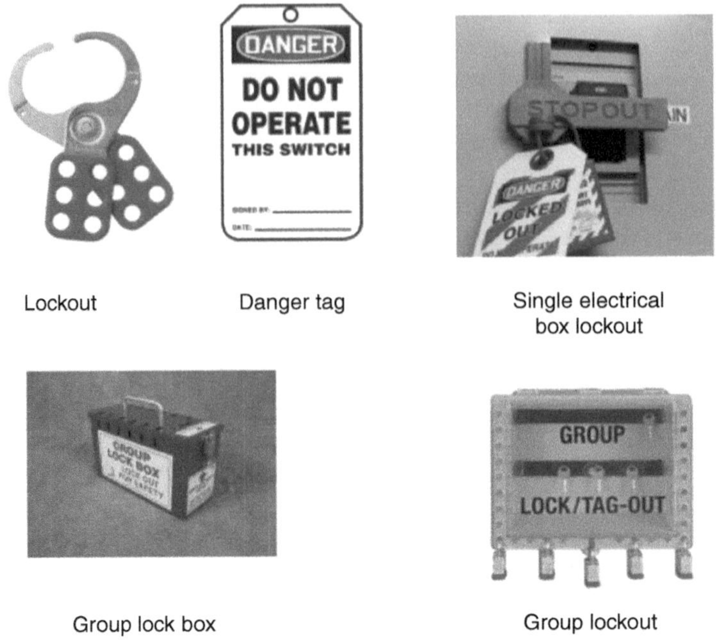

FIGURE 16.4 Component for lockout and tagout.

2. Entry to the confined space is limited or restricted. Entry that is allowed only after the space has been prepared for entry following certain procedures outlined later. This is necessary to prevent asphyxiation of the employees who need to perform some task in this space.
3. Asphyxiation can be caused by:
   a. lack of oxygen
   b. presence of inert gas
   c. presence of toxic vapor

16.4.10.2 Need for an Entry Procedure for a Confined Space

It is necessary for the oil-processing plant to have a procedure for confined space entry to prevent any loss of life.

This procedure must be followed every time workers have to enter any confined space for any work.

This discussion is limited to the entry procedure that is appropriate for storage tanks maintained under a nitrogen atmosphere to protect the oil.

16.4.10.3 What is a Tank Entry Procedure?

This procedure provides the necessary steps that must be followed by plant personnel when it becomes necessary to enter a nitrogen-blanketed tank for inspection, repair, or cleaning.

16.4.10.4 Who Should Be Familiar With the Procedure?

All plant personnel who are associated with the process—such as operators, process mechanics, supervisors, and the department managers—must be trained on the tank entry procedure. This does not mean that they all will need to enter the vessel, but their familiarity with the procedure is important. For example, the production manager may not ever enter the vessel, but he must be fully aware of the procedure in to prevent any mishap.

16.4.10.5 Why Must It Be Followed?

Adherence to the procedure must be treated as mandatory. The interior of a tank may be unsuitable to support life for lack of oxygen, and can asphyxiate any individual who enters the confined space without taking proper precautions. This can result in loss of human life.

16.4.10.6 Equipment Needed for Tank Entry

Following is a list of equipment that is needed for the preparation of a tank that is maintained under a nitrogen atmosphere.

1. An air blower with a capacity of 200–300 ft.^{3}/min of airflow and a flexible discharge hose with 6–12 in. diameter is needed. The hose must be long enough to reach the center of the tank floor. The capacity of the blower will vary with the size of the tank.
2. A mechanical pulley device to rescue the person entering the tank in case of an emergency.
3. A portable oxygen meter with a sampling tube that is long enough to reach various corners of the tank to measure oxygen content.
4. A resuscitator to provide artificial respiration to the victim if needed.
5. A safety kit equipped with a fully charged oxygen canister, a gas mask, and a safety harness.
6. A two-way radio or walkie–talkie.
7. A thermometer to check the inside temperature of the tank.

16.4.10.7 Signs to Be Displayed on a Nitrogen-Blanketed Tank

A nitrogen-blanketed tank must have the following items:

1. A permanent sign on the side of the tank reading "NITROGEN ENVIRONMENT" and "TANK ENTRY PERMIT REQUIRED."

   The manhead on the tank must also bear the same sign, "NITROGEN ENVIRONMENT" and "TANK ENTRY PERMIT REQUIRED."

   A *union* on the nitrogen supply line located between the isolation valve and the tank that can be easily disconnected.

16.4.10.8 Preparation for Tank Entry

The following steps are required to prepare a nitrogen-blanketed tank for entry:

1. Empty the tank.
2. Close the isolation valve and *disconnect the union* on the nitrogen supply line to the tank.
3. If the tank uses steam heat, close the manual steam shutoff valve before the steam control valve. This should provide protection against steam leakage through the control valve. However, make sure that the steam is shut off.
4. Follow the same procedures described in step 3 for a tank heated by hot water or a thermal system.
5. Put a chain and a padlock on the wheel of the manual steam shutoff valve or hot water supply valve. The safety coordinator should hold the key until the tank is ready to be put back into service. The lock can be removed only by the safety coordinator.
6. Remove the fuse or flip the circuit breaker for the tank agitator motor.
7. Lock out the electrical box containing the circuit breaker (or fuse) for the mixer motor. The key must be held by the safety coordinator. In addition, every person who needs to enter the tank to do any work must have a separate lock and a key. Every individual entering the tank must put a lock on the electrical box and carry the key before entering the tank. The lock can be removed only by the individual exiting the tank, and the last lock can be removed only by the safety coordinator when the tank is ready to be used again. Normally the supervisor or a manager of the department is in charge of the safety locks and the keys. The same person holds the keys to all the locks on the shutoff valves as described earlier.
8. Open and insert the discharge hose of the air blower into the tank through the open man head, reaching the center of the tank floor, and then turn it on. Open the hatch at the top of the tank. If there is no hatch on the top, the vacuum vent valve on the tank will allow the air to exhaust from the tank.
9. Leave the air blower on for approximately 24 h. The amount of time required for blowing air depends on the size of the tank.
10. Calibrate the portable oxygen meter and make sure it is working properly. The portable oxygen meter must have a motorized sampling probe so that, with the help of a connecting hose, it can draw a sample from inside the tank to measure the percentage of oxygen.
11. Check the percentage of oxygen by extending the sampling tube to several locations in the tank. If the oxygen content at these points is above *19.5%*, the atmosphere in the tank can support human life. If, not continue to blow air until the oxygen content exceeds 19.5%.
12. Check the temperature of the tank. It must be less than 110°F. Generally, after 24 h of air blowing, the tank will reach ambient temperature.
13. Keep the blower operating at low speed when the oxygen content is above 19.5%, the temperature in the tank is almost ambient, and the repair crew is at work inside the vessel.
14. With an explosimeter check inside of the tank for any explosive gas and make sure there is no explosive gas inside the tank.

Note:

It is now safe to enter the tank.

16.4.11 The Tank Entry Permit Must be Filled out and Signed by two Persons

The safety coordinator and process manager (supervisor) must sign the tank entry permit as shown on the sample tank entry permit.

16.4.11.1 Tank Entry Permit

1. A signed tank entry permit certifies that the tank is safe to enter.
2. The safety coordinator and the department supervisor jointly make the preentry tank inspection as indicated further and sign the tank entry permit with the date and time. The signed and dated permit is posted on the outside of the tank in a protective clear envelope.
3. The tank entry permit expires at the end of the shift or when the last worker exits the tank even before the end of the shift.
4. A new tank entry form must be filled to resume tank entry if the job is to be continued. A sample tank entry permit is shown:

Questionnaire on the Tank Entry Permit	Yes	No
Is the tank floor clean for working inside?		
Is the oxygen content in the tank above 19.5%?		
Is the nitrogen line disconnected?		
Is the steam valve closed and locked?		
Is the hot-water heating valve locked?		
Is the electrical supply disconnected?		
Is the electrical box locked?		
Is the tank temperature below 110°F?		
Is the air blower on?		
Has the resuscitator been inspected and certified?		
Is the safety kit in working order?		
Is the safety harness operational?		
Signed: Safety Coordinator	Date	Time
Signed: Process Manager/Supervisor	Date	Time

16.4.12 Entering the Tank

Once the tank entry permit is posted on the tank, personnel may enter the tank using the following procedure:

1. Keep the air blower on at low speed to allow fresh air into the tank for the comfort of the personnel entering the tank.
2. A safety person must be present at the tank entry point before anyone enters the tank. He checks the safety harness, eye protection, and gloves on every individual entering the tank.

CAUTION:

Remember! The tank entry permit expires at the end of the shift. A new permit must be issued and posted before work can be resumed.

3. Each person must put the lock on the electrical box before entering the tank and keep the key in possession.
4. One or more persons can enter the tank to do the necessary repair job. Each person must wear a safety harness, goggles, or face shield, and protective gloves.
5. The plant must install some mechanical device to pull the person out of the tank in case he or she is overcome or hurt for any reason.
6. The resuscitation kit, safety kit, and one of the two-way radios must be kept near the tank and within the reach of the safety person.
7. The second two-way radio must be with the safety coordinator or department foreman, whoever is in charge at the time.
8. The safety person frequently talks to the person or persons inside the tank to make sure they are alert and well.
9. If the safety person observes that the person or persons inside are not alert or are nonresponsive or hurt, he or she must do the following:
   a. Call the safety coordinator on the two-way radio and ask him or her to come immediately with additional help.
   b. Get immediate help from others to pull the person out of the tank using the pulley and the safety harness.
   c. The safety person must prepare to enter the tank by putting on the safety mask and harness.
   d. The safety person must not enter the tank until the additional help arrives.
10. Once the help arrives, the safety person, already wearing the safety mask and harness, enters the tank to rescue his or her colleagues. The safety coordinator or the additional help must be beside the tank. This person will have control over the safety harness and the two-way radio to call the safety coordinator.
11. The victim is brought outside and immediately given oxygen or resuscitation, if needed.
12. Medical attention is provided immediately to evaluate the condition of the individual or individuals affected.

16.5 SPECIAL NOTES

1. A nitrogen environment does not support life. A person can become brain-dead in 5–8 min if exposed to any nonlife-supporting gas, such as nitrogen. The person becomes disoriented and euphoric at some point before

becoming brain-dead. Nitrogen is an odorless gas. Although 79.1% of the air we breathe is made up of nitrogen (20.9% oxygen), a concentration of oxygen less than 19.5% cannot support human life. Therefore, all of the aforementioned precautions are critical to prevent any loss of life.

2. All operating personnel must be given full training on the safety procedure for tank entry.
3. There must be periodic refresher training for all employees.
4. All new hires must go through this training as a condition for employment.
5. The resuscitator must be inspected according to the instructions supplied by the manufacturer.
6. The oxygen canister in the safety kit must be recharged after every time it is used.
7. The safety harness must be checked to make sure it is not damaged.
8. The oxygen meter must be calibrated, and the internal cell and the battery, as well as the sensor must be replaced according to the instructions provided by the manufacturer.

Chapter 17

Regulatory Agencies and Their Roles in a Vegetable Oil Plant

17.1 INTRODUCTION

There are food safety regulatory agencies almost in every country, such as United States, Europe, Japan, China, India, Latin America, The Middle East, Australia, India, some Eastern European countries, some Latin American countries and Africa that are involved in the safety of the food product made distributed. There are also agencies that are responsible for the safety of the workers for health and prevention of accidents. Some of these food safety agencies are affiliated with specific religious sectors that have their strict guidelines for foods. Some of these agencies even have infrastructures set up to monitor and certify food-manufacturing facilities and also supervise to ensure that the manufacturers follow the prescribed guidelines. Vegetable oil production facility comes under their jurisdiction only if the plant processes meat fat, dairy products, or use ingredients that are supplied by other manufacturers and the finished product packaging bear their certified emblems. These agencies do not have any regulating authority but they can influence their clients so the product manufacturer is encouraged to follow their guidelines to get their business.

17.2 AGENCIES OVERSEEING FOOD INDUSTRY

17.2.1 United States

Following agencies are connected to the US Food Industries. They perform various functions.

- Occupational Safety and Health Administration (OSHA)
- Environmental Protection Agency (EPA)
- National Fire Protection Agency (NFPA)
- US Food and Drug Administration (US FDA)
- US Department Of Agriculture (USDA)
- Rabbinical Assembly
- National Institute of Oil Processors Association (NIOP)
- National Oilseed Processors' Association (NOPA)

Practical Guide to Vegetable Oil Processing. http://dx.doi.org/10.1016/B978-1-63067-050-4.00017-9
Copyright © 2017 AOCS Press. Published by Elsevier Inc. All rights reserved.

17.2.2 Europe

European agencies that overlook and guide the edible oil industry are:

- The Federation of Oils, Seeds and Fats Associations Ltd. (FOSFA)
- FEDIOL
- EFSA European Food Safety Authority
- Food Safety Authority (FSA) in several EU member countries
- Rapid Alert System for Food and Feed (RASFF)

> **Note**
>
> There are regulatory agencies in almost every developed and developing country in the world. The information is too numerous to be included here in this chapter.

17.2.2.1 Occupational Safety and Health Administration

In response to dangerous conditions across the nation and as a culmination of decades of reform the Occupational Safety and Health Act of 1970 was signed into Law by President Richard M. Nixon in 1971.

The function of OSHA is to:

1. Conduct workplace inspections for ensuring whether employers meet minimum requirements of the OHSA standards, and providing a safe and healthful workplace for workers.
2. Create an environment conducive for the participation of the industry members engaging in occupational health and safety practices.
3. Harmonize national, regional, and international laws and standards on occupational health and safety.
4. Develop comprehensive database for work related accidents and diseases through:
   a. information
   b. guidelines
   c. recording
   d. notification and reporting system
5. Promote research and development.
6. Improve occupational health and safety skills and human resources in public and private sector.
   a. Establish adequate funding arrangements to fund OSHA activities.

17.2.2.2 Role of OSHA in an Oil Plant

OSHA rules come into play to protect the workers wherever there are:

1. workers performing duties
2. machinery and equipment to be operated

3. building facility used by the operation
4. heating, ventilation, and air conditioning are used

Since the inception of OSHA, industrial accidents and deaths on the job from accidents have declined significantly. On the job deaths statistics:

- 11/100,000 in 1972
- 3.6/100,000 in 2009
- 3.3/100,000 in 2014

17.3 ENVIRONMENTAL PROTECTION AGENCY

The EPA is a government organization formed in 1970 under the Executive Branch. The agency was formed to regulate the environmentally unsafe practices that had potentially adverse effect to the environment.

Today our environment is much cleaner because of the inception of the agency. The EPA is engaged in the following programs:

1. Air, Noise, and Radiation Programs
2. Water and Waste Management Programs
3. Solid Waste Emergency Response Programs
4. Legal and Enforcement Counseling
5. Pesticides and Toxic Substances Programs
6. Research and development on all environment related programs
7. Major developments in environment area

The significant acts implemented by the EPA are:

1. Federal Water Pollution Control Act.
2. Safe Drinking Water Act.
3. Amended Clean Air Act.
4. Environmental Pesticide Control Act.
5. Compensation and Liability Act, or the EPA superfund. The statute, which was passed in 1980, aims to provide the EPA with the funds necessary to clean up myriads of abandoned hazardous waste dumpsites all across the United States.

17.3.1 Role of EPA in a Food Plant

The EPA is involved where the facility has:

1. Personnel working—to ascertain workers' safety from exposure to dangerous chemicals or conditions.
2. Emission of gas or exhaust.
3. Effluent as industrial waste or even water.
4. Any form of environmental pollution.

17.4 NATIONAL FIRE PROTECTION ASSOCIATION

The National Fire Protection Association (NFPA) is an international nonprofit organization established in 1896. It is the world's leading advocate of fire prevention and an authoritative source on public safety. It has a worldwide membership over 70,000 from nearly 100 nations. The Association's mission is to reduce the worldwide burden of fire and other hazards on the quality of life by providing and advocating:

1. establishing consensus codes,
2. setting standards, and
3. conducting research, training, and education.

17.4.1 NFPA's Role in an Oil Plant

NFPA's role is very extensive, such as:

1. Planning for construction of the building and the facility.
2. Instituting protection codes applicable for:
   a. the building
   b. cooling towers
   c. storage of dry materials, including raw materials, finished goods even wooden pallets
   d. special requirements for potentially combustible material
3. Proper ventilation of vapors and fumes to prevent fire hazards.
4. Any heating and exhaust fumes from combustible materials.
5. Fuel storage and delivery system.
6. Storage for any potentially explosive materials, such as compressed hydrogen gas.
7. Establishing emergency evacuation route and procedure.
8. Fire-drill protocol.

17.5 US DEPARTMENT OF AGRICULTURE

It is part of the Executive Branch of the U S Government. The USDA's goals are:

1. protecting public health
2. preventing foodborne illness
3. enhancing public education and outreach
4. investing in innovative technology, processes and tools to protect public health

As a department of the government services, it performs numerous functions. It provides the following services:

1. Agricultural Marketing Service (NRCS)
2. Agricultural Research Service (ARS)
3. Animal and Plant Health Inspection Service (APHIS)
4. Economic Research Service (ERS)

5. Farm Service Agency (FSA)
6. Food and Nutrition Service (FNS)
7. Food Safety and Inspection Service (FSIS)
8. Foreign Agricultural Service (FAS)
9. Forest Service (USFS)
10. Grain Inspection, Packers and Stockyards Administration (FGIS)
11. National Agricultural Statistics Service (NASS)
12. National Institute of Food and Agriculture (NIFA)
13. Natural Resources Conservation Service (NRCS)
14. Risk Management Agency (RMA)
15. Rural Development (RD)
16. National Appeals Division (NAD)
17. Departmental Management (DM)

As part of the Food and Nutrition Service the agency releases *The Dietary Guidelines for Americans* every 5 years. The *Dietary Guidelines* is required under the 1990 National Nutrition Monitoring and Related Research Act, which states that every 5 years, the US Department of Agriculture (USDA) and Health and Human Services (HHS) must jointly publish a report containing nutritional and dietary information and guidelines for the general public. The *2015–2020 Dietary Guidelines for Americans* is the 8th edition released since 1980 and remains the current edition until the *2020–2025 Dietary Guidelines for Americans* is released. The *Dietary Guidelines for Americans* is the cornerstone of Federal nutrition policy and nutrition education activities.

17.6 ROLE OF USDA AT AN EDIBLE OIL PLANT

The USDA gets directly involved in an oil processing plant if the plant processes animal fats in addition to vegetable oils. In such a case, the Rabbinical Assembly gets involved if the plant also processes vegetable oils and manufactures foods bearing Kosher symbol on the product packaging.

All seeds for crushing are inspected and graded by USDA inspectors before they arrive at the crushing plant.

USDA Agricultural Research Service (ARS) is known to assist with research work on crops, grain handling, milling, grain research on fermentation, oil extraction and processing research, biodiesel production, and product development research. The technology developed by the USDA Research group has pioneered the work in the area of vegetable oils that has put the industry in today's position. The organization performs technology transfer from USDA ARS to US and overseas industries.

17.7 US FOOD AND DRUG ADMINISTRATION

The US Food and Drug Administration (USFDA or FDA) is the oldest comprehensive consumer protection agency in the US federal government. Its goes back to the appointment of Lewis Caleb Beck in the Patent Office around 1848 to carry out chemical analyses of agricultural products. The USDA inherited the aforementioned function in 1862.

FDA's modern regulatory functions began with the passage of the Pure Food and Drugs Act in 1906. A quarter-century later this law was enforcing prohibition of interstate commerce in adulterated and misbranded food and drugs. FDA received its present name in 1938.

The scope of FDA's regulatory authority is very broad. The agency's responsibilities are closely related to those of several other government agencies. Often frustrating and confusing for consumers in determining the appropriate regulatory agency to contact

The traditionally recognized product categories that fall under FDA's regulatory jurisdiction are listed later. However, this is not a complete list by any means.

- foods
- cosmetics
- drugs
- biologics
- medical devices
- electronic products that give off radiation
- veterinary products
- tobacco products

Function of the FDA has evolved over the years. In the 1970s, Food Safety was mostly about basic sanitation, filth, and chemical safety. FDA's food safety role was mostly about inspections and court cases against the violators. At that time, when FDA decided it needed to propose a regulation—to put a limit on PCB's in fish or to address the safety of caffeine—it was written and published with little or no review by anyone else. It was more or less autonomous.

Foodborne illness caused by microbiological pathogens emerged as a public health concern and a threat to public confidence in the food supply in the 1980s. FDA started to focus on food safety and manner of working began a major shift in the late 1980s and 1990s. Seafood safety was a particular concern at that time, and FDA responded by issuing the seafood hazard analysis and critical control points (HACCP) rule in 1995. HACCP is a management system in which food safety is addressed through the analysis and control of biological, chemical, and physical hazards from raw material production, procurement and handling, to manufacturing, distribution, and consumption of the finished product. This rule, combined with parallel HACCP reforms at USDA for meat and poultry, really launched the modern era of food safety regulation.

It established the principle that food safety depends primarily on food companies taking responsibility for preventing hazards in their operations, not on government taking enforcement actions after problems occur.

It also established the precedent of FDA providing technical guidance and assistance to seafood processors—through a tome that explained in clear steps so they could know what works in preventing specific hazards. With the seafood

HACCP rule, FDA also found itself working more closely with an array of external constituencies and partners, including the seafood industry and the states. By mid-90s, the executive orders mandating Office of Management and Budget (OMB) review of new regulations were in full effect. OMB assists the President in overseeing the preparation of the Federal budget and in supervising its administration in Federal agencies. The OMB also oversees and coordinates the Administration's procurement, financial management, information, and regulatory policies.

The Food Safety Modernization Act (FSMA) is the most sweeping reform of our food safety laws in more than 70 years. President Obama signed FSMA into a law on January 4, 2011. It aims to ensure the US food supply is safe by shifting the focus from responding to contamination to preventing it. Through FSMA the FDA implemented the most comprehensive changes in food safety, "Mandatory Food Recall."

FSMA has shifted the burden of delivering safe food to the manufacturers throughout its supply chain, including risk management and product recall. The number and areas of provisions in the FSMA is too extensive to list. The FDA updated the FSMA Rules and Guidance for Industries were updated on June 15, 2016. Some of these provisions are still under final review but most of them have reached the final stage.

17.8 RABBINICAL ASSEMBLY

There are two dominant rabbinical assemblies in USA that oversee the processing, production, and distribution of foods at every level:

1. raw material
2. preparation
3. processing
4. product shipping and handling

The two major rabbinical assemblies in USA are Circle U with a symbol, Ⓤ and Circle K with a symbol, Ⓚ.

Although the Rabbinical Assembly does not have any regulatory power, it does determine what food can be purchased and consumed by the Jewish community in general and especially the Orthodox Jews.

The organizations make sure the food meets the Kosher standards. "Kosher" is a Hebrew word that literally means "fit" or "proper." When used in relation to food products, "Kosher" means that the food item in question meets the dietary requirements of "Jewish law."

Jewish dietary law is a comprehensive set of guidelines that are designed for making food safe and good for human health.

Kosher foods are divided into three categories:

1. meat
2. dairy
3. pareve

A brief discussion on the aforementioned items would be appropriate to discuss some of the following items.

17.8.1 Meat

Meat includes products from kosher animals and birds that have been slaughtered and processed under Kosher guidelines. Even if the meat is Kosher the fat from the Kosher animal may not be so. This is because the fat from certain parts of the animal can be Kosher but not from all parts of the animal.

The fat from the animal can be Kosher only if it comes from the acceptable parts of the animal's body, processed in a Kosher plant, packaged in Kosher packaging and distributed appropriately following the Kosher manufacturing guidelines. The entire manufacturing process has to be certified Kosher by the Rabbinical Assembly.

17.8.2 Dairy

All foods derived from, or containing, milk are classified as dairy, including milk, butter, yogurt, and all cheese—hard, soft, and cream. Even a trace amount of dairy can cause a food to be considered dairy. The Kosher requirements are:

1. They must come from Kosher animals.
2. All ingredients must be Kosher and free of meat derivatives. Conventional rennet, gelatin, etc., are of animal origin and cannot be used in Kosher dairy.
3. All ingredients and additives must be produced, processed and packaged on Kosher equipment.

17.8.3 Pareve

Foods that are neither meat nor dairy are called pareve.

17.9 ROLE OF RABBINICAL ASSEMBLY IN AN OIL PLANT

If the food plant wants to manufacture, package and distribute products with Kosher symbol on the label, they must comply with the Jewish Food Laws that constitute Kosher status for food. This applies to:

1. All incoming raw materials have to be Kosher.
2. Packaging materials must be Kosher.
3. All process and handling equipment must be Kosher.
4. Storage and Distribution system and practices according to the Kosher requirement.
5. Additives, such as emulsifiers etcetera must be Kosher.
6. Enzymes used for degumming crude oil cannot be of animal origin.

17.10 NATIONAL INSTITUTE OF OILSEED PRODUCTS

The National Institute of Oilseed Products (NIOP) is an international trade association with the principal objective of promoting the general business welfare of:

1. persons,
2. firms, and
3. corporations engaged in the buying, selling, processing, shipping, storage, and use of vegetable oils and raw materials.

The Headquarter of NIOP is located in Washington DC, USA.

If a food plant in USA or elsewhere is buying edible oils from USA or selling edible oils to USA, the company has to abide by the trading rules set forth in the "NIOP Trading Rule Book." NIOP has been serving for over 80 years, has served nearly 120 member firms in 14 countries.

The NIOP Trading Rules include:

1. types of sales
2. shipment
3. tank cars, barges, and containers
4. performance of quality specifications
5. quality specifications/survey procedures/prior cargoes
6. vegetable oil specifications
7. seeds and nuts
8. meals
9. payment/letters of credit/insurance and other financial obligations
10. force majeure/default/bankruptcy and insolvency
11. arbitration of disputes

17.11 NATIONAL OILSEED PROCESSORS ASSOCIATION

The National Oilseed Processors Association (NOPA), a national trade association, represents the US soybean, canola, flaxseed, safflower seed, and sunflower seed, crushing industries. Organized in 1930, NOPA's membership also includes associate members that are consumers of vegetable oil or oilseed meal, including independent oil refiners, mixed feed manufacturers and meat and poultry customers and end-users of oils and meals.

NOPA is a member-driven trade association. It represents its members' interests in the areas of:

- trading rules,
- international trade matters, and
- federal legislative and regulatory matters, and biotechnology policies.

NOPA advocates for an efficient global supply chain system on behalf of the US oilseed processing industry, by providing leadership through education, information and market-based solutions to:

- policymakers,
- trade negotiators,

- growers customers,
- suppliers, and
- global oilseed organizations.

17.12 FEDERATION OF OILS, SEEDS AND FATS ASSOCIATIONS

The Federation of Oils, Seeds and Fats Associations Ltd (FOSFA) is a professional international contract issuing and arbitral body, such as NIOP in USA, concerned exclusively with the world trade in oilseeds, oils, and fats. The Headquarter is located in London, UK.

FOSFA is much larger than NIOP. It has 1103 members in 89 countries. These members include:

- producers,
- processors,
- shippers and dealers,
- traders,
- brokers and agents,
- superintendents,
- analysts,
- ship-owners,
- ship brokers,
- tank storage,
- companies,
- arbitrators,
- consultants,
- lawyers, and
- insurers and others, providing services to traders.

FOSFA has an extensive range of standard forms of contracts covering goods shipped as:

- Cost, Insurance and Freight (CIF)
- Cost and Freight (C&F)
- Free On Board (FOB)

FOSFA provides the above for:

- soybeans,
- sunflower seeds,
- rapeseed and others,
- vegetable and marine oils and fats,
- refined oils and fats, from all origins worldwide,
- methods of transportation, and
- terms of trade.

Internationally, 85% of the global trade in oils and fats is traded under FOSFA contracts. There is an extensive list of forms for each type of trade. These forms can be purchased from FOSFA.

Like NIOP in USA, the FOSFA contracts incorporate a dispute procedure involving arbitration by experienced individuals from within the trade.

17.13 FEDIOL

FEDIOL represents the interests of the EU vegetable oil and protein-meal industry. It represents the interest of public and private organizations to ensure a favorable business environment for European Institutions, as well as international bodies and stakeholders, such as:

- suppliers
- customers
- civil societies

The agency is engaged in areas that deals with:

- Formulating, as well as voicing common positions to contribute to a constructive regulatory framework within the EU.
- Developing professional principles and good manufacturing practices, ensuring the quality and safety of products.
- Helping the industry to diversify its activities and sources of supply at EU and international level.
- Strengthening the competitiveness of the EU oilseed industry.
- Representation the EU vegetable oil and protein-meal industry to maintain its positive image.

FEDIOL does not have any legislative function but it recommends to EU with position papers on matters pertaining foods and feeds, based on scientific justification.

17.14 EUROPEAN FOOD SAFETY AUTHORITY

European Food Safety Authority (EFSA) was set up in 2002 to be a source of scientific advice and communication on risks associated with the food chain following a series of food crises in the late 1990s, one of the significant incidents was incoming palm oil was adulterated with diesel oil. EU established the agency under the General Food Law—Regulation 178/2002.

The General Food Law created a European food safety system in which responsibility for risk assessment (science) and for risk management (policy) are kept under separate agencies. EFSA has a duty to communicate on its scientific findings to the public.

The organization is funded by European Union but it works independently of the of the European legislative and executive institutions (Commission,

Council, Parliament) and the EU Member States. However, the agency is legally bound to share information with the EU, EC, and the member states. Its Headquarter is located in Brussels, Belgium.

EFSA produces scientific opinions and advice, which form the basis for European food policies and legislation.

This agency is somewhat similar to FDA in USA and oversees the following areas but does not develop policies for food safety or regulations:

- food and feed safety
- nutrition
- animal health and welfare
- plant protection
- plant health

Most of EFSA's work is undertaken in response to requests for scientific advice from the European Commission, the European Parliament and EU member states. The agency also carries out scientific work on its own initiative, in particular to examine emerging issues and new hazards and to update the EC and the member states on the assessment methods and approaches. This is known as "self-tasking."

Recommendations from EFSA becomes a law through the European Commission, it is then implemented and enforced by the Food Safety Authorities of individual member countries.

The European Commission is the EU's executive body. It represents the interests of the European Union as a whole (not the interests of individual countries). The term "Commission" refers to both the College of Commissioners and to the institution itself.

The European Commission has its headquarters in Brussels, Belgium, and some services also in Luxembourg. The Commission has Representations in all EU Member States and 139 Delegations across the globe.

17.15 FOOD SAFETY AUTHORITY

There are Food Safety Authorities (FSA) in several EU countries that come under the individual governments. These offices are responsible for protecting consumers and the industries from any harmful situation that affects food and feed. These agencies have the authority to establish policies and enforce them and they work in close communication with EFSA, FEDIOL, and EC. Some of these Food Standard Authorities are listed here:

- Austria: AGES (Österreichische Agentur für Gesundheit und Ernährungssicherheit GmbH), http://www.ages.at/ages/ages-oesterreichische-agentur-fuer-gesundheit-und-ernaehrungssicherheit/
- Belgium: AFSCA (Agence Fédérale belge pour la Sécurité de la Chaîne Alimentaire), http://www.favv-afsca.fgov.be

- Denmark: Ministry of Food, Agriculture and Fisheries, http://www.fvm.dk
- Finland: National Food Agency Finland, http://www.evira.fi/portal/en/
- France: Anses (Agence nationale de sécurité sanitaire de l'alimentation, de l'environnement et du travail), http://www.anses.fr/fr
- Germany: Bundesministerium für Ernährung, Landwirtschaft und Verbraucherschutz, http://www.verbraucherministerium.de
- Greece: Hellenic Food Authority, http://www.efet.gr/
- Ireland: FSAI (Food Safety Authority of Ireland), http://www.fsai.ie
- Sweden: Livsmedelsverket (National Food Agency), http://www.slv.se/en/
- The Netherlands: VWA (Voedsel en Waren Autoriteit),
- United Kingdom: FSA (Food Standards Agency), http://www.food.gov.uk

17.16 RAPID ALERT SYSTEM FOR FOOD AND FEED

RSAFF was created in 1979. The agency facilitates:

- Transmittal and sharing information efficiently between its members (EU-28 national food safety authorities, Commission, EFSA, ESA, Norway, Liechtenstein, Iceland, and Switzerland) when there is any sign of safety concern related to food discovered in any member state.
- It provides a round-the-clock service to ensure that urgent notifications are sent, received, and responded to collectively and efficiently.
- RASFF has saved the EU members and other countries from food safety risks on numerous occasions.
- The agency has averted many situations on food safety before they could cause any harm to the European consumers.
- The agency has established a very robust communication system between the Agency Headquarter in Brussels and the agencies in the member countries, EFSA, FSA, FEDIOL, and EC that works round the clock 7 days a week.

The RASFF allows national food and feed control authorities in the EEA to share information about measures taken in response to serious risks detected in relation to food or feed. This exchange of information helps Member States to act more rapidly and in a coordinated manner in response to health threats caused by food or feed.

Vital information exchanged through RASFF can lead to products being recalled from the market. Its service has been invaluable to ensure food safety in the EU and beyond.

An edible oil company must have internal or external resources, which can provide the necessary information about the regulatory and trade authorities in the respective parts of the world when it comes to import or export edible oils.

Index

D

E

G

H

M

N

O